T0320444

Solid Waste-Based Materials for Environmental Remediation

This book provides a multifaceted examination of solid waste management methods, the preparation, properties, and application of solid waste materials in the remediation of various environmental media, as well as the combination of solid waste materials and artificial intelligence.

Based on the latest research results and cutting-edge technologies from around the world, the contributors combine the design principles of solid waste materials with application examples, including a complete system, clear routes, and illustrations. They integrate the idea of ecological civilization, the concept of sustainable development, and engineering innovation thinking, providing a reliable reference for resource recycling and contributing to global low-carbon emission.

The book is suitable for teachers and students, as well as researchers, industrial technicians, and managers involved in solid waste resources and environmental remediation.

Guanyi Chen is a distinguished professor at Tianjin University of Commerce and Tianjin University, China. His research interests are solid waste/biomass thermo-chemical conversion technology; recycling of solid waste; biochar preparation and its environmental use. He has published more than 400 articles in journals such as ES&T, WR, *Nature Communications*, *Applied Energy*, and *Biomass and Bioenergy*.

Ning Li is an associate professor at the School of Environmental Science and Engineering, Tianjin University, China. His research interests are environmental catalysis and advanced oxidation process; recycling of solid waste. He has published several articles in journals such as ES&T and WR.

Zhanjun Cheng is a professor at the School of Environmental Science and Engineering, Tianjin University, China. His research interests are clean and efficient use of solid waste resources; pyrolysis reaction kinetics; catalysis science and technology. He has published several articles in journals such as JACS, ES&T, and ACB.

Solid Waste-Based Materials for Environmental Remediation

Edited by
Guanyi Chen, Ning Li, and Zhanjun Cheng

CRC Press
Taylor & Francis Group
Boca Raton London New York

CRC Press is an imprint of the
Taylor & Francis Group, an **informa** business

Cover image designed by Production Perig, Shutterstock No.1703479210

First edition published 2025
by CRC Press
2385 NW Executive Center Drive, Suite 320, Boca Raton FL 33431

and by CRC Press
4 Park Square, Milton Park, Abingdon, Oxon, OX14 4RN

CRC Press is an imprint of Taylor & Francis Group, LLC

ISBN: 978-1-032-86390-0 (hbk)
ISBN: 978-1-032-87922-2 (pbk)
ISBN: 978-1-003-53540-9 (ebk)

DOI: 10.1201/9781003535409

Typeset in Minion
by Newgen Publishing UK

Contents

JINGLAN WANG, ZHANJUN CHENG, XIAOQIANG CUI, JIAN LI,
LAN MU, BEIBEI YAN, NING LI, AND GUANYI CHEN

NING LI, WENJIE GAO, LAN LIANG, YANGLI CUI,
ZHANJUN CHENG, BEIBEI YAN, ESLAM SALAMA,
MONA OSSMAN, AND GUANYI CHEN

ZIJIE XIAO, BOWEN YANG, YINGZI ZENG, HONGTAO SHI, AND XIAOCHI FENG

Contributors

Guanyi Chen
School of Environmental Science
and Engineering
Tianjin University/Tianjin Key Lab
of Biomass/Wastes Utilisation
Tianjin, China
School of Mechanical Engineering
Tianjin University of Commerce
Tianjin, China

Chen Chen
School of Mechanical Engineering
Tianjin University of Commerce
Tianjin, China

Xiaoshi Cheng
College of Environment
Hohai University/Key Laboratory
of Integrated Regulation and
Resource Development on
Shallow Lakes
Nanjing, China

Zhanjun Cheng
School of Environmental Science
and Engineering
Tianjin University/Tianjin Key Lab
of Biomass/Waste Utilisation
Tianjin, China

Xiaoqiang Cui
School of Environmental Science
and Engineering
Tianjin University/Tianjin Key Lab
of Biomass/Waste Utilisation
Tianjin, China

Yangli Cui
School of Environmental Science
and Engineering
Tianjin University/Tianjin Key Lab
of Biomass/Wastes Utilisation
Tianjin, China

Xiaochi Feng
School of Civil and Environmental
Engineering
Harbin Institute of Technology
(Shenzhen)/State Key Laboratory
of Urban Water Resource and
Environment
Shenzhen, China

Wenjie Gao
School of Environmental Science
and Engineering
Tianjin University/Tianjin Key Lab
of Biomass/Waste Utilisation
Tianjin, China

Gaoqi Han
School of Environmental Science
and Engineering
Tianjin University
Tianjin, China

Rui Han
School of Environmental Science
and Engineering
Tianjin University
Tianjin, China

Chao He
Faculty of Engineering and Natural
Sciences
Tampere University
Tampere, Finland

Chao Jia
Department of Environmental
Science and Engineering
Fudan University
Shanghai, China

Jian Li
School of Environmental Science
and Engineering
Tianjin University/Tianjin
Key Lab of Biomass/Waste
Utilisation
Tianjin, China

Ning Li
School of Environmental Science
and Engineering
Tianjin University/Tianjin Key Lab
of Biomass/Waste Utilisation
Tianjin, China

Lan Liang
School of Environmental Science
and Engineering
Tianjin University/Tianjin Key Lab
of Biomass/Waste Utilisation
Tianjin, China

Jingyang Luo
College of Environment
Hohai University/Key Laboratory
of Integrated Regulation and
Resource Development on
Shallow Lakes
Nanjing, China

Yuting Luo
College of Environment
Hohai University/Key Laboratory
of Integrated Regulation and
Resource Development on
Shallow Lakes
Nanjing, China

Jizhong Meng
Civil Engineering, College of
Science and Engineering
University of Galway
Ireland

Lan Mu
School of Mechanical
Engineering
Tianjin University of Commerce
Tianjin, China

Mona Ossman
City of Scientific Research and
Technological Applications

New Borg El-Arab
Alexanderia, Egypt

Eslam Salama
City of Scientific Research and
  Technological Applications
New Borg El-Arab
Alexanderia, Egypt

Hongtao Shi
School of Civil and Environmental
  Engineering
Harbin Institute of Technology
  (Shenzhen)/State Key
  Laboratory of Urban
  Water Resource and
  Environment
Shenzhen, China

Liming Sun
State Key of Laboratory of Soil and
  Sustainable Agriculture
Institute of Soil Science, Chinese
  Academy of Science
Nanjing, China

Qingbai Tian
School of Environmental Science
  and Engineering
Shandong University
Qingdao, China

Jinglan Wang
School of Environmental Science
  and Engineering
Tianjin University/Tianjin Key
  Lab of Biomass/Waste
  Utilisation
Tianjin, China

Shaobin Wang
School of Chemical Engineering
  and Advanced Materials
The University of Adelaide
Adelaide, Australia

Zijie Xiao
School of Civil and Environmental
  Engineering
Harbin Institute of Technology
  (Shenzhen)/State Key Laboratory
  of Urban Water Resource and
  Environment
Shenzhen, China
Department of Chemical
  Engineering
KU Leuven
Leuven, Belgium

Runze Xu
College of Environment
Hohai University/Key Laboratory
  of Integrated Regulation and
  Resource Development on
  Shallow Lakes
Nanjing, China

Xing Xu
School of Environmental Science
  and Engineering
Shandong University
Qingdao, China

Beibei Yan
School of Environmental Science
  and Engineering
Tianjin University/Tianjin Key Lab
  of Biomass/Waste Utilisation
Tianjin, China

Han Yan
School of Environmental Science
  and Engineering
Tianjin University
Tianjin, China

Bowen Yang
School of Civil and Environmental
  Engineering
Harbin Institute of Technology
  (Shenzhen)/State Key Laboratory
  of Urban Water Resource and
  Environment
Shenzhen, China

Fengbo Yu
State Key of Laboratory of Soil and
  Sustainable Agriculture
Institute of Soil Science, Chinese
  Academy of Science
Nanjing, China

Yingzi Zeng
School of Civil and Environmental
  Engineering
Harbin Institute of Technology
  (Shenzhen)/State Key
  Laboratory of Urban
  Water Resource and
  Environment
Shenzhen, China

Xiaotong Zhao
Queen's University Belfast
Belfast, Northern Ireland

Xiangdong Zhu
State Key of Laboratory of
  Soil and Sustainable
  Agriculture
Institute of Soil Science, Chinese
  Academy of Science
Nanjing, China

CHAPTER 1

Intelligent Classification and Management of Solid Waste

Runze Xu, Xiaoshi Cheng, Yuting Luo, Chao He, and Jingyang Luo

1.1 INTRODUCTION

The disposal of common goods after consumption, known as solid waste, has become an increasingly pressing global issue due to factors such as shifting consumption habits, rapid urban development, and dwindling disposal choices. According to the World Bank, approximately 2.01 billion tonnes of solid waste are generated annually. Of that amount, 33% of garbage is improperly managed, and by 2050, that amount will increase to 3.4 billion tonnes.[1] The problem of "waste siege" is growing. The continuous discharge of solid waste into the environment has raised serious concerns about human health. The burning of informal waste pollutes the air, and leaching harmful chemicals from solid waste into surface water poses serious health risks. Despite the potential risks to human health and the environment caused by solid waste exposed to the atmosphere, it is being recognized as one of the most renewable resources due to its ability to be converted into fuel, materials, energy, and higher-value byproducts.[2] Maximizing the efficiency of the use of solid waste and minimizing the hazards associated with it requires a systematic approach to waste management and the use of treatment and resource utilization technologies. Despite the fact that the "Waste Classification" policy and the "Zero-Waste

City" program were proposed to support solid waste reduction and recovery, the actual implementation of these policies and programs has not been the same in Chinese cities due to regional differences in solid waste generation.[3] Furthermore, China's municipal solid waste has a very complex composition and characterization due to different living habits and material needs across the country. Research on the spatiotemporal patterns of solid waste generation is crucial for efficient management and policy formulation.[4]

Advancements in artificial intelligence techniques, such as machine learning, are now being applied to predict trends related to solid waste, offering insights into future waste generation and aiding in the development of effective management policies. Intelligent technology creates computer systems and programs capable of emulating human attributes including problem-solving, learning, perception, understanding, reasoning, and environmental awareness. Machine learning models, including artificial neural networks, fuzzy logic, expert systems and genetic algorithms are adept at addressing ill-defined issues, establishing complex mappings, and predicting outcomes. Based on the diverse selection of machine learning models, there have been studies covering the diverse applications of machine learning in solid waste-related fields. Artificial neural networks have emerged as the predominant machine learning technique, accounting for over 50% of usage, followed by support vector machines at 14%.[5] Notably, artificial neural networks implemented in 19 of the 24 identified sub-fields (*for example*, predicting waste generation, biogas production, and incineration pollutants emissions).

This chapter aims to introduce and compare the algorithms of intelligent technologies and discuss their current status, challenges, and future prospects in the intelligent classification and management of solid waste. Initially, the typical intelligent technology algorithms and improvement pathways utilized in the classification and management of solid waste are introduced. The applications of intelligent technology in waste generation, waste collection and transportation, waste sorting, and waste characterization, as well as waste disposal and recycling are then reviewed. Furthermore, this discussion underscores the challenges and opportunities inherent in applying intelligent technology to solid waste categorization and management. Thus, this chapter endeavors to provide a comprehensive and practical insight into the deployment of intelligent technologies in the realm of solid waste management.

1.2 INTELLIGENT TECHNOLOGY

Intelligent technology based on machine learning algorithms is a multidisciplinary field that covers computer science, statistics, approximate theory, and mathematics.[6] Intelligent technology such as artificial neural networks, support vector machines, K-nearest neighbors, decision trees, random forests, adaptive network fuzzy inference systems, and deep learning has been widely used to solve complex problems in the classification and management of solid waste. The detailed description of these common intelligent technologies has been provided in many previous reviews.[7, 8]

The workflow of the intelligent machine learning methods typically involves three phases: (1) organizing datasets and preprocessing data; (2) comparing proper machine learning algorithms and optimizing parameters, and (3) testing and evaluating performance.[9] For a single machine learning algorithm, a complete establishment process involves the following cycles: (i) model training, where internal information and features are learned from the prepared training dataset through training algorithms; (ii) model validation, determining the optimal hyperparameters for the machine learning algorithm using the validation dataset; and (iii) model testing, evaluating the generalization ability of the final model based on a testing dataset.[7]

The simulation performance of machine learning models largely hinges on data quality, data normalization method, model depth, optimization algorithms, training algorithms and the model structure (Figure 1.1). With the arising of more complicated issues, more dedicated and efficient model structures and algorithms are required. The following points summarize key strategies for enhancing intelligent technology (Figure 1.2):

(1) Except for daily or online monitoring data related to solid waste, datasets collected from lab-scale experiments normally are scarce and incomplete due to equipment limitation, experiment costs and labor shortage.[10] While generating more training data can improve model performance, filling gaps with synthesized data and utilizing mathematical techniques to enhance datasets are also beneficial, for instance, linearly interpolating incomplete data to fill in missing data and embedding adequate noise into the prepared datasets to augment datasets.[11] In addition, several data augmentation methods such as rotating, flipping, mirroring, and adjusting contrast can be applied to increase the amount of image data.[12]

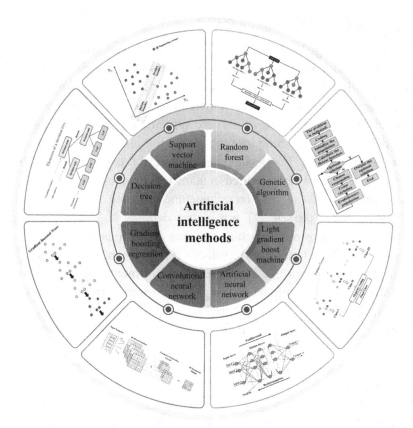

FIGURE 1.1 The illustration of the common intelligent technologies applied in the classification and management of solid waste.

(2) Data normalization, also called standardization, is essential for the data preprocessing procedure. Sigmoid or tanh functions usually are chosen as transfer functions for hidden layers. When inputs of hidden layers are excessively large (absolute value), sigmoid and tanh functions both have very low first-order derivative values. This over-saturated effect can significantly decelerate the learning rate of machine learning models by causing the gradients to flatten, almost approaching zero.[13] Normalizing data brings input values within a narrower range centered around zero, which ensures that the gradients are sufficiently steep for efficient training of machine learning models.[11]

(3) Becoming deeper is an inevitable development trend of some types of machine learning models (such as deep neural networks) due to

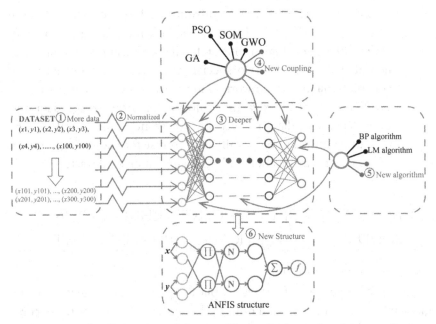

FIGURE 1.2   The illustration of the possible methods used to improve the performance of intelligent technology.

the emerging intricate problems and higher requirements proposed by the public and engineers. Deep learning models possess superior feature learning capabilities compared to shallow neural networks, enabling them to tackle more intricate issues.[13] The flexible structure and powerful learning ability of deep learning make it particularly effective for tasks like forecasting time-series data, as well as for carrying out visualization and classification tasks.

(4) Coupling different kinds of optimization algorithms with machine learning models is a feasible way to improve model performance. Some algorithms such as the genetic algorithm can optimize the setting of hyperparameters and accelerate the training process[14, 15] while other algorithms such as grey wolf optimization can find the most appropriate input variables.[16]

(5) The Levenberg Marquart algorithm has acceptable training speed and training performance for machine learning models. However, when the structure of models becomes deeper, new training algorithms are required due to the weakness of the gradient descent algorithm.[17]

(6) Fuzzy neural network and adaptive network-based fuzzy inference systems are two types of neural network models based on fuzzy logic, which show higher interpretability and accuracy than normal neural networks.[16, 18] Two special structures of neural networks, convolutional neural networks and recurrent neural networks, have excellent ability in extracting information from image data and time-dependent data, respectively. These novel structural designs of neural networks hold significant promise for addressing complex issues in the classification and management of solid waste.[19, 20]

1.3 APPLICATION OF INTELLIGENT TECHNOLOGY IN SOLID WASTE CLASSIFICATION AND MANAGEMENT

1.3.1 Waste Generation

Solid waste comprise of food waste, plastic, paper, rubber, textile, wood, glass, metals, ash, diapers, and others. The sources of solid waste are commercial, institutional, residential, and industrial.[21] The rapid increase in population resulted in an increase in the quantity of solid waste generation. Selecting and implementing waste management strategies and pollution control technologies has become exceedingly challenging. This is due to the complex and nonlinear nature of solid waste generation, which is influenced by numerous factors[22] such as temporal dynamics, socio-economic variables, origins of waste, and environmental conditions (Figure 1.3). To address the intricacies of waste disposal and recycling, a range of machine learning algorithms have been employed to examine and resolve the issues associated with solid waste generation (Table 1.1).

1.3.1.1 Time Scales

Waste generation projections account for the complex interaction between socio-economic elements, technical advances, regulatory frameworks, and environmental issues across different time scales. In short-term prediction, factors to consider include seasonal variations, consumption trends during holidays and events, and temporary policy adjustments. Medium-term projections involve extending current patterns and predicting socio-economic and environmental changes. Long-term waste projections account for cultural, economic, and environmental changes.[23]

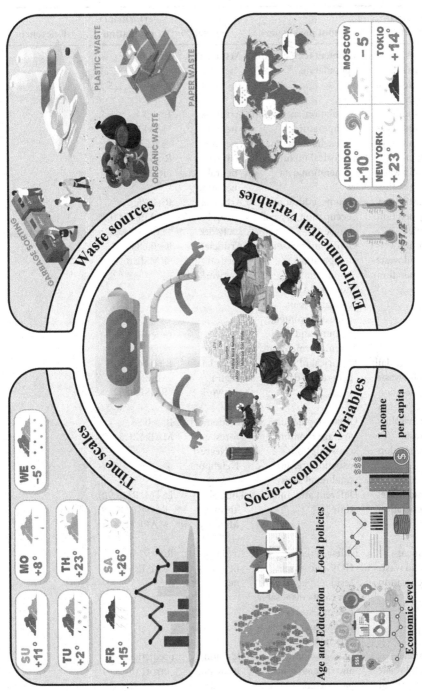

FIGURE 1.3   The main influencing variables of municipal solid waste generation.

TABLE 1.1   Overview of machine learning application in solid waste generation

Research object	Input parameters	Model	Performance evaluation	Reference
Weekly yard waste generation rate	Ambient temperature, population	Artificial neural network	R^2: 0.71; MAPE: 18.72%; MSE: 27,425	25
Monthly solid waste generation	Population, maximum temperature, household income	Artificial neural network	R^2: 0.74; MAPE: 4.6%; MSE: 0.26	59
Monthly solid waste generation	Historical monthly generation data	Artificial neural network	R^2: 0.739; RMSE: 0.019	30
Plastic waste generation rate	Income, education, occupation, type of houses	Artificial neural network	R^2: 0.75; MAPE: 34.07%	47
Monthly solid waste generation	GDP, rain, maximum temperature, population, household size, educated man, educated women, income, and the unemployment rate	Artificial neural network	R^2: 0.68; RMSE: 0.12	27
Average daily solid waste generation per month	Historical MSW generation data	Artificial neural network	R^2: 0.747; MSE: 0.0004; RMSE: 0.02	44
Daily solid waste generation	Urban population, average monthly consumption expenditure, and total retail sales	Random forest and K Nearest Neighbors	$R^2 > 0.96$ MAE: 121.5–125.0	46
Solid waste generation	Different area and time	Long Short-Term Memory (LSTM),	LSTM: R^2 (0.92); RMSE (935.08); MAPE (114.36)	60
Annual solid waste generation	Different regions, GDP, permanent resident population, per capita disposable income, and general public budget expenditure	Artificial neural network	R^2: 0.916, RMSE: 59.3	34
Annual solid waste generation	Socio-economic and demographic parameters	Decision trees and neural networks	prediction error: 16–23%	29

TABLE 1.1 (Continued)

Research object	Input parameters	Model	Performance evaluation	Reference
Solid waste generation	Gross domestic product, urban population, and energy consumption	Grey (1, 1), linear regression, and artificial neural network	MAPE: 0.0143; RMSE: 450.84; MAE: 228.53; R^2: 0.931	53
Annual solid waste generation	Population; area; degree of urbanization; purchase power index; deprivation index and other 8 variables	Artificial neural network	R^2:0.73; Error: 10^{-7}–10^{-3}	35

Researchers have implemented artificial neural network modeling and prediction in waste generation in terms of daily,[24] weekly,[25] monthly,[26] quarterly,[27] and annual[27] data. For example, Ali Abdoli et al.[26] constructed an artificial neural network model to predict monthly solid waste generation by using population, highest temperature, and household income as input parameters and achieved model performance with an R^2 of 0.74. These studies have verified that artificial neural network modeling is appropriate for waste generation prediction on all short-, medium-, and long-term time scales with satisfying results. Azadi and Karimi-Jashni[28] used feed forward artificial neural network models with monthly temperature and historical generation rates to predict seasonal waste generation for 20 cities in Iran, and noted that artificial neural network models generally out-perform multiple linear regression models in the projection of mean seasonal municipal waste generation rates. Feed forward artificial neural network models were also utilized by Kannangara et al.[29] to predict annual regional municipal solid waste generation and diversion rates in Canada.

Given the periodic properties of solid waste generation, the temporal variations of waste amount can be predicted by relying solely on historical data, without the need to factor in socio-economic influences. For instance, Singh and Satija[30] established an artificial neural network model for predicting monthly solid waste generation with historical monthly data as input parameters and achieved an R^2 value of 0.74. Sunayana et al.[31] forecasts the monthly solid waste generation in Nagpur (India) for the year 2023 with maximum absolute error of 6.34% using non-linear autoregressive models. It had been estimated that the minimum waste generation from the year 2017 to 2023 will increase by approximately 5345 tonnes.

1.3.1.2 Sociocultural Variables

Sociocultural considerations in solid waste generation often lead to various assumptions and generalizations grounded in socially recognized norms.[32] For example, aspects related to the local population's attitude towards waste, the link between governance and political stability, or the communication gaps between the community and the solid waste management authorities must be addressed.[33] The sociocultural descriptors (namely, age, demographic conditions and education) were mainly discussed. Wu et al.[34] utilized socioeconomic indicators in developing the artificial neural network models to investigate the impact of regional differences on the solid waste generation rates. Oliveira et al.[35] also selected socioeconomic and demographic data to build artificial neural network models for packaging waste estimation, demonstrating the significantly higher explanatory power of artificial neural network than regression models. Wan et al.[36] developed a district-level predictive approach for solid waste generation using machine learning, with a case study for New York City. The forecast model showed a significant decline in accuracy from years prior to 2020, the year of the pandemic outbreak with an anomalous variation in solid waste generation. And the factors related to residents' employment status can improve the model performance, both before and after the epidemic, after identifying and analyzing four proxy variables relevant to the socio-economic effects of the pandemic.

Variables such as population age, population growth, population density per square meter, number of households per dwelling show a positive correlation to solid waste generation.[31] According to Sinha and Prabhudev,[37] younger population groups (*namely*, less than 25 years old) have a different contribution to solid waste generation when compared to older people. And this age group generates fewer solid waste than the 25–45 age group. In terms of education, it is observed that it is not regularly included as a predictor variable given the difficulties in terms of data accessibility.[38] However, from the review by Kolekar et al.[39] and studies by Keser et al.[40]

and Grazhdani,[41] it is determined that education influences solid waste generation significantly, especially in developing countries. Abdella Ahmed et al.[42] investigates the effect of different living styles on the type of generated solid waste. A deep learning time series forecasting network was used for predicting solid waste generation. The analyzed results indicated that the average solid waste was 0.42, 0.65, and 0.86 kg/person/day for poor, social, and privileged zones, respectively.[42]

1.3.1.3 Economic Variables

It is essential to analyze the influence of the economic sector and consumption patterns related to the generation process of solid waste. These consumption patterns can be described through economic growth, industrial processes, and economic activities, amongst others.[38]

Economic variables are a highly relevant aspect in the design, planning and execution of activities related to the management of solid waste. This is particularly true for economic factors, as a country's level of economic development significantly affects solid waste generation. Generally, the greater the economic growth and consumption within a society, the higher the production of solid waste.[43]

Economic data of cites or regions has been effectively applied as the input variables of machine learning models,[44] yielding strong predictive results for waste generation. According to Kaza et al.,[1] high-income countries manage solid waste more effectively than middle- or low-income countries. This also suggests a marked distinction between urban and rural areas within low-income countries, with the former typically managing waste more efficiently. For example, in terms of solid waste generation, by 2016, high-income countries contributed 34% of total solid waste generation versus 29% for lower-middle income countries or 5% for low-income countries.

From this perspective, the most relevant variables studied in this dimension are the following: gross domestic product per capita,[45] income per capita, gross national product and economic and tax policies.[46] Nguyen et al.[46] compared six types of machine learning models to predict the generation of solid waste from selected residential areas of Vietnam. Random forest and K-Nearest Neighbors algorithms showed good predictive ability of the training data (80% of the data), with an R^2 value > 0.96 and a MAE of 121.5–125.0 for the testing data (20% of the data). The model simulation results showed that the urban population, average monthly consumption expenditure, and total retail sales were the most influential variables for solid waste generation. Yang et al.[22] found that a deep neural network model performs best among all machine learning models based on the empirical analysis of provincial panel data from 2008 to 2019 in China, and unravel the correlation between solid waste production and socioeconomic features (for example, total regional gross domestic product, population density). It was observed that an increase in the urban population and the concentration of wholesale and retail industries tend to escalate solid waste production in economically developed regions, and vice versa.

1.3.1.4 Waste Sources

Predicting solid waste generation across different sources requires a nuanced understanding of the specific factors, including economic, technological, regulatory, and social dynamics. To predict the amount of waste generated from various sources, it is essential to evaluate the unique features of each source. The composition and quantity of solid waste is influenced by lifestyle, culinary habits, and purchasing behaviors. Production of packaging, electronic trash, and other disposable items is linked to economic growth and advancements in industry.

Machine learning models have also been applied to predict the generation amount of solid waste from different sources, such as household package waste,[35] plastic waste,[47] medical waste,[48] yard waste,[25] and hotel waste.[49] For example, Kumar et al.[50] used income, education, occupation, and house type as input parameters to construct an artificial neural network model and predicted the plastic waste generation rate with an R^2 of 0.75. Relevant studies have indicated that artificial neural network modeling is also effective for predicting the generation and fractions of various waste sources. Azarmi et al.[49] forecasted the waste generation rates of the accommodation sector in North Cyprus based on an artificial neural network, which showed the highest prediction performance, specifically, MAE (1.378) and the highest R^2 value (0.998). The analyzed waste was categorized into recyclable, general waste and food residue. And the total waste generated during the lean season reached 2010.5 kg/d, in which large hotels accounted for the largest fraction (66.7%), followed by medium-sized hotels (19.4%) and guesthouses (2.6%). Interestingly, 45% of the waste was generated by British tourists, while the least waste was generated by African tourists (7.5%). The artificial neural network predicted that small and large hotels would produce 5.45 and 22.24 tonnes of waste by the year 2020, respectively.[49]

1.3.1.5 Environmental Variables

The generation and accumulation of solid waste affects not only the exposed populations but also the environment, making the latter factor significant in the prediction of waste generation. Consequently, characterizing environmental aspects such as energy consumption, along with climatic elements like temperature and precipitation, is crucial.[27, 51, 52, 53] Environmental and climatic characteristics and variations are important in forecasting solid waste generation, as certain assumptions and generalizations may vary based on seasonal weather patterns or thermal fluctuations.[27]

Regarding climatic variables, temperature is considered the most relevant variable ahead of precipitation and humidity.[25] According to Han et al.,[54] temperature has a strong correlation with solid waste generation, as it affects consumption patterns. For this reason, Abbasi et al.[27] argues that a characterization in seasonal terms is important as long as site-specific characteristics allow it. Along the same lines, precipitation and, indirectly, humidity also affect solid waste generation due to the added moisture content since solid waste is exposed to weather conditions prior to municipal collection and when transported to the final disposition sites.[38] Wenjing Lu[55] constructed a database of MSW generation and feature variables covering 130 cities across China and based on a gradient boosting machine algorithm to predict solid waste generation. The annual precipitation, population density and annual mean temperature were selected as key influencing factors with weights of 13%, 11% and 10%, respectively. The gradient boosting machine shows good performance with $R^2 = 0.939$. Model prediction on solid waste generation in Beijing and Shenzhen indicates that waste generation in Beijing will increase gradually over the next 3–5 years, while that in Shenzhen is expected to grow rapidly in the next three years.

In terms of environmental variables, energy consumption is directly associated with the utilization of fossil fuels.[56] As such, social metabolism, understood as a system open to the influx of matter and energy and the expulsion of waste, has undergone significant transformations in solid waste generation following the industrial revolution.[57] For example, Khan et al.[57] examined municipal solid waste in eight of China's eastern coastal regions. Results found that solid waste generation is increasing in Shandong, Guangdong, Zhejiang, and Fujian provinces, but declining in other eastern coastal cities, provinces, and special zones. And Cárdenas-Mamani et al.[58] carried out a study in Lima (Peru) in which energy consumption was related to daily household activities (namely, cooking, heating and air conditioning systems, amongst others), evidencing the link between economic status and energy consumption level (namely, high income districts have a higher average energy consumption).

1.3.2 Waste Collection and Transportation

Logistics and transportation of refuse are fundamental to waste management. Moreover, the logistics and transportation system for the waste is a critical hub connecting the source of the waste and its treatment.[61] However, the current waste logistics and transportation systems suffer from several shortcomings. One primary concern is the high cost, especially during collection, with transportation expenses comprising about 70–80% of total waste management cost. Additionally, staffing challenges contribute to inefficiencies, manifesting as disorganized collection schedules and inadequate vehicles.[62] Consequently, solutions based on artificial intelligence have been developed and implemented to optimize waste logistics and transportation processes.[5] This involves optimizing waste transportation and logistics from four perspectives: transportation distance, transportation cost, transportation time, and efficiency.[63]

1.3.2.1 Collection

A machine learning model was combined with a road sweeper, which is a popular machine that helps preserve the cleanliness of cities during solid waste transportation. The convolutional neural network algorithm was added to the road sweeper-operated system to save power energy. The experimental results demonstrated that the waste identification system could conserve over 80% of the electrical power consumed by conventional cleaning systems whilst also extending the lifespan of the brushes used. Furthermore, the Internet of Things (IoT) technology was extensively utilized in waste collection. To establish an automatic question answering system for waste classification, Jiang et al. applied the convolutional and recurrent neural network model to train a waste image classification and natural language processing.[64] The model achieved an accuracy rate of 95%, precision rates between 0.784 and 0.907, recall rates from 0.791 to 0.898, and F1 scores ranging from 0.787 to 0.902. Additionally, Baras et al. invented a smart recycling bin equipped with a convolutional neural network model,[65] which attained an accuracy rate of 93.4%.

1.3.2.2 Transportation Vehicle Routing

Akdaş et al.[66] introduced a method for vehicle routing using an ant colony optimization algorithm. Initially, they collected data from 110 points in a specific city in Turkey, converted the point coordinates, and imported them into a database. These points were visualized on a map to create a distance matrix. The ant colony optimization algorithm then calculated the shortest route based on this matrix. Researchers found that the 10th iteration of the ant colony optimization algorithm can reduce the transportation distance by 13% to reach the optimal solution. In addition, other studies have shown that using Dijkstra and Tabu search algorithms can also reduce the distance of waste transportation.[67] The Dijkstra algorithm first calculated the shortest path between two points, while the Tabu search algorithm determined the quickest route. In a subsequent experiment involving 200 waste collection points in a specific area of Turkey, the total transportation distance was 16,106 meters. After optimization for 80 waste collection points, the transportation distance was reduced to 5,497 meters. Rızvanoğlu et al.[67] confirmed that Dijkstra and Tabu search algorithms could reduce transportation distance by 28%. Overall, the application of the ant colony optimization algorithm led to an average reduction of 13% in waste transport distances. Furthermore, the combined use of the Dijkstra-Tabu search algorithm achieved a 28% decrease in these distances.

Amal et al.[68] introduced genetic algorithms for optimizing vehicle routing. Initially, they employed a geographic information system to determine the vehicle's route, which was then refined using a genetic algorithm. The optimal solution was visualized using ArcGIS and Python scripts. In a subsequent experiment conducted in a Tunisian city, ten iterations of the genetic algorithm decreased the running time from 15.2 hours to 10.91 hours, amounting to a 28.22% reduction in vehicle running time.[68] In addition, parallel annealing algorithms are also used to optimize vehicle collection paths.[69] Using the parallel annealing algorithm to optimize the waste collection path in Xuanwu District, Beijing, Zhang et al.[69] it was found that the optimized scheme of the parallel annealing algorithm can reduce the time by 12% compared with the original scheme. Nonetheless, the research has certain limitations, such as assuming a constant vehicle speed, which is impractical. In conclusion, the genetic and parallel annealing algorithms have demonstrated the capability to reduce transportation times by 28.22% and 12%, respectively, although the latter's optimization is constrained by the assumption of a fixed vehicle speed.

1.3.2.3 Transportation Route Optimization

In terms of transportation route optimization, a geographic information system integrated with an artificial neural network model, was applied to optimize the waste collection route according to the volumes of recyclables and garbage in various scenarios. Vu et al.[25] showed that compared with single and dual-compartment trucks, the dual-compartment trucks can save 10.3–16.0% in travel distance and slightly reduce emissions. Purkayastha et al.[70] also used the artificial neural network model to assess the allocation of garbage collection bins. It could help countries locating collecting bins and enhancing the collection efficiency of garbage. Ferrão et al.[71] simulated the optimal routes for efficient MSW transportation based on the constructive genetic algorithm and tabu search and applied in a municipality located at Southern Brazil, with circa 40000 inhabitants. The implementation achieved a remarkable 25.44% reduction in daily mileage of the vehicles, resulting in savings of 150.80 km/month and 1,809.60 km/ year. Additionally, it reduced greenhouse gas emissions (including fossil CO_2, CH_4, N_2O, total CO_2e and biogenic CO_2) by an average of 26.15%. Moreover, it saved 39 min of daily working time.

Tirkolaee et al.[72] presented the simulated annealing algorithm for generating initial values based on a random algorithm. The simulated annealing algorithm is used for optimization based on obtaining the initial value. An area in Iran with 330 square kilometers and 43 recycling nodes is optimized using the simulated annealing algorithm.[72] The simulated annealing algorithm reduced the total cost by 13.3%. Akhtar et al.[73] proposed an improved backtracking search algorithm on capable vehicle routing problems modeled under smart bins, using the backtracking search algorithm to optimize based on the original route. At the same time, the data provided by the smart garbage bin is used to find the optimal range to reduce the number of garbage bins, thereby minimizing the distance. After four days of simulation experiments, Akhtar et al.[73] found that the efficacy of waste collection increased by 36.78%. Algorithms and models can be developed further if more constraints and uncertainties are considered. Furthermore, Nowakowski et al.[74] proposed a harmony search algorithm for optimizing vehicle collection routes. After applying the harmony search algorithm to optimize vehicle routes for collecting electronic waste in a region of Poland, Nowakowski et al.[74] found that the harmony search algorithm could increase the number of collection points visited by 5.4%. In summary, the backtracking search algorithm can increase efficiency by 36.78%, and the harmony search algorithm can increase the number of access points by 5.4%.

1.3.3 Waste Sorting

It is worth noted that the classifying or sorting of MSW is a critical step toward its utilization and determining its potential harm.[75, 76] Solid waste can generally be classified into two categories: organic and inorganic categories. Specifically, organic waste includes food scraps and yard debris, while inorganic waste comprises plastics, metals, cans, paper, glass, and more. The waste composition differs from country to country, influenced by factors such as level of economic development, cultural norms, geographical location, energy sources, and climate. As countries urbanize and their populations grow wealthier, the proportion of inorganic waste like plastics, paper, and aluminum tends to increase, while the relative amount of organic waste decreases.[77] On the other hand, in low- and middle-income countries, organic waste often constitutes a significant portion of the urban waste stream, ranging from 40 to 85% of the total. Meanwhile, the shares of paper, plastic, glass, and metal tend to rise in the waste streams of middle- and high-income countries.[78]

Most of the current studies for solid waste classification are based on imaging techniques (Table 1.2). Abdulkareem et al[79] presented a two-layer intelligent decision system for waste sorting based on fused features of deep learning models as well as a selection of an optimal deep waste-sorting model based on multi-criteria decision making. This system can assist urban decision-makers in prioritizing and selecting an AI-optimized sorting model. To swiftly and automatically generate datasets for stacked construction waste, Ji et al[80] built an acquisition and detection platform to automatically collect different groups of RGB-D images for instances labeling. Two automatic annotation methods for real stacked construction solid waste datasets based on semi-supervised self-training and RGB-D fusion edge detection were proposed, and datasets under real-world conditions yield better models training results.[80] Huang et al[81] propose the ResMsCapsule network, which is a trash picture categorization model based on the capsule network. By combining the residual network and multi-scale module, the ResMsCapsule network can improve the performance of the basic capsule network greatly. Extensive experiments using the publicly available dataset TrashNet show that the ResMsCapsule method has a simpler network structure and higher garbage classification accuracy.

TABLE 1.2 Summary of some convolutional neural network-based models used in classification of solid wastes

Model	Classification	Data size	Accuracy (%)	Advantage	Disadvantage	References
GoogLeNet, Inception v3, Xception, ResNet50	Metal, glass, cardboard, and general waste	1451 samples	98	Introducing a two-layer intelligent decision system for waste sorting based on fused features of deep learning models as well as a selection of optimal deep waste-sorting models	Lack of practical use cases	79
ResNet101 and feature pyramid network	Bricks, concrete, wood, and rubber	298 samples	97.74	Enabling the quick creation of new datasets to adapt to changing working conditions	Limit amount of data	80
Convolutional neural network	Pure metal, pure glass, metal and mixed waste, glass and mixed waste, glass and metal	2059 samples	98	Identifying the occurrence of glass and metal in consumer trash bags with high accuracy	Lack of test in realistic settings	83
Convolutional neural network (VGGNet, DenseNet and NASNetLarge)	Paper, glass, metal, plastic, textile, and organic waste	5000 images	96.5 and 94	Combining different candidate classifiers to improve accuracy	Limit amount of data	84
Convolutional neural network (Improved Dense Net)	Cardboard, glass, metal, paper, plastic, trash	10108 images	99.6	Optimizing fully connected layer of convolutional neural networks	Lack of comparison experiments	85
Convolutional neural network (Improved ResNet)	1. Paper, glass, metal, plastic, textile, and organic waste; 2. Organic, inorganic, medical waste	5904 images	94 and 98	Comparing the classification accuracy of multiple models	Limit amount of data	86

Autoencoder + convolutional neural network (AlexNet, GoogLeNet and ResNet) + Ridge Regression + Support Vector Machine	Organic or recyclable waste	25077 waste images	99.95	High accuracy	High complexity	87
Convolutional neural network (AlexNet)+ Multilayer Perceptron	Recyclable or the others	Manually collect (5000 images)	98.2 (fixed orientation) 91.6 (random orientations)	Higher accuracy than using convolutional neural networks alone	Lack of comparison experiments	88

The classification accuracy of the ResMsCapsule network is 91.41%, and the number of parameters is only 40% of that of ResNet18, which is better than other image classification algorithms.

In addition, there are also studies that classify based on other pathways and devices, such as entropy weighting and detectors. Xi et al[82] combined the analytic hierarchy process and ANN models to assess the solid waste separation capability based on 18 selected indicators of solid waste separation in 15 cities in China. The entropy weight method was used in the analytic hierarchy process to optimize and determine the indicators and then evaluate their weights, which showed that the general public budget expenditure had the highest weight (0.5239). This implied that the solid waste separation capability could be mainly influenced by government financial support. The ANN based on scan optimization and machine learning methods was established ($R^2 = 0.9992$) to predict the missing indicators. Using trash bags containing various amounts of glass and metal, in addition to common waste found in households, Funch et al[83] used a combination of sound recording and a beat-frequency oscillation metal detector as inputs to a machine learning algorithm to identify the occurrence of glass and metal in trash bags. A custom-built test rig was developed to mimic a real waste collection truck, which was used to test different sensors and build the datasets. Convolutional neural networks were trained for the classification task, achieving accuracies of up to 98%.

1.3.4 Waste Characterization

Efficient collection, treatment and disposal of solid waste are dependent on accurate prediction of waste characteristics, which are largely affected by various technical, socioeconomic, legal, environmental, political, and cultural elements. Due to the interdependence between these elements as well as insufficient data and associated uncertainties, unconventional modelling techniques were expected to account for these factors. The majority of the studies that explored the applications of AI in waste management analyzed the prediction of solid waste characteristics (Figure 1.4).

As the most effective approach for reducing the mass and volume of solid waste (especially bulky waste), incineration could also help with energy recovery.[89] The biogenic share of thermally utilized solid waste is usually classified as renewable energy and thus excluded from national CO_2 inventories. Lan et al[90] developed a novel approach to directly classify the carbon

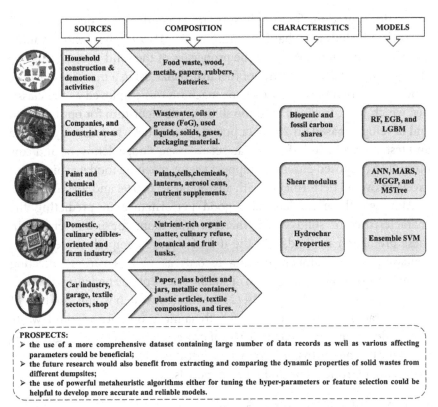

SOURCES	COMPOSITION	CHARACTERISTICS	MODELS
Household construction & demotion activities	Food waste, wood, metals, papers, rubbers, batteries.		
Companies, and industrial areas	Wastewater, oils or grease (FoG), used liquids, solids, gases, packaging material.	Biogenic and fossil carbon shares	RF, EGB, and LGBM
Paint and chemical facilities	Paints,cells,chemieals, lanterns, aerosol cans, nutrient supplements.	Shear modulus	ANN, MARS, MGGP, and M5Tree
Domestic, culinary edibles-oriented and farm industry	Nutrient-rich organic matter, culinary refuse, botanical and fruit husks.	Hydrochar Properties	Ensemble SVM
Car industry, garage, textile sectors, shop	Paper, glass bottles and jars, metallic containers, plastic articles, textile compositions, and tires.		

PROSPECTS:
➤ the use of a more comprehensive dataset containing large number of data records as well as various affecting parameters could be beneficial;
➤ the future research would also benefit from extracting and comparing the dynamic properties of solid wastes from different dumpsites;
➤ the use of powerful metaheuristic algorithms either for tuning the hyper-parameters or feature selection could be helpful to develop more accurate and reliable models.

FIGURE 1.4   An overview of solid waste sources, composition, characteristics and predictive models.

group and predict carbon content using the hyperspectral imaging spectra of solid waste in conjunction with state-of-the-art tree-based machine learning models, including a random forest, an extreme gradient boost, and a light gradient boost machine. All of the classifiers and regressors were able to achieve an accuracy above 0.95 and an R^2 of 0.96 in the test set, respectively.[90] In addition, the dynamic properties of Municipal Solid Waste are site-specific and need to be evaluated separately in different regions. Alidoust et al[91] evaluated the shear modulus using 153 cyclic triaxial tests. A comparison of the performance of developed models indicated that the ANN model outperformed the other models. More specifically, for ANN, MARS, MGGP, and M5 Tree models, the corresponding values of R^2 were equal to 0.9897, 0.9640, 0.9617, and 0.8482 for the training dataset, while the values for the testing dataset for ANN, MARS, MGGP, and M5 Trees are 0.9812, 0.9551, 0.9574, and 0.8745. Hydrochar with premium gasoline properties is used for fuel combustion for strength. The properties of fuel hydrochar, including C char (carbon content), HHV (higher heating value), and yield, are mainly based on the properties of the solid waste. Velusamy et al[92] predicted the properties of fuel hydrochar using a machine learning model. An ensemble support vector machine was employed as the classifier, which was combined with the slime mode algorithm for optimization and developed based on 281 data points. The model was primarily trained and tested on a fusion of three datasets: sewage sludge, leftovers, and cow dung. The proposed ensemble support vector machine model achieved an excellent overall performance with an average R^2 of 0.94 and RMSE of 2.62.

Although these models have good performance in predicting the characteristics of solid waste, there is still room for improvement in these prediction methods: (1) the use of a more comprehensive dataset containing large numbers of data records as well as various affecting parameters on the damping ratio of solid waste could be beneficial towards developing more reliable models; (2) the future research would also benefit from extracting and comparing the dynamic properties of solid waste from different dumpsites; (3) the use of powerful metaheuristic algorithms either for tuning the hyper-parameters or feature selection could be helpful to develop more accurate and reliable models.

1.3.5 Waste Disposal and Recycling

Solid waste is typically disposed of and recycled via sanitary landfills, thermochemical processes (namely, incineration, gasification and pyrolysis), composting, and anaerobic digestion.[61, 93] Several machine learning algorithms have been used in these procedures to look into issues with their recycling and disposal.

1.3.5.1 Sanitary Landfill

Sanitary landfill is the most widely used disposal process due to lower investment and easy operation. By utilizing socioeconomic and meteorological data (for example, average temperature, humidity, and maximum wind speed) as well as socioeconomic data (for example, weekly earning and unemployment rates), Recurrent Neural Networks (RNNs) could anticipate the overall amount of landfill waste.[94] In terms of landfill area, the modified adaptive neural fuzzy inference system succeeded in minimizing the land requirements for solid waste disposal by up to 43%.[95] The populations of 0–14, 15–64 and 65+ age groups were identified as the most important variables, which indicates the direct relationship between people and the activities that generate solid waste.[95]

Meanwhile, sanitary landfill has obvious side-effects in generating leachate and gas. Leachate management is one of the foremost concerns related to the landfill environmental impacts.[96] Therefore, the characteristics of landfill leachate such as Chemical Oxygen Demand (COD), distribution feature, and heavy metals are important prediction targets based on machine learning algorithms.[96, 97] Variables including rainfall depth, number of days after waste deposition, thickness of top and bottom compacted clay liners, and thickness of top cover over the lysimeter could be utilized to accurately predict the leachate COD load through ANNs (R > 0.98).[96] To predict the underground distribution of leachate, a deep network for multi-view fusion was applied to invert the real resistivity distribution of the medium caused by leachate. The average RMSE of synthetic models reached 0.98.[98] An efficient integration of geoelectrical tomographic data could also enable machine learning techniques for mapping leachate dispersion.[99] Landfilling fly ash from municipal solid waste incineration is a significant difficulty due to the leaching danger from heavy metals. Heavy metal leaching potentials could be evaluated using explainable machine learning techniques such extreme gradient boosting.[100] The leaching concentrations of six heavy metals (Cr, Cd, Cu, Ni, Pb, and Zn) were determined by measuring the pH, concentrations of soluble chlorine, free calcium oxide, soluble lead, soluble calcium, and acid neutralizing capacity.[100] Furthermore, the RNN model could be employed to predict lead removal from industrial sludge leachate using cement kiln dust based on dosage, contact time, and temperature.[101]

The odors and methane caused by landfills also need to be treated from the environmental and safe perspective. Landfills will release odorous gases from their working surface, and the emission rates are crucial for odor and health risk assessment.[102] Meteorological parameters (for example, temperature, humidity, atmospheric pressure) and waste properties (for example, contents of protein, lipid, carbohydrate, ash, and moisture) could be utilized to construct ANN models for predicting the emission rate of odorous gases. The temperature, atmospheric pressure, protein and lipid contents are parameters sensitive to emission rates, and meteorological parameters have significant impacts on the modelling uncertainty.[102] On the other hand, the methane released from landfills has high potency as a renewable energy source.[103] ANN models had been built to perform a multi-objective optimization to obtain the optimal power generation conditions for landfills. The biogas inlet pressure, temperature, ignition time, and the equivalence ratio were input variables whereas the output was the produced power, specific fuel consumption and nitrogen oxides (NOx) emissions.[103] The optimized operating parameters provided by the ANN models could maximize the landfill power generation by at least 1 MW for each genset.[103] Fallah et al.[104] also built an ANN model for predicting methane generation rate based on 12 meteorological parameters including maximum, mean, and minimum daily temperature, dew point, maximum and minimum daily relative humidity, air pressure and wind speed. However, Mehrdad et al.[105] reported that the support vector machine model was superior to both the adaptive neuro-fuzzy inference system and ANN models for predicting methane generation. The support vector machine model was able to capture 90% and 82% of the variation in methane emission from landfills with and without leachate recirculation, respectively.[105]

1.3.5.2 Incineration, Gasification and Pyrolysis

Incineration has the advantages of quick treatment and effective energy recovery while reducing waste volume and bulk by 90% and 70%–80%, respectively.[106] However, hazardous gas pollutants such as NOx, toxic heavy metals, acidic gases, and organic pollutants may be produced during incineration. The application of waste incineration is significantly impacted by flue gas treatment. There are some studies that focused on the emission of pollutants during incineration, such as dioxin and carbon monoxide.[106-108]

The accurate prediction of the NOx emissions is extremely important for pollutant control in municipal solid waste incineration process. Wang et al.[107] proposed a method based on a modular neural network and adaptive ensemble stochastic configuration network for predicting NOx concentration based on 18 feature variables including primary air flow, secondary air flow, activated carbon accumulation and average furnace temperature. By integrating 10 input variables (namely, six temperature variables, inlet flow rate flue gas, increment of steam in boiler, combustion chamber draft, and flue gas oxygen content), Huang et al.[106] established a Takagi–Sugeno–Kang fuzzy neural network for predicting NOx emissions and the combustion efficiency of incineration processes. The adaptive large-scale multi-objective competitive swarm optimization algorithm could be utilized to reduce the NOx emissions (15.6%) and improve the combustion efficiency (10%).[106] Dioxin as a persistent organic pollutant produced from incineration processes needs to be considered and minimized due to its high toxicity.[108] Xia et al.[108] proposed a fuzzy tree broad learning system for online predicting the dioxin emission, whereas deep learning models including RNN and CNN trained by online dioxin emissions data also could be utilized to predict real-time dioxin emissions.[109] Temperature, air-flow, and time dimension were identified as important variables for dioxin prediction.[109] Furthermore, incineration processes will generate incin-eration fly ash which contains heavy metals constituting approximately 0.5% of the weight of fly ash.[110] The gradient boosting regression model succeeded in predicting the heavy metal immobilization rate of fly ash by geopolymers. The feature categories influencing heavy metal immobiliza-tion rate are ranked in order of importance as heavy metal properties > geopolymer raw material properties > curing conditions > alkali activator properties.[110]

Except for incineration, gasification and pyrolysis are cost-effective ways to recover energy from solid waste by producing a wider range of products such as liquid transport fuels, bulk chemicals, and biochar.[111] The machine learning methods have been mainly applied to predict the thermogravimetric characteristic[112] and syngas characteristics[113] during gasification and pyrolysis processes. For example, the ANN model was developed for generating differential thermogravimetric analysis curves for hemicellulose, cellulose, and lignin in biomass.[112] Moreover, the ANN model showed better performance than the stepwise linear regression model for syngas characteristics such as low heating value, dry gas ratio, volume fraction of H_2 and CO, with $R^2 = 0.807 - 0.939$.[113] Results of import-ance analysis showed that flow rates of the work gas – N_2, feedstock type, flow rates of the work gas – steam, and input power are the most critical parameters for low heating value, gas yield, and volume fraction of CH_4 and H_2, respectively. Input power and specific energy requirements are the most important factors affecting volume fractions of H_2 (25.7–57.3 vol%).[113]

1.3.5.3 Composting
Composting could be considered a cost-effective option for solid waste disposal due to their high level of organic matter. After composting, solid waste becomes hygienic and odorless humus, realizing the key aspects of harmlessness, waste reduction, and recycling. Machine learning

algorithms can help model the complex processes that occur during composting, such as maturity prediction, parameters control, and optimization.[7] For instance, Li et al.[114] established Extra Trees models for predicting seed germination index ($R^2 = 0.928$) and T value ($R^2 = 0.957$) which indicates green waste compost maturity. Results of Shapley Additive exPlanations (SHAP) analysis showed that duration had a significant positive effect on compost decay and was the most influential factor for both seed germination index and T value in the context of garden waste compost decay. Similarly, Wan et al.[115] applied a fusion model with the highest R^2 of 0.977 and 0.986 for the multi-task prediction of seed germination index and C/N ratio. Singh and Uppaluri[116] utilized several machine learning models to simulate compost generation rate as a function of climatic parameters and organic waste content. The gradient boosting model had better prediction performance ($R^2 = 0.99$) than the k-nearest neighbor, the random forest and the autoregressive integrated moving average model.[116]

To optimize starting conditions for composting, Ding et al.[117] utilized stacking models and SHAP to predict composting maturity and identify key parameters. They found that optimal starting conditions should be maintained in the mesophilic state (temperature: 30–45°C, moisture content: 55–65%, pH: 6.3–8.0), and nutrients (total nitrogen >2.3%, total organic carbon >35%) and should be adjusted in the thermophilic state. Interestingly, Ding et al.[118] further employed machine learning models to explore key microbial genera and to optimize composting systems. The SHAP method identified *Bacillus, Acinetobacter, Thermobacillus, Pseudomonas, Psychrobacter,* and *Thermobifida* as prominent microbial genera. By preparing microbial agents to target the identified key genera, the composting quality score was 76.06 for the treatment and 70.96 for the control.[118] Similar to fly ash from incineration, composting also has potential risk of heavy metals. Therefore, heavy metal immobilization during composting needs to be predicted and optimized. Guo et al.[119] found that gradient boosting regression showed the best performance for predicting both heavy metal bioavailability variations and immobilization. Feature importance analysis revealed that the heavy metal initial bioavailability factor, total phosphorus, and composting duration were the determinant factors for heavy metal bioavailability variations (together contributing >75%). After genetic algorithm optimization, the maximum immobilization rates of Cu, Zn, Cd, As, and Cr were 79.53, 31.30, 14.91, 46.25, and 66.27%, respectively, superior to over 90% of the measured data.[119]

1.3.5.4 Anaerobic Digestion

Anaerobic digestion is also a recycling technology for solid waste containing high levels of biodegradable organic fractions. Anaerobic digestion uses microorganisms to degrade organic matter under anaerobic conditions, which can produce highly energetic methane and valuable volatile fatty acids.[120] Machine learning models mainly focused on predicting the production of biogas, methane, and lipase enzymes, as well as the concentration of volatile fatty acids. The most important application of machine learning models for anaerobic digestion is to understand the interplay between the input and output variables for a wide range of parametric scenarios, thus enabling process optimization. For example, Zhan and Zhu[121] optimized the methane yield by ANN coupling with a genetic algorithm. The optimal conditions of C/N ration (24.46), total solid (5.03%) and biochar (8.73%) showing an improvement of 20.6%.[121] In contrast, Zhan et al.[122] found that response surface methodology had better performance than ANN models in predicting methane content and daily methane yield. The optimal conditions for maximum daily methane yield obtained by response surface methodology were a C/N ratio of 22.73, total solids of 2.27%, and hydraulic retention time of 11.45 days, under which the validation trials showed methane content of $58.37 \pm 0.25\%$ and daily methane yield of 184.36 ± 0.51 mL CH_4/g VS.[122] Gan et al.[123] utilized a random forest model to predict and optimize both biogas production as well as COD removal. Optimal conditions for the anaerobic digester, including a temperature of 41.4°C, an organic loading rate of 0.87 kg COD/m^3/day, and a dilution ratio of 1:10, led to a 14.9% improvement in biogas production and a 17.5% increment in COD removal efficiency. Feature importance analysis revealed that the dilution ratio significantly influenced biogas production, whereas organic loading rate played a dominant role in COD removal.[123]

1.3.6 Challenges

As mentioned above, deep learning has a potential to substantially increase the efficiency throughout the whole life cycle of solid waste disposal. Cases included more accurate prediction of generation, collection, transportation, classification, characterization, disposal and recycling of solid waste. Despite rapid progress in this area of research, intelligent technologies for managing solid waste are largely still in the experimental and developmental stages. The transition of these technologies into practical applications is met with considerable challenges and necessitates concerted efforts for further advancement:

1. A significant obstacle to implementing intelligent technologies in waste management is the scarcity of data. Large datasets are fundamental to the development of such technologies.[124] Intelligent technology is inherently data-driven, requiring comprehensive information to function effectively. Data of numerical, textual, media, and other types is required to be handled by intelligent technology. In addition, the absence of high-quality public datasets oriented to the classification and management of solid waste makes it difficult to allow meaningful performance evaluation based on a unified standard. This would distract research efforts from the core line of the continuous improvement of intelligent treatment of solid waste. Moreover, to obtain and transfer these data, investment should be devoted not only to industrial equipment for solid waste, but also in cutting-edge Internet technology infrastructure, such as wireless sensor networks and IoT devices.

2. Intelligent technologies based on machine learning algorithms are often criticized for their "black box" properties. The internal workings of data-driven models are opaque, and estimating the relative importance of each input variable can be challenging, leading to uncertainties in the practical application of intelligent technologies for solid waste management. Consequently, their lack of transparency poses significant hurdles to broader adoption. Notably, several interpretable methods such as gradient-weighted class activation mapping and SHapley Additive exPlanations have been developed to clarify the mechanisms inside the intelligent technology. These methods can make the "black box" become more transparent and interpretable.

3. The attempts to integrate intelligent technology for solid waste are hampered by the multiplicity of different machine learning algorithms and their quick transformation.[125] While numerous studies have applied various machine learning models with success in this domain, the absence of comparative analyses makes it challenging to offer clear guidance for further research or practical applications. Therefore,

more comprehensive and detailed model evaluation work needs to be conducted. Furthermore, there appears to be no continuous line of forward-looking research in specific fields. In other words, despite the handful of studies, the actual progress may not be as substantial as one would expect.[126]

4. A significant barrier to the deployment of intelligent systems in waste management is the scarcity of workers skilled in this area. Intelligent technology needs relevant skills such as programming, data analysis, probability, and statistics. Currently, industries generate a huge amount of waste that needs to be tackled smartly by hiring workers specializing in intelligent technology or training existing workers.[5] Arranging training and workshops for the workers in the waste management industry will help to enhance their skills to effectively employ various intelligent techniques for efficient management of solid waste.

1.4 CONCLUSION

In this chapter, we reviewed the state-of-the-art intelligent technology applied in the classification and management of MSW, exploring various aspects such as waste generation, collection and transportation, sorting, characterization, disposal and recycling. The exploration of machine learning modelling and its applications to solid waste has provided valuable insights into the potential of sustainable and intelligent management of solid waste. Throughout this chapter, the artificial neural network and its variants (for example, convolutional neural network and recurrent neural network) has been demonstrated to be the most powerful approach for simulation and classification tasks in all fields related to the solid waste. Following model training, validation and testing using both experimental data and practical data, machine learning-based intelligent technologies have emerged as robust tools for establishing intelligent classification and management systems for solid waste. Future research in the field should concentrate on enhancing modeling techniques, improving practical data collection, and developing IoT technology. Additionally, a deeper comprehension of solid waste characteristics and their behavior during disposal and recycling could lead to more precise models. Due to the promising results from existing machine learning classification and dynamical simulation, it is essential for modelers to remain updated with the rapid advancements in these fields. Integrated computational biology and artificial intelligence will further drive significant enhancements in the intelligent classification and management of MSW.

REFERENCES

1. Kaza, S.; Yao, L.; Bhada-Tata, P.; Van Woerden, F. What a waste 2.0: A global snapshot of solid waste management to 2050. World Bank Publications: 2018.

2. Khawer, M. U. B.; Naqvi, S. R.; Ali, I.; Arshad, M.; Juchelková, D.; Anjum, M. W.; Naqvi, M. Anaerobic digestion of sewage sludge for biogas & biohydrogen production: State-of-the-art trends and prospects. *Fuel* 2022, 329, 125416.

3. Kang, Y.; Yang, Q.; Wang, L.; Chen, Y.; Lin, G.; Huang, J.; Yang, H.; Chen, H. China's changing city-level greenhouse gas emissions from municipal solid waste treatment and driving factors. *Resour. Conserv. Recy.* 2022, 180, 106168.

4. Halkos, G.; Petrou, K. N. The relationship between MSW and education: WKC evidence from 25 OECD countries. *Waste Manage.* 2020, 114, 240–252.

5. Abdallah, M.; Abu Talib, M.; Feroz, S.; Nasir, Q.; Abdalla, H.; Mahfood, B. Artificial intelligence applications in solid waste management: A systematic research review. *Waste Manage.* 2020, 109, 231–246.

6. Asnicar, F.; Thomas, A. M.; Passerini, A.; Waldron, L.; Segata, N. Machine learning for microbiologists. *Nat. Rev. Microbiol.* 2023, 22, 191–205.

7. Huang, L. T.; Hou, J. Y.; Liu, H. T. Machine-learning intervention progress in the field of organic waste composting: Simulation, prediction, optimization, and challenges. *Waste Manage.* 2024, 178, 155–167.

8. Khan, M.; Chuenchart, W.; Surendra, K. C.; Kumar Khanal, S. Applications of artificial intelligence in anaerobic co-digestion: Recent advances and prospects. *Bioresour. Technol.* 2023, 370, 128501.

9. Sun, L.; Li, M.; Liu, B.; Li, R.; Deng, H.; Zhu, X.; Zhu, X.; Tsang, D. C. W. Machine learning for municipal sludge recycling by thermochemical conversion towards sustainability. *Bioresour. Technol.* 2024, 394, 130254.

10. Del Rio-Chanona, E. A.; Cong, X.; Bradford, E.; Zhang, D.; Jing, K. Review of advanced physical and data-driven models for dynamic bioprocess simulation: Case study of algae-bacteria consortium wastewater treatment. *Biotechnol. Bioeng.* 2019, 116 (2), 342–353.

11. Del Rio-Chanona, E. A.; Fiorelli, F.; Zhang, D.; Ahmed, N. R.; Jing, K.; Shah, N. An efficient model construction strategy to simulate microalgal lutein photo-production dynamic process. *Biotechnol. Bioeng.* 2017, 114 (11), 2518–2527.

12. Yurtsever, M.; Yurtsever, U. Use of a convolutional neural network for the classification of microbeads in urban wastewater. *Chemosphere* 2019, 216, 271–280.

13. Goodfellow, I.; Bengio, Y.; Courville, A. *Deep Learning.* Cambridge, MA, USA: MIT Press 2016.

14. Jacob, S.; Banerjee, R. Modeling and optimization of anaerobic codigestion of potato waste and aquatic weed by response surface methodology and artificial neural network coupled genetic algorithm. *Bioresour. Technol.* 2016, 214, 386–395.

15. Karri, R. R.; Sahu, J. N. Modeling and optimization by particle swarm embedded neural network for adsorption of zinc (II) by palm kernel shell based activated carbon from aqueous environment. *J. Environ. Manage.* 2018, 206, 178–191.

16. Dehghani, M.; Seifi, A.; Riahi-Madvar, H. Novel forecasting models for immediate-short-term to long-term influent flow prediction by combining ANFIS and grey wolf optimization. *J. Hydrol.* 2019, 576, 698–725.

17. Schmidhuber, J. Deep learning in neural networks: An overview. *Neural Networks* 2015, 61, 85–117.

18. Asadi, M.; Guo, H.; McPhedran, K. Biogas production estimation using data-driven approaches for cold region municipal wastewater anaerobic digestion. *J. Environ. Manage.* 2019, 253, 109708.

19. Gidon, A.; Zolnik, T. A.; Fidzinski, P.; Bolduan, F.; Papoutsi, A.; Poirazi, P.; Holtkamp, M.; Vida, I.; Larkum, M. E. Dendritic action potentials and computation in human layer 2/3 cortical neurons. *Science* 2020, 367 (6473), 83–87.

20. Zhang, X.; Zhuo, Y.; Luo, Q.; Wu, Z.; Midya, R.; Wang, Z.; Song, W.; Wang, R.; Upadhyay, N. K.; Fang, Y.; Kiani, F.; Rao, M.; Yang, Y.; Xia, Q.; Liu, Q.; Liu, M.; Yang, J. J. An artificial spiking afferent nerve based on Mott memristors for neurorobotics. *Nat. Commun.* 2020, 11 (1), 51.

21. Abdulyekeen, K. A.; Umar, A. A.; Patah, M. F. A.; Daud, W. M. A. W. Torrefaction of biomass: Production of enhanced solid biofuel from municipal solid waste and other types of biomass. *Renew. Sustain. Energ. Rev.* 2021, 150, 111436.

22. Yang, L.; Zhao, Y.; Niu, X.; Song, Z.; Gao, Q.; Wu, J. Municipal solid waste forecasting in China based on machine learning models. *Front. Energ. Res.* 2021, 9 , 763977.

23. Xu, A.; Chang, H.; Xu, Y.; Li, R.; Li, X.; Zhao, Y. Applying artificial neural networks (ANNs) to solve solid waste-related issues: A critical review. *Waste Manage.* 2021, 124, 385–402.

24. Nabavi-Pelesaraei, A.; Bayat, R.; Hosseinzadeh-Bandbafha, H.; Afrasyabi, H.; Berrada, A. Prognostication of energy use and environmental impacts

for recycle system of municipal solid waste management. *J. Clean. Prod.* 2017, 154, 602–613.

25. Vu, H. L.; Ng, K. T. W.; Bolingbroke, D. Time-lagged effects of weekly climatic and socio-economic factors on ANN municipal yard waste prediction models. *Waste Manage.* 2019, 84, 129–140.

26. Ali Abdoli, M.; Falah Nezhad, M.; Salehi Sede, R.; Behboudian, S. Long-term forecasting of solid waste generation by the artificial neural networks. *Environ. Prog. Sustain. Energ.* 2012, 31 (4), 628–636.

27. Abbasi, M.; Rastgoo, M. N.; Nakisa, B. Monthly and seasonal modeling of municipal waste generation using radial basis function neural network. *Environ. Prog. Sustain.* Energ. 2019, 38 (3), e13033.

28. Azadi, S.; Karimi-Jashni, A. Verifying the performance of artificial neural network and multiple linear regression in predicting the mean seasonal municipal solid waste generation rate: A case study of Fars province, Iran. *Waste Manage.* 2016, 48, 14–23.

29. Kannangara, M.; Dua, R.; Ahmadi, L.; Bensebaa, F. Modeling and prediction of regional municipal solid waste generation and diversion in Canada using machine learning approaches. *Waste Manage.* 2018, 74, 3–15.

30. Singh, D.; Satija, A. Prediction of municipal solid waste generation for optimum planning and management with artificial neural network – case study: Faridabad City in Haryana State (India). *Int. J. Syst. Assur. Eng.* 2018, 9, 91–97.

31. Sunayana; Kumar, S.; Kumar, R. Forecasting of municipal solid waste generation using non-linear autoregressive (NAR) neural models. *Waste Manage.* 2021, 121, 206–214.

32. Khan, A. H.; López-Maldonado, E. A.; Khan, N. A.; Villarreal-Gómez, L. J.; Munshi, F. M.; Alsabhan, A. H.; Perveen, K. Current solid waste management strategies and energy recovery in developing countries – state of art review. *Chemosphere* 2022, 291, 133088.

33. Kundariya, N.; Mohanty, S. S.; Varjani, S.; Hao Ngo, H.; W. C. Wong, J.; Taherzadeh, M. J.; Chang, J. S.; Yong Ng, H.; Kim, S. H.; Bui, X. T. A review on integrated approaches for municipal solid waste for environmental and economical relevance: Monitoring tools, technologies, and strategic innovations. *Bioresour. Technol.* 2021, 342, 125982.

34. Wu, F.; Niu, D.; Dai, S.; Wu, B. New insights into regional differences of the predictions of municipal solid waste generation rates using artificial neural networks. *Waste Manage.* 2020, 107, 182–190.

35. Oliveira, V.; Sousa, V.; Dias-Ferreira, C. Artificial neural network modelling of the amount of separately-collected household packaging waste. *J. Clean. Prod.* 2019, 210, 401–409.

36. Wan, S.; Nik-Bakht, M.; Ng, K. T. W.; Tian, X.; An, C.; Sun, H.; Yue, R. Insights into the urban municipal solid waste generation during the COVID-19 pandemic from machine learning analysis. *Sustain. Cities Soc.* 2024, 100, 105044.

37. Sinha, R.; Prabhudev, B. Impact of socio-cultural challenges in solid waste management. *Int. J. Eng. Res. Technol.* 2016, 4 (27), 1–3.

38. Izquierdo-Horna, L.; Kahhat, R.; Vázquez-Rowe, I. Reviewing the influence of sociocultural, environmental and economic variables to forecast municipal solid waste (MSW) generation. *Sustain. Prod. Consump.* 2022, 33, 809–819.

39. Kolekar, K. A.; Hazra, T.; Chakrabarty, S. N. A review on prediction of municipal solid waste generation models. *Procedia Environ. Sci.* 2016, 35, 238–244.

40. Keser, S.; Duzgun, S.; Aksoy, A. Application of spatial and non-spatial data analysis in determination of the factors that impact municipal solid waste generation rates in Turkey. *Waste Manage.* 2012, 32 (3), 359–371.

41. Grazhdani, D. Assessing the variables affecting on the rate of solid waste generation and recycling: An empirical analysis in Prespa Park. *Waste Manage.* 2016, 48, 3–13.

42. Abdella Ahmed, A. K.; Ibraheem, A. M.; Abd-Ellah, M. K. Forecasting of municipal solid waste multi-classification by using time-series deep learning depending on the living standard. *Results Eng.* 2022, 16, 100655.

43. Namlis, K.-G.; Komilis, D. Influence of four socioeconomic indices and the impact of economic crisis on solid waste generation in Europe. *Waste Manage.* 2019, 89, 190–200.

44. Ali, S. A.; Ahmad, A. Forecasting MSW generation using artificial neural network time series model: A study from metropolitan city. *SN Appl. Sci.* 2019, 1 (11), 1338.

45. Beigl, P.; Lebersorger, S.; Salhofer, S. Modelling municipal solid waste generation: A review. *Waste Manage.* 2008, 28 (1), 200–214.

46. Nguyen, X. C.; Nguyen, T. T. H.; La, D. D.; Kumar, G.; Rene, E. R.; Nguyen, D. D.; Chang, S. W.; Chung, W. J.; Nguyen, X. H.; Nguyen, V. K. Development of machine learning-based models to forecast solid waste generation in residential areas: A case study from Vietnam. *Resour. Conserv. Recy.* 2021, 167, 105381.

47. Kumar, A.; Agrawal, A. Recent trends in solid waste management status, challenges, and potential for the future Indian cities – a review. *Curr. Res. Environ. Sustain.* 2020, 2, 100011.

48. Golbaz, S.; Nabizadeh, R.; Sajadi, H. S. Comparative study of predicting hospital solid waste generation using multiple linear regression and artificial intelligence. *J. Environ. Health Sci. Eng.* 2019, 17, 41–51.

49. Azarmi, S. L.; Oladipo, A. A.; Vaziri, R.; Alipour, H. Comparative modelling and artificial neural network inspired prediction of waste generation rates of hospitality industry: The case of North Cyprus. *Sustainability* 2018, 10 (9), 2965.

50. Kumar, A.; Samadder, S.; Kumar, N.; Singh, C. Estimation of the generation rate of different types of plastic waste and possible revenue recovery from informal recycling. *Waste Manage.* 2018, 79, 781–790.

51. Cervantes, D. E. T.; Martínez, A. L.; Hernández, M. C.; de Cortázar, A. L. G. Using indicators as a tool to evaluate municipal solid waste management: A critical review. *Waste Manage.* 2018, 80, 51–63.

52. Johnson, N. E.; Ianiuk, O.; Cazap, D.; Liu, L.; Starobin, D.; Dobler, G.; Ghandehari, M. Patterns of waste generation: A gradient boosting model for short-term waste prediction in New York City. *Waste Manage.* 2017, 62, 3–11.

53. Chhay, L.; Reyad, M. A. H.; Suy, R.; Islam, M. R.; Mian, M. M. Municipal solid waste generation in China: Influencing factor analysis and multi-model forecasting. *J. Mater. Cycles Waste.* 2018, 20, 1761–1770.

54. Han, Z.; Liu, Y.; Zhong, M.; Shi, G.; Li, Q.; Zeng, D.; Zhang, Y.; Fei, Y.; Xie, Y. Influencing factors of domestic waste characteristics in rural areas of developing countries. *Waste Manage.* 2018, 72, 45–54.

55. Lu, W.; Huo, W.; Gulina, H.; Pan, C. Development of machine learning multi-city model for municipal solid waste generation prediction. *Front. Environ. Sci. Eng.* 2022, 16 (9): 119.

56. Shah, A. V.; Srivastava, V. K.; Mohanty, S. S.; Varjani, S. Municipal solid waste as a sustainable resource for energy production: State-of-the-art review. *J. Environ. Chem. Eng.* 2021, 9 (4), 105717.

57. Khan, S.; Anjum, R.; Raza, S. T.; Bazai, N. A.; Ihtisham, M. Technologies for municipal solid waste management: Current status, challenges, and future perspectives. *Chemosphere* 2022, 288, 132403.

58. Cárdenas-Mamani, Ú.; Kahhat, R.; Vázquez-Rowe, I. District-level analysis for household-related energy consumption and greenhouse gas emissions: A case study in Lima, Peru. *Sustain. Cities Soc.* 2022, 77, 103572.

59. Ali Abdoli, M.; Falah Nezhad, M.; Salehi Sede, R.; Behboudian, S. Long-term forecasting of solid waste generation by the artificial neural networks. *Environ. Prog. Sustain.* 2012, 31 (4), 628–636.

60. Niu, D.; Wu, F.; Dai, S.; He, S.; Wu, B. Detection of long-term effect in forecasting municipal solid waste using a long short-term memory neural network. *J. Clean. Prod.* 2021, 290, 125187.

61. Xia, W.; Jiang, Y.; Chen, X.; Zhao, R. Application of machine learning algorithms in municipal solid waste management: A mini review. *Waste Manage. Res.* 2022, 40 (6), 609–624.

62. Andeobu, L.; Wibowo, S.; Grandhi, S. Artificial intelligence applications for sustainable solid waste management practices in Australia: A systematic review. *Sci. Total Environ.* 2022, 834, 155389.

63. Fang, B.; Yu, J.; Chen, Z.; Osman, A. I. I.; Farghali, M.; Ihara, I.; Hamza, E. H. H.; Rooney, D. W. W.; Yap, P.-S. Artificial intelligence for waste management in smart cities: A review. *Environ. Chem. Lett.* 2023, 21 (4), 1959–1989.

64. Jiang, C.; Su, Q.; Zhang, L.; Huang, B. Automatic question answering system based on convolutional neural network and its application to waste collection system. *J. Circuit. Syst. Comp.* 2021, 30 (1), 2150013.

65. Baras, N.; Ziouzios, D.; Dasygenis, M.; Tsanaktsidis, C.; IEEE in A cloud based smart recycling bin for waste classification, 9th International Conference on Modern Circuits and Systems Technologies (MOCAST), Electr Network, 7–9 Sept. 2020; Electr Network, 2020.

66. Akdaş, H. Ş.; Demir, Ö.; Doğan, B.; Bas, A.; Uslu, B. Ç. In Vehicle route optimization for solid waste management: A case study of Maltepe, Istanbul, 2021 13th International Conference on Electronics, Computers and Artificial Intelligence (ECAI), 1–3 July 2021; pp 1–6.

67. Rızvanoğlu, O.; Kaya, S.; Ulukavak, M.; Yesilnacar, M. I. Optimization of municipal solid waste collection and transportation routes, through linear programming and geographic information system: A case study from Sanliurfa, Turkey. *Environ. Monit. Assess.* 2020, 192 (1), 9.

68. Amal, L.; Le Hoang, S.; Chabchoub, H. SGA: Spatial GIS-based genetic algorithm for route optimization of municipal solid waste collection. *Environ. Sci. Pollut. Res.* 2018, 25 (27), 27569–27582.

69. Zhang, S.; Mu, D.; Wang, C. A solution for the full-load collection vehicle routing problem with multiple trips and demands: An application in Beijing. *IEEE Access* 2020, 8, 89381–89394.

70. Purkayastha, D.; Majumder, M.; Chakrabarti, S. Suitability index assessment for collection bin allocation using analytical hierarchy process (AHP) cascaded to artificial neural network (ANN). *Detritus* 2020, 9, 38–49.

71. Ferrão, C. C.; Moraes, J. A. R.; Fava, L. P.; Furtado, J. C.; Machado, E.; Rodrigues, A.; Sellitto, M. A. Optimizing routes of municipal waste collection: An application algorithm. *Manage. Environ. Qual.: Int. J.* 2024, ahead-of-print.

72. Tirkolaee, E. B.; Abbasian, P.; Soltani, M.; Ghaffarian, S. A. Developing an applied algorithm for multi-trip vehicle routing problem with time windows in urban waste collection: A case study. *Waste Manage. Res.* 2019, 37, 4–13.

73. Akhtar, M.; Hannan, M. A.; Begum, R. A.; Basri, H.; Scavino, E. Backtracking search algorithm in CVRP models for efficient solid waste collection and route optimization. *Waste Manage.* 2017, 61, 117–128.

74. Nowakowski, P.; Szwarc, K.; Boryczka, U. Combining an artificial intelligence algorithm and a novel vehicle for sustainable e-waste collection. *Sci. Total Environ.* 2020, 730, 138726.

75. Wei, J.; Li, H.; Liu, J. Curbing dioxin emissions from municipal solid waste incineration: China's action and global share. *J. Hazard. Mater.* 2022, 435, 129076.

76. Liang, Y.; Song, Q.; Wu, N.; Li, J.; Zhong, Y.; Zeng, W. Repercussions of COVID-19 pandemic on solid waste generation and management strategies. *Front. Environ. Sci. Eng.* 2021, 15 (6), 115.

77. Kamaruddin, M. A.; Yusoff, M. S.; Rui, L. M.; Isa, A. M.; Zawawi, M. H.; Alrozi, R. An overview of municipal solid waste management and landfill leachate treatment: Malaysia and Asian perspectives. *Environ. Sci. Pollut. Res.* 2017, 24 (35), 26988–27020.

78. Jammeli, H.; Ksantini, R.; Ben Abdelaziz, F.; Masri, H. Sequential artificial intelligence models to forecast urban solid waste in the city of Sousse, Tunisia. *IEEE Trans. Eng. Manage.* 2023, 70 (5), 1912–1922.

79. Abdulkareem, K. H.; Subhi, M. A.; Mohammed, M. A.; Aljibawi, M.; Nedoma, J.; Martinek, R.; Deveci, M.; Shang, W.-L.; Pedrycz, W. A manifold intelligent decision system for fusion and benchmarking of deep waste-sorting models. *Eng. Appl. Artif. Intel.* 2024, 132, 107926.

80. Ji, T.; Li, J.; Fang, H.; Zhang, R.; Yang, J.; Fan, L. Rapid dataset generation methods for stacked construction solid waste based on machine vision and deep learning. *PLoS One* 2024, 19 (1), e0296666.

81. Huang, L.; Li, M.; Xu, T.; Dong, S. Q. A waste classification method based on a capsule network. *Environ. Sci. Pollut. Res.* 2023, 30 (36), 86454–86462.

82. Xi, H.; Li, Z.; Han, J.; Shen, D.; Li, N.; Long, Y.; Chen, Z.; Xu, L.; Zhang, X.; Niu, D.; Liu, H. Evaluating the capability of municipal solid waste separation in China based on AHP-EWM and BP neural network. *Waste Manage.* 2022, 139, 208–216.

83. Funch, O. I.; Marhaug, R.; Kohtala, S.; Steinert, M. Detecting glass and metal in consumer trash bags during waste collection using convolutional neural networks. *Waste Manage.* 2021, 119, 30–38.

84. Huang, G. L.; He, J.; Xu, Z.; Huang, G. A combination model based on transfer learning for waste classification. *Concurr. Comp.-Pract. Exp.* 2020, 32 (19), e5751.

85. Mao, W.-L.; Chen, W.-C.; Wang, C.-T.; Lin, Y.-H. Recycling waste classification using optimized convolutional neural network. *Resour. Conserv. Recy.* 2021, 164, 105132.

86. Vo, A. H.; Hoang Son, L.; Vo, M. T.; Le, T. A novel framework for trash classification using deep transfer learning. *IEEE Access* 2019, 7, 178631–178639.

87. Toğaçar, M.; Ergen, B.; Cömert, Z. Waste classification using AutoEncoder network with integrated feature selection method in convolutional neural network models. *Measurement* 2020, 153, 107459.

88. Chu, Y.; Huang, C.; Xie, X.; Tan, B.; Kamal, S.; Xiong, X. Multilayer hybrid deep-learning method for waste classification and recycling. *Comput. Intel. Neurosci.* 2018, 2018, 5060857.

89. Zhou, Q.; Yang, J.; Liu, M.; Liu, Y.; Sarnat, S.; Bi, J. Toxicological risk by inhalation exposure of air pollution emitted from China's municipal solid waste incineration. *Environ. Sci. Technol.* 2018, 52 (20), 11490–11499.

90. Lan, D.-Y.; He, P.-J.; Qi, Y.-P.; Wu, T.-W.; Xian, H.-Y.; Wang, R.-H.; Lü, F.; Zhang, H.; Long, J.-S. Machine learning and hyperspectral imaging-aided forecast for the share of biogenic and fossil carbon in solid waste. *ACS Sustain. Chem. Eng.* 2023, 11 (10), 4020–4029.

91. Alidoust, P.; Keramati, M.; Hamidian, P.; Amlashi, A. T.; Gharehveran, M. M.; Behnood, A. Prediction of the shear modulus of municipal solid waste (MSW): An application of machine learning techniques. *J. Clean. Prod.* 2021, 303, 127053.

92. Velusamy, P.; Srinivasan, J.; Subramanian, N.; Mahendran, R. K.; Saleem, M. Q.; Ahmad, M.; Shafiq, M.; Choi, J.-G. Optimization-driven machine learning approach for the prediction of hydrochar properties from municipal solid waste. *Sustainability* 2023, 15 (7), 6088.

93. Guo, H.-N.; Wu, S.-B.; Tian, Y.-J.; Zhang, J.; Liu, H.-T. Application of machine learning methods for the prediction of organic solid waste treatment and recycling processes: A review. *Bioresour. Technol.* 2021, 319, 124114.

94. Vu, H. L.; Ng, K. T. W.; Richter, A.; An, C. Analysis of input set characteristics and variances on k-fold cross validation for a Recurrent Neural Network model on waste disposal rate estimation. *J. Environ. Manage.* 2022, 311, 114869.

95. Younes, M. K.; Nopiah, Z. M.; Basri, N. E.; Basri, H.; Abushammala, M. F.; Younes, M. Y. Landfill area estimation based on integrated waste disposal options and solid waste forecasting using modified ANFIS model. *Waste Manage.* 2016, 55, 3–11.

96. Azadi, S.; Amiri, H.; Rakhshandehroo, G. R. Evaluating the ability of artificial neural network and PCA-M5P models in predicting leachate COD load in landfills. *Waste Manage.* 2016, 55, 220–230.

97. Bagheri, M.; Bazvand, A.; Ehteshami, M. Application of artificial intelligence for the management of landfill leachate penetration into groundwater, and assessment of its environmental impacts. *J. Clean. Prod.* 2017, 149, 784–796.

98. Sun, X.; Qian, X.; Nai, C.; Xu, Y.; Liu, Y.; Yao, G.; Dong, L. LDI-MVFNet: A multi-view fusion deep network for leachate distribution imaging. *Waste Manage.* 2023, 157, 180–189.

99. Piegari, E.; De Donno, G.; Melegari, D.; Paoletti, V. A machine learning-based approach for mapping leachate contamination using geoelectrical methods. *Waste Manage.* 2023, 157, 121–129.

100. Liu, Z.; Lu, M.; Zhang, Y.; Zhou, J.; Wang, J. Identification of heavy metal leaching patterns in municipal solid waste incineration fly ash based on an explainable machine learning approach. *J. Environ. Manage.* 2022, 317, 115387.

101. Coruh, S.; Gurkan, E. H.; Kilic, E. Modelling of lead removal from battery industrial wastewater treatment sludge leachate on cement kiln dust by using Elman's RNN. *Int. J. Global Warm.* 2017, 13 (1), 92–102.

102. Xu, A.; Li, R.; Chang, H.; Xu, Y.; Li, X.; Lin, G.; Zhao, Y. Artificial neural network (ANN) modeling for the prediction of odor emission rates from landfill working surface. *Waste Manage.* 2022, 138, 158–171.

103. Alrbai, M.; Abubaker, A. M.; Darwish Ahmad, A.; Al-Dahidi, S.; Ayadi, O.; Hjouj, D.; Al-Ghussain, L. Optimization of energy production from biogas fuel in a closed landfill using artificial neural networks: A case study of Al Ghabawi Landfill, Jordan. *Waste Manage.* 2022, 150, 218–226.

104. Fallah, B.; Ng, K. T. W.; Vu, H. L.; Torabi, F. Application of a multi-stage neural network approach for time-series landfill gas modeling with missing data imputation. *Waste Manage.* 2020, 116, 66–78.

105. Mehrdad, S. M.; Abbasi, M.; Yeganeh, B.; Kamalan, H. Prediction of methane emission from landfills using machine learning models. *Environ. Prog. Sustain.* 2021, 40 (4), e13629.

106. Huang, W.; Ding, H.; Qiao, J. Large-scale and knowledge-based dynamic multiobjective optimization for MSWI process using adaptive competitive swarm optimization. *IEEE Trans. Syst. Man Cy.: Syst.* 2024, 54 (1), 379–390.

107. Wang, R.; Li, F.; Yan, A. Modular stochastic configuration network based prediction model for NOx emissions in municipal solid waste incineration process. *Eng. Appl. Artif. Intel.* 2024, 127, 107315.

108. Xia, H.; Tang, J.; Yu, W.; Qiao, J. Online measurement of dioxin emission in solid waste incineration using fuzzy broad learning. *IEEE Trans. Ind. Inform.* 2024, 20 (1), 358–368.

109. Wen, C.; Lin, X.; Ying, Y.; Ma, Y.; Yu, H.; Li, X.; Yan, J. Dioxin emission prediction from a full-scale municipal solid waste incinerator: Deep learning model in time-series input. *Waste Manage.* 2023, 170, 93–102.

110. Guo, L.; Xu, X.; Wang, Q.; Park, J.; Lei, H.; Zhou, L.; Wang, X. Machine learning-based prediction of heavy metal immobilization rate in the

solidification/stabilization of municipal solid waste incineration fly ash (MSWIFA) by geopolymers. *J. Hazard. Mater.* 2024, 467, 133682.

111. Ascher, S.; Watson, I.; You, S. Machine learning methods for modelling the gasification and pyrolysis of biomass and waste. *Renew. Sustain. Energ. Rev.* 2022, 155, 111902.

112. Kartal, F.; Dalbudak, Y.; Özveren, U. Prediction of thermal degradation of biopolymers in biomass under pyrolysis atmosphere by means of machine learning. *Renew. Energ.* 2023, 204, 774–787.

113. Chu, C.; Boré, A.; Liu, X. W.; Cui, J. C.; Wang, P.; Liu, X.; Chen, G. Y.; Liu, B.; Ma, W. C.; Lou, Z. Y.; Tao, Y.; Bary, A. Modeling the impact of some independent parameters on the syngas characteristics during plasma gasification of municipal solid waste using artificial neural network and stepwise linear regression methods. *Renew. Sustain. Energ. Rev.* 2022, 157, 112052.

114. Li, Y.; Xue, Z.; Li, S.; Sun, X.; Hao, D. Prediction of composting maturity and identification of critical parameters for green waste compost using machine learning. *Bioresour. Technol.* 2023, 385, 129444.

115. Wan, X.; Li, J.; Xie, L.; Wei, Z.; Wu, J.; Wah Tong, Y.; Wang, X.; He, Y.; Zhang, J. Machine learning framework for intelligent prediction of compost maturity towards automation of food waste composting system. *Bioresour. Technol.* 2022, 365, 128107.

116. Singh, T.; Uppaluri, R. V. S. Application of ANN and traditional ML algorithms in modelling compost production under different climatic conditions. *Neural Comput. Appl.* 2023, 35 (18), 13465–13484.

117. Ding, S.; Huang, W.; Xu, W.; Wu, Y.; Zhao, Y.; Fang, P.; Hu, B.; Lou, L. Improving kitchen waste composting maturity by optimizing the processing parameters based on machine learning model. *Bioresour. Technol.* 2022, 360, 127606.

118. Ding, S.; Jiang, L.; Hu, J.; Huang, W.; Lou, L. Microbiome data analysis via machine learning models: Exploring vital players to optimize kitchen waste composting system. *Bioresour. Technol.* 2023, 388, 129731.

119. Guo, H. N.; Liu, H. T.; Wu, S. Simulation, prediction and optimization of typical heavy metals immobilization in swine manure composting by using machine learning models and genetic algorithm. *J. Environ. Manage.* 2022, 323, 116266.

120. Xu, R.; Fang, S.; Zhang, L.; Cheng, X.; Huang, W.; Wang, F.; Fang, F.; Cao, J.; Wang, D.; Luo, J. Revealing the intrinsic drawbacks of waste activated sludge for efficient anaerobic digestion and the potential mitigation strategies. *Bioresour. Technol.* 2021, 345, 126482.

121. Zhan, Y.; Zhu, J. Response surface methodology and artificial neural network-genetic algorithm for modeling and optimization of bioenergy

production from biochar-improved anaerobic digestion. *Appl. Energ.* 2024, 355, 122336.

122. Zhan, Y.; Zhu, J.; Schrader, L. C.; Wang, D. Modeling and optimization of bioenergy production from co-digestion of poultry litter with wheat straw in anaerobic sequencing batch reactor: Response surface methodology and artificial neural network. *Appl. Energ.* 2023, 345, 121373.

123. Gan, E. Y. T.; Chan, Y. J.; Wan, Y. K.; Tiong, T. J.; Chong, W. C.; Lim, J. W. Examining the synergistic effects through machine learning prediction and optimisation in the anaerobic co-digestion (ACoD) of palm oil mill effluent (POME) and decanter cake (DC) with economic analysis. *J. Clean. Prod.* 2024, 437, 140666.

124. Nnaji, C.; Karakhan, A. A. Technologies for safety and health management in construction: Current use, implementation benefits and limitations, and adoption barriers. *J. Build. Eng.* 2020, 29, 101212.

125. Yigitcanlar, T.; Cugurullo, F. The sustainability of artificial intelligence: An urbanistic viewpoint from the lens of smart and sustainable cities. *Sustainability* 2020, 12 (20), 8548.

126. Gangsar, P.; Tiwari, R. Signal based condition monitoring techniques for fault detection and diagnosis of induction motors: A state-of-the-art review. *Mech. Syst. Signal Process.* 2020, 144, 106908.

Preparation Methods of Solid Waste-Based Materials

Qingbai Tian, Jizhong Meng, and Xing Xu

2.1 INTRODUCTION

With the remarkable development of urbanization and industrialization all over the world, environmental problems have risen to prominence as an increasingly urgent issue. For example, various pollutants – including phosphorus, nitrogen, heavy metal ions or related compounds, along with trace organic matter such as pharmaceutical residues, antibiotics, and microplastics – are contaminating water and soil environments, leading to significant environmental and human health ramifications. In response, several effective water/wastewater or soil remediation methods have been developed, including adsorption/ion exchange, bioremediation, Advanced Oxidation Processes (AOPs), and carrier/membrane treatments. These highly efficient treatments require some specialized environmentally friendly functional materials. Recent studies have widely used a series of materials, such as activated carbon, graphene, metal oxides, carbon nanotubes, Metal-Organic Framework (MOFs) and other materials in environmental remediation. Yet, these functional materials pose challenges due to their costly and complex production methods.

With the rapid developments of industrial and agricultural production, the production of various solid waste, including agricultural waste,

DOI: 10.1201/9781003535409-2

industrial waste, domestic waste, and so forth, have increased greatly in recent decades.[1, 2] In particular, human demand has led to a dramatic expansion of agroforestry and has facilitated a rapid increase in all types of agricultural and forestry waste. For example, the global production of agricultural waste reached 20.3 billion tonnes in 2019.[3] In addition, the full utilization of these biomass resources has gained increasing attention. Improper management not only constitutes a significant waste of resources but also exacerbates a range of environmental and biological issues.[3-5] To promote the effective usage of these solid waste, mainly agricultural residues, methods beyond incineration have been explored, such as transformation into functional materials through pyrolysis, gasification, surface modification, and other processes.[5-7] For example, a large amount of literature has reported that various biomass (namely, corn stalk, wheatgrass stalk, corncob, peanut shell, rice husk, bagasse, wood chip, and the like) can be converted into a variety of activated carbons, adsorbents or catalysts (Figure 2.1).[2, 4, 5, 8-13] Furthermore, biochar, derived from biomass-based waste, boasts advantageous properties such as abundant pore structures,

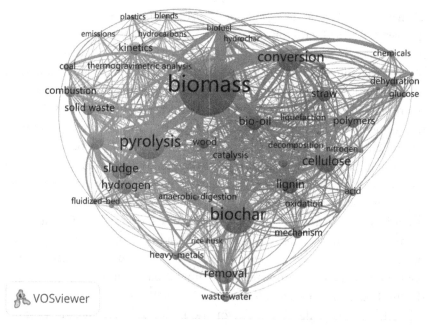

FIGURE 2.1 Co-occurred keywords in published researches on the solid waste-based materials (mainly biomass-based materials) [13]. Reprinted with permission from Copyright 2024 Elsevier.

low cost, ease of production, and versatility.[1,5] These materials have exhibited efficient adsorption capacity for various heavy metal ions, as well as inorganic nitrogen and phosphorus pollutants and organic antibiotics.[2, 7, 8, 12, 14, 15] Recent works also reported that these solid waste-based materials can serve as the activators for various oxidants, including hydrogen peroxide (H_2O_2), ozone (O_3), and PerSulfate (PS). A series of Reactive Oxygen Species (ROS) and non-radical electron transfer pathways can be triggered via these AOPs for effectively removing organic pollutants from water or soil.[9-11]

A number of physical, thermochemical and biochemical methods provide key means for the preparation of waste solid-based materials for environmental use. A series of pretreatment procedures, including physical, chemical, and biological techniques, can be efficiently used in the treatment of the solid waste for further modification to achieve specific purposes. This solid waste, primarily biomass-based, can be transformed into porous materials through thermochemical treatments such as conventional pyrolysis, microwave-assisted pyrolysis,[16-18] hydrothermal carbonization, and ion thermal carbonization. In addition, advancements in modification methods such as structural pore creation, metal/non-metal doping, functionalization, and acid treatments have enabled the enhancement of solid waste-based materials for diverse applications.

The purpose of this chapter is to systematically summarize the latest progress and future prospects in the synthesis of solid waste-based materials. Initially, several pretreatment procedures of solid waste were reviewed. Following this, various thermochemical treatment methods for the manufactures of solid waste-based materials were highlighted. Additionally, the chapter explores the subsequent modifications of these materials, such as acid-base treatments and metal/non-metal incorporation. Promising instances of solid waste-based materials are showcased to analyze fabrication conditions and further elucidate their respective preparatory approaches/strategies. Ultimately, this chapter identifies the limitations and forthcoming opportunities of these synthetic routines.

2.2 PRETREATMENT OF SOLID WASTE

Pretreatment can effectively change the structure of solid waste, especially the structures of biomass-based waste, so as to reduce the stability of carbohydrates in the biomass and facilitate the subsequent pyrolysis or loading processes.[19-21] Therefore, this process is considered an important link in the preparation of solid waste-based materials and directly affects the cost of solid waste-based materials (mainly biomass-based materials).[22, 23] At present, the existing pretreatment technologies mainly include physical, chemical, physical chemical and biological methods.[22, 23]

Physical processes, such as drying, washing, milling, extrusion, ultrasound and microwave radiation, have been intensively used for the pretreatment of various solid waste-based materials.[20, 21, 24-26] The drying and washing can only remove the surface moisture and impurities. Other physical pretreatments, including milling, extrusion, and ultrasound, use mechanical energy or wave energy to change the structure of solid waste.[23] These techniques are sometimes classified for the purpose of reducing particle size, but there are problems such as their high energy consumption and greater wear on the instruments.[23] Among the many pre-treatment techniques that have been proven to be very effective and intensively used, mechanical grinding can significantly improve the particle size of pre-treated materials and promote the modification efficiency of subsequent processes.[23, 26, 27] In the grinding process, the materials are subjected to strong mechanical stress, which has a relatively large impact on the physical and chemical properties of the raw materials.[20] Pant and co-workers reported that, as a widely accepted strategy for effectively altering the structural characteristics of biomass-based solid waste, the milling of this solid waste is often integrated with other pretreating techniques in industrial applications.[20] Due to the complexity of the cross-linking between lignocellulosic components in biomass-based solid waste, it is necessary to break the cross-linking before hydrolysis. The pretreatment process helps to (i) weaken the action of the bonds within the cellulose and reduce the crystallinity of the cellulose, (ii) remove/destroy some lignin and hemicellulose, and (iii) increase the internal surface area to improve the structural modification of the fiber by subsequent treatment.

Chemical pretreatment generally uses acid, alkali and other chemical reagents to treat various solid waste before pyrolysis and metal/non-metal loading.[27, 28] Chemical pretreatment is considered to be the most effective means of treating biomass-solid waste, because hemicellulose and lignin in the biomass-solid waste can is first dissolved and destroyed during the chemical pretreatment.[27, 29] For example, various alkali solutions (sodium carbonate, calcium hydroxide and sodium hydroxide, and the like) can cleave the glycosidic bond of lignocellulose through esterification reactions, remove the hemicellulose and lignin, and further improve the porous structures in the biomass-based waste.[28] Although alkali pretreatment can efficiently facilitate the structural improvement of solid waste, some problems such as the residual neutralization, and secondary pollution still need to be solved.[28] Acid pretreatment is reported to be more effective than alkali pretreatment in treating the biomass-based solids because the acid pretreatment process can better dissolve the hemicellulose in the biomass, destroy the internal crystalline components, and improve the accessibility of subsequent treatment for the cellulose. However, acid pretreatment has many problems, including (i) the inability to effectively remove lignin, (ii) the production of formic acid, acetic acid and other inhibitors during the treatment process, and (iii) the corrosion to the equipment due to the acidic solution. Organic solvents, namely, ethanol, methanol, formic acid, acetic acid, propanol, can also enable the efficient removal of lignin for better subsequent hydrolysis.[28, 30] This has the advantage of a high hydrolysis rate of cellulose, green safety and a reusable solvent. Topakas and his colleagues tried to use isobutanol as an effective organic solvent to remove lignin from beechwoods.[29] The organic solvent with better permeability can promote the expansion of cellulose, and reduce the crystallinity of cellulose, thus promoting the conversion of lignocellulose.[28] Biomass-based waste can also be chemically pretreated by adding engineered nanoparticles or natural/man-made active substances (namely, graphene, carbon nanotubes, montmorillonite and clay) to improve the material properties (for example, surface area, functional groups).[26, 27, 31] For example, Bacterial Cellulose (BC) was mixed with graphene oxide and graphene for constructing the carbon aerogel through synergies of hydrogen bonding and electrostatic interaction during freeze-casting and carbonization (Figure 2.2a).[31] The aerogels were alternately wound by layered structures (graphene oxide and graphene) and fibrous structure (BC nanofibers), exhibiting ultra-light weight and excellent mechanical strength (Figure 2.2b-e). The piezoresistive sensor prepared by the aerogel can detect physiological signals in real time, effectively distinguish subtle sound signals and realize versatile applications (Figure 2.2f and 2g). However, in practical applications, there are also shortcomings such as high manufacturing efficiency and unfriendliness to the environment.

Biological pretreatment includes both anaerobic and aerobic processes, and specific enzymes such as lipase, carbohydrase, and peptidase can be added to the pretreatment systems. It is mainly used for the pretreatment

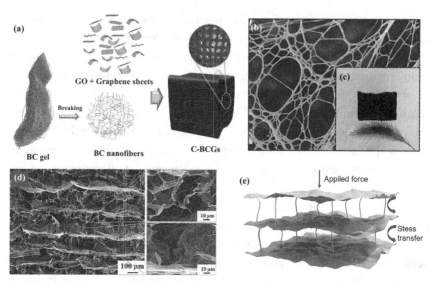

FIGURE 2.2 (a) Fabrication scheme of the carbon aerogel derived from BC nanofibers, graphene oxide and graphene; (b) SEM image of BC nanofibers, (c) Actual image of the carbon aerogel; (d) SEM images inside the carbon aerogel and (e) Microstructure scheme of the carbon aerogel; (f) Signal identification of the sounds in entire piano; (g) Comparison of acoustic spectra and current spectra[31]. Reprinted with permission from Copyright 2024 Elsevier.

of carbon-based solid waste. Biological pretreatment can improve biomass feedstock and be used for large-scale production of various kinds of engineered biochar. Anaerobic digestion is an effective biochar pretreatment technique, and the anaerobic pretreatment of various biomass-based waste (namely, bagasse, straw, sludge and animal manures) to produce biochar has been realized on a large scale. Compared with biochar prepared by pyrolysis alone, the parameters such as cation exchange capacity, specific surface area, pore structure and the surface negative charge of biochar obtained by integrated anaerobic digestion and pyrolysis were improved. Another approach to biological pretreatment is to directly utilize biomass rich in high concentrations of metals or other mineral components,[32] or to irrigate biomass crops by a solution with the target elements involved so that the desired elements can naturally accumulate in the biomass. Compared to other pretreatment processes, biological pretreatment is a relatively mild process, and this pretreatment process will not cause the formation of inhibitory matter, which is a relatively environmentally friendly and efficient strategy.[23, 27]

Numerous pretreatment methodologies exist for solid waste, particularly those derived from biomass. However, each method typically targets specific aspects of biomass solid waste structure, achieving optimal results only in particular domains. By combining various pretreatments, their respective strengths and weaknesses can be synergistically balanced, significantly enhancing the overall efficacy. Such an integrated approach can mitigate the limitations inherent in single pretreatments and reduce the number of processing steps required. While the combined pretreatments can improve the treatment efficiency of biomass-based solid waste, they necessitate more precise control throughout the process and can result in increased costs.

2.3 PREPARATION OF SOLID WASTE-BASED MATERIALS

The fabrication of eco-friendly materials from solid waste encompasses a diversity of methods, such as conventional pyrolysis, microwave-assisted pyrolysis, hydrothermal carbonization, and ion thermal carbonization. Materials synthesized through these varied approaches demonstrate a spectrum of structures and physicochemical properties, offering broad prospects for environmental applications.

2.3.1 Conventional Pyrolysis

The pyrolysis process is based on a non-oxidizing thermal decomposition reaction, which can thermally degrade various solid waste (mainly organic solid materials) into carbon-rich by-products at a range of temperatures (300–800°C).[33-35] These non-oxidizing pyrolysis conditions are mainly maintained via keeping (i) a vacuum atmosphere; (ii) an inert environment fed with nitrogen, argon or helium; (iii) feeding with reactive gases, such as steam or carbon dioxide.[33-36] Therefore, the properties of the resulting pyrolysis by-products of solid waste are often closely related to their pyrolysis conditions such as inert gas type, temperature, heating rate, flow rate of feeding gas and operation time.[35, 36]

At present, traditional pyrolysis techniques are very mature for the pyrolysis treatment of solid waste.[35] By modulating the operating conditions in pyrolysis processes, the yields and performances of the target products derived from solid waste can be optimized. The pyrolysis of solid waste produces three unique products: the remaining solid residues (for example, biochar), condensable liquids (for example, bio-oil) and the non-condensable gases (for example, syngas).[37] Since there is no oxygen in the pyrolysis process, the solid waste (mainly carbon-based waste) does not burn, so the pyrolysis process can be an efficient means to obtain various carbon-based solid materials and bio-oils from solid waste. Moreover, lower pyrolysis temperatures and longer residence times lead to higher yields of carbon-based solid materials (for example, biochar). For example, prolonging the reaction time can promote the polymerization of carbon-based solid waste for the production of more solid materials.[35] Lee and colleagues reported that single-use plastic masks can be converted into non-graphitized toner through acid treatment and long-term pyrolysis.[24] The extended pyrolysis could significantly promote the cross-linking in the plastic mask-derived materials and the formation of polyaromatic hydrocarbons in the non-graphitized toner. In contrast, shorter residence times at higher temperatures usually result in the formation of more liquids.[35,37]

Slow pyrolysis reactions in traditional pyrolysis processes often require long residence times (5–30 minutes, or even 24–36 hours), low temperatures (250–600°C), and low heating rates (0.1–1.0°C/s).[24] Slow pyrolysis facilitates the secondary reaction of solid waste by extending the vapor residence time, thereby maximizing the yield (30–60%) of the remaining solid waste, such as biochar, as shown in Figure 2.3.[34] The yield of these solid waste-based materials is closely associated to the physicochemical characteristics

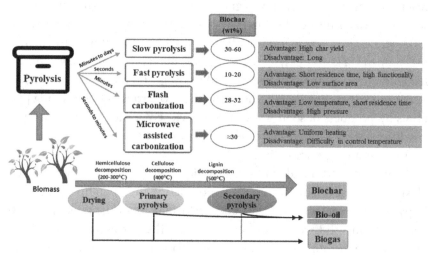

FIGURE 2.3 Pyrolysis process including conventional pyrolysis and microwave assisted pyrolysis of solid waste (mainly carbon-based solid waste) [34]. Reprinted with permission from Copyright 2022 Elsevier.

and structural properties of the materials and the conditions of the pyrolysis process, especially the parameters such as pyrolysis temperatures, rates of heating, residence times and the types of the pyrolysis reactor. In general, slow pyrolysis has a longer pyrolysis time, so it is possible to treat a wider range of solid waste, including larger solid waste or smaller particles and debris. Rapid pyrolysis is characterized by extremely high temperature, fast heating rates, and short residence times. The rapid pyrolysis can avoid the secondary reactions due to the short steam residence time and high heating rate. Since there are no secondary reactions, the whole rapid pyrolysis is an endothermic process. Rapid pyrolysis maximizes the yield of bio-oil. Flash pyrolysis is a transient thermal cracking reaction that occurs under low pressure and near vacuum conditions. The products produced by flash pyrolysis are basically the same as those produced by fast pyrolysis. The temperature of flash pyrolysis is between 800 and 1000°C. In general, flash pyrolysis is often used for the pyrolysis of biomass solid materials with particle size less than 0.2 mm. As for microwave-assisted pyrolysis, it is a relatively new pyrolysis method that can serve as one of the attractive alternatives for accelerating and optimizing pyrolysis reactions. Compared with conventional pyrolysis processes, the quick chemical reactions of microwave pyrolysis can be carried out effectively due to its excellent heat transfer characteristics. However, the ratio of the resulting solid residue after microwave-assisted pyrolysis is about 30%, which is similar to flash pyrolysis but lower than slow pyrolysis. Microwave assisted pyrolysis will be discussed in detail in the following section.

Basically, various carbon-based materials derived from solid waste have a range of physiochemical and mechanical characteristics, mainly depending on factors such as the material, pyrolysis preparation and production operations, which alter the suitability of carbon-based materials for various environmental applications. Slow pyrolysis is more suitable for the fabrication of solid waste-based materials (mainly carbon-based materials) with mass of 30–60%.

2.3.2 Microwave-Assisted Pyrolysis

Microwaves are defined as the regions of the electromagnetic spectrum from 0.01 to 1 m, as shown in Figure 2.4a, with frequencies ranging from 0.3 to 300 GHz.[17, 38] Recently, the Federal Communications Commission (FCC) have designated the frequencies of 915 MHz and 2450 MHz for industrial, scientific, and medical purposes.[18, 38] Microwave-assisted

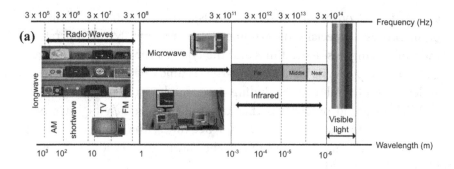

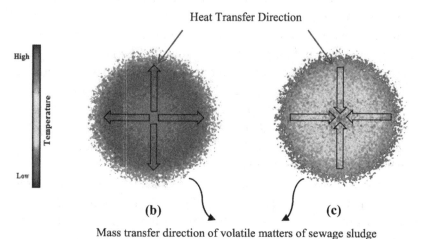

FIGURE 2.4 (a) Scheme of the Electromagnetic spectra [38] Reprinted with permission from Copyright 2024 Elsevier; Distributions of temperature in the (b) microwave-assisted pyrolysis and (c) conventional pyrolysis [43] Reprinted with permission from Copyright 2019 Elsevier.

pyrolysis is an effective technique to prepare various porous materials from homogeneous waste such as sludges, plastic scraps, tires, food waste, and agricultural waste.[39-41] Microwaves can penetrate the solid waste materials and can be converted into the heat inside the materials.[41, 42] Since heat loss affects the surface of the materials, heat is first generated inside the material and transferred to the outside, which is different from the conventional pyrolysis with heat transferring from the outside to the inside (Figure 2.4b and 2.4c).[43] Given the poor thermal conductivities of some solid waste, a temperature gradient can be formed from the core of these solid waste to the outside. Compared with conventional pyrolysis, the activation energy of microwave radiation is reduced by 40 ~ 150 kJ/mol on average.[18, 38, 40, 44] The lower activation energy can further facilitate the chemical cracking in the solid waste, which reduces the operational costs and improves pyrolysis efficiency.

Microwave-assisted pyrolysis of various solid waste usually results in three broad categories of products, primarily dependent on the parameters (namely, reactor types, feedstocks types, chemical additives, waste sizes, microwave output power, residence time, catalyst types, microwave types) that affect the microwave-assisted pyrolysis reactions.[18, 38, 41, 42] The solid residue (namely, biochar), liquid (namely, bio-oil) and gas are the three stages of products generated during the microwave-assisted pyrolysis.[45-47] An obvious advantage of microwave assisted pyrolysis is that the process is highly likely to produce gaseous and liquid fuels. The obtained solid waste residues also have some environmental application values. For example, the conversion of plastic-based waste can be realized in a much shorter time via microwave pyrolysis as compared to conventional heating techniques.[48, 49] Microwave treatments of plastic waste typically results in a range of multiphase products, such as carbon-based solid materials, bio-oils, and gases.[48-50] Among them, the yields of carbon-based solids are always easily affected by the pyrolysis temperatures.[48, 50] The higher the temperature of microwave pyrolysis, the higher the degree of thermal pyrolysis, resulting in lower solid yield. Thomas and co-workers first reported that the microwaves could assist the activation of iron-based catalyst particles to effectively degrade crushed plastic waste to produce hydrogen and Multi-Walled Carbon NanoTubes (MWCNTs).[50] Unlike conventional heating methods, the iron-based catalysts are able to absorb microwave energy more effectively and activate the surrounding plastic particles via heat transfer and thermal convection. The entire process takes just 30–90 seconds in a single step to convert a variety of real-life plastic waste into clean fuel hydrogen, with a yield of up to 55.6 mmol/g of plastic. In addition, the solid residues left after the degradation of plastic waste were based on substantial amounts of filamentous carbons with ~92 wt.% of MWCNTs, which also have high environmental application values. As a result, microwave assisted pyrolysis can facilitate the conversion of plastic-based waste into the high-value environmental materials and fuels.

Microwave heating technology facilitates the production of various types of solid waste-based materials which, in most instances, efficiently serve numerous environmental applications. These applications range from the adsorption of diverse pollutants to roles in batteries, electrode materials, composite membrane materials, catalysts, and catalyst carriers.[51-54] Nonetheless, the advancement and synthesis of more effective environmentally functional materials via microwave-assisted pyrolysis necessitates meticulous control over processing conditions, as well as raw material pretreatment and subsequent post-treatment steps, to enhance their specific application performances.[18, 42, 46, 48] Furthermore, recent literature has also reported that compared to traditional pyrolysis methods, the solid residues (such as biochar) obtained by microwave assisted pyrolysis have very low microporous and mesoporous volumes, and have stronger leaching capacity for organic matter and heavy metals, and consequently, their efficacy in environmental applications may be compromised.[51, 53]

2.3.3 Hydrothermal Carbonization

Some solid waste (mainly organic solid waste) can be efficiently converted to carbon-based materials via HydroThermal Carbonization (HTC) by using a specific aqueous solution as the medium.[55, 56] During the hydrothermal carbonization, hydrolysis of the carbon-based materials is the first step, followed by a series of steps such as dehydration, decarboxylation, recondensation/polymerization, aromatization, and carbonization.[8]

Most components in biomass-based solid waste, including hemicellulose, carbohydrates, glucose, fats, and lipids can be hydrolyzed at a temperature as low as approximately 180°C.[55] Due to the homogeneity and long chain structure of cellulose, its hydrolysis requires a relatively long reaction process and requires a certain high temperature (230°C).[55] Compared with conventional pyrolysis (slow pyrolysis), the biomass-based materials produced by hydrothermal carbonization have a higher carbon content. In addition, these materials produced by the HTC process are more acidic than those produced by slow pyrolysis. The HTC technique is more energy intensive than other methods because it requires the use of materials with a high water content.[34] Hence, HTC carbonization is considered to be an inert slow pyrolysis with low temperature. In addition, the liquid vapor pressure rises sharply by rapidly raising the temperature of the liquid (such as water) in the HTC process, which promotes the reduction of its surface tension, resulting in the production of carbon materials with unique structures. In addition, liquids also exhibit different liquid polarity, viscosity and molecular diffusivity in the subcritical states, which is conducive to the dissolution of some polar and non-polar compounds within solid waste, especially biomass-based solid waste.[55] Consequently, HTC is frequently characterized as a comparatively swift chemical reaction process capable of effectively transforming biomass-based solid waste, such as straw, coconut husks, and rice husks, into high-value products suited for environmental applications.[55, 57-59]

The process of preparing biochar based on hydrothermal carbonization is strongly dependent on the reaction conditions of the carbonization process. For example, the hydrothermal temperatures (150–350°C) play the very essential roles in the hydrothermal carbonization processes.[55] As mentioned above, biomass-based solid waste are able to be activated more vigorously at higher temperatures, which can promote the hydrolysis of hemicellulose and cellulose.[55] Algal biomass carbonization produces solid fuels at lower temperatures (180–270°C). In contrast, higher temperatures are beneficial to the carbonization of biomass and increase the calorific values of the solid fuels obtained at temperatures >270°C.[55] The HTC time is important for the determining the structures of products. Zhang and colleagues prepared a series of tobacco straw-based carbon materials by regulating the carbonization time, and analyzed the influence of different carbonization times (1–12 h) on the structural properties of the prepared carbon-based materials.[60] It is found that a longer HTC time is conducive to the formation of spherical nanocarbons. Additionally, the HTC process can introduce more oxygen-containing functional groups on the surface of carbon materials. For example, during the hydrothermal carbonization process, the chemical compositions and morphologies of rice husk-based materials were changed significantly, and more oxygen-containing functional groups were formed in as-prepared rice husk-based biochar. The HTC technique can also be used to prepare carbon-based materials doped with various heteroatoms to improve the heat resistance and electrochemical properties of biochar. However, it must be pointed out that although hydrothermal carbonization has the advantages of simple preparation, low cost, low pollution, high carbon yield, and so forth, compared to conventional pyrolysis processes, the solid waste-based materials (mainly biomass-based materials) prepared by HTC usually show smaller surface area and lower porosity. Therefore, it is often necessary to combine other treatment processes to further improve the pores and surface areas of the materials for better environmental applications.

2.3.4 Ionothermal Carbonization

Recent studies have found that porous carbon materials with high porosity can be efficiently prepared by ionothermal carbonization.[55, 61, 62] Ionic liquids are essentially salts in a liquid state at high temperatures, such as KCl, or KOH when liquified, thereby acting as ionic liquids. These substances are known for their chemical resistance, low melting points, and high thermal stability. Meanwhile, ionic liquids are considered to be environmentally friendly solvents due to their unique properties, including (i) being odorless, non-combustible, and possessing exceedingly low vapor pressure, making them suitable for high vacuum systems while minimizing environmental pollution through volatilization; (ii) displaying excellent solubility for both organic and inorganic substances, enabling homogeneous reaction conditions and reduced equipment sizes; (iii) offering broad operating temperature ranges (-40 ~ 300°C), along with robust chemical and thermal stabilities, and ease of separation for recycling; (iv) exhibiting Lewis and Franklin acidity with adjustable acid strength. Due to these characteristics, ionic liquids are increasingly used as the templates, solvents and ostomies in the productions of various biochar. For example, Yu and co-workers observed that ionothermal carbonization could help to fabricate the nitrogen-doped layered porous carbon using polyherbal as a biomass precursor.[62] White et al. used ionic liquids to prepare biochar from carbohydrates and found that biochar yields can be improved with significant layered pores due to the pore formation of iron species in ionic liquids.[61] Consequently, ionothermal carbonization proves to be an effective method for generating carbon materials with a desirable balance of mesoporous and microporous features.

2.4 MODIFICATION OF SOLID WASTE-BASED MATERIALS

2.4.1 Acid-Base Treatment

Acid modification is one of the most common methods to change the surface composition and structural properties of solid waste, especially biomass waste.[63-66] This is because acid modification can promote the dissolution of soluble substances present in biomass waste, improve the porosity of biochar and expand its specific surface area. In addition, acid modification often introduces oxygen-containing functional groups (for example, $-SO_3H$, $-OOH$, and $-P=O$), which play the essential roles in the adsorption of heavy metals, or can also serve as potential catalytic sites. At present, the most common acid modifiers mainly include various inorganic acids (for example, H_2SO_4, HNO_3, HCl, H_3PO_4) and weak organic acids (namely, citric acid, acetic acid). Acid modification of biomass-based materials can be achieved by two means: (i) Mix biomass materials with an acid modifier, and then pyrolyze at a certain temperature; (ii) The biomass material is first pyrolyzed to produce biochar, and then treated with acid modifier. Kan et al, reported that the modification of waste printed circuit boards by HNO_3 could efficiently remove the residual Cu species, and produce high populations of pore structures for the adsorption of antibiotic contaminants.[65] However, acid modification also has some problems. For example, the collection and treatment cost of acid waste liquid is high, and the acid modification process also produces some toxic substances, such as Volatile Organic Chemicals (VOCs), and Polycyclic Aromatic Hydrocarbons (PAHs). These substances may remain in the biochar and be released when used for soil or water restoration, causing secondary pollution or damage to the microbial community.

Among all these acid modifiers, it is reported that phosphoric acid (H_3PO_4) has been the most intensively used for the modification of biomass-based solid waste.[66] The advantages of H_3PO_4 as an acid modifier are that H_3PO_4 is easy to recover and has a good pore-making ability to obtain higher biochar yield, pore structure and specific surface area.[64, 65, 67-69] In addition, less PAHs and VOCs can be produced during the preparation of the biomass-based carbon by H_3PO_4 activation, so H_3PO_4-modified biochar exhibits better environmental friendliness and can be better applied to water and soil remediation. At present, the H_3PO_4 activation process has been widely used in industries to fabricate various biomass-based carbon materials with high populations of pore structures and specific surface areas from various crop waste and processing residues.

The activation process of H_3PO_4 involves a series of complex and step-by-step chemical reactions, including surface hydrolysis, dehydration, crosslinking, aromatization, and pore formation.[68, 69] Specifically, the following steps are used to form rich pore structures in biomass-based materials. (i) Initially, the ionization of H_3PO_4 induces the expansion and dispersion of cellulose and other polymers in the biomass into a colloidal form, facilitating hydrolysis into monosaccharides or polysaccharides. This step is conducive to the formation of graphite microcrystalline structures and enhances pore production. (ii) Biomass-based solid waste subsequently undergo catalytic dehydration, improving carbon yield. (iii) H_3PO_4 further cross-linked with the polymer, results in an increase in porous structures within the biochar during high-temperature pyrolysis, such as phosphate esters and pyrophosphates. (iv) The carbon skeleton undergoes aromatization and aromatic ring condensation reactions to promote the formation of microcrystalline structures. (v) H_3PO_4 provides a place for the new carbon atom aggregation of the precursor. Finally, the excess inorganic components such as H_3PO_4 are removed by washing, and the available adsorption regions and pore structures in the carbon-based materials are also improved. Zhou and colleague prepared a H_3PO_4-modified biochar derived from the waste coffee grounds.[67] Two steps were involved for the modification procedures, including (i) biochar obtained from slow pyrolysis for 120 min (300°C), and (ii) subsequent H_3PO_4 impregnation with the biochar. The surface area of H_3PO_4-modified waste coffee grounds was almost 40 times higher than that of the original waste, exhibiting superior adsorption capacity towards 2,4-dichlorophenoxyacetic acid with adsorbed amount of 323.76 mg/g. Similar excellent adsorption results were also reported for other H_3PO_4-modified waste, for example, palm kernel shells,[64] peanut shells,[68, 69] corncobs,[69] since H_3PO_4 could remarkably facilitate the formation of various mesopores and micropores during slow pyrolysis, creating large populations of active sites for the adsorption of pollutants.

In summary, acid modification can effectively facilitate the production of the pore structures and specific surface areas of solid waste-based materials (mainly biomass-based materials), and enrich the oxygen-containing groups on their surfaces. Since acid modification has some negative effects on the soil/water environment, these effects should be minimized when acid modification of biomass materials is carried out. Considering that the biochar prepared by H_3PO_4 activation has a significantly lower impact on the environment, H_3PO_4 activation has been implemented in engineering applications and has been widely adopted.

2.4.2 Non-Metal Loading

In addition to the structural characteristics (rich pore structure) of solid waste-based materials, some surface functional groups (such as OH, -COOH, -NH_3 and -SO_3 groups) and heteroatoms (for example, N, O, S, P), will also affect their performance in environmental applications and be even more critical than their own pore structure in some aspects.[70, 71] The introduction of -OH and -COOH functional groups in biochar can significantly improve the adsorption capacity of CO_2 and the removal capacities of various heavy metal pollutants in soil/water environments.[70] These functional groups or heteroatoms can be doped on the surface of engineered biochar by various modifications. For example, in order to increase oxygen-containing functional groups (such as carboxylic groups and phenolic hydroxyl groups), surface modifications via H_2O_2 and HNO_3 oxidation is commonly used. The amination reaction on the surface of solid waste can be achieved by high temperature ammonia treatment to increase the amino functional group. Sulfonation on the surface of solid waste can be achieved by sulfuric acid. In addition to single heteroatom doping or the introduction of functional groups, double or triple doping treatments are currently used to increase the abundance of surface-active sites. In environmental applications, these multiple dopants or introductions can achieve synergies to further improve performances. For example, N and S co-doping in biochar is often better than N single-doping for supercapacitors, or for activating persulfate for contaminant degradation.[72]

At present, the commonly used organic modified reagents are melamine, thiourea, dicyandiamide, glutaraldehyde, polyethylene imide, methanol, urea and so on.[73-75] The purpose of these modifiers is mainly to introduce the required functional groups or introduce common nitrogen, sulfur and other heteroatoms. Organic reagent modification plays an essential role in regulating the compositions and chemical properties of solid waste-based materials (mainly biochar), which can cause many property changes of biochar. Firstly, the populations of surface-active functional groups or non-metallic heteroatoms of biochar can be significantly increased by the modification of organic reagents. Xu et al. reported that through the modification of ethylenediamine and triethylamine, the amide and quaternary amino functional groups on the surface of various stalks can be increased to 3–10% of the mass of the stalks.[73, 75] In addition, organic reagents can change the elemental compositions of biochar, affecting the polarity, aromatics and hydrophilic/hydrophobic properties of the biochar surface. Wang et al. developed a novel electrochemical advanced oxidation process for a Janus cathode catalyzed by S, N self-doping derived from Ginkgo leaf.[74] The co-doped graphite N, pyridine N and thiophene S played an important role in H_2O_2 generation and in situ activation to produce reactive oxygen species, and the catalytic membrane can achieve high efficiency enrichment and rapid degradation of pollutants.

2.4.3 Metal Loading

Many metal structures with very high adsorption or catalytic properties (metal oxides, metal nanoparticles, or metal monatomic sites) cannot be directly used alone in water/wastewater treatments. This is because the agglomerations of these metal structures will occur, resulting in a great reduction in adsorption or catalytic properties, which will greatly hinder the environmental applications of these metal structures. Therefore, recent studies have now focused on fixing these highly active metal structures onto various carriers.[76,77] For example, Xu and colleagues adsorbed fluoride or phosphate ions in water by loading zirconia onto grapefruit peel, and achieved excellent adsorption results.[77] Moreover, the zirconia loaded on these porous biomass carriers does not easily fall off, which is conducive to subsequent continuous operation. The Layered Double Hydroxides (LDHs) are a class of metal hydroxides consisting of two or more metal elements, which are layered with anions and water molecules overlapped in the main layer or between layers.[76, 78] At present, integrating the biochar with LDH represents a great advantage in water or soil treatments, providing an effective method for the removal of various contaminants. For example, Xiao and colleagues fabricated a CoNi-LDH/biochar composite for the adsorption of lopinavir (Figures 2.5a-h), which exhibited the maximum adsorption capacity of 832.9 mg/g.[76] Various biochar can also be used to fix a range of transition metals and their compounds (namely, $Fe^{(0)}$, Fe_3C, Fe_3O_4, $CoFeO_4$, $FeMnO_4$, Mn_2O_3, $NiCo_2O_4$, MnO_2, CuO, Co_3O_4) for advanced oxidation treatment of contaminants in water. These transition metals and their derivatives have strong magnetic properties, which can be endowed with certain magnetic properties after being introduced into biochar and are also conducive to the catalytic separation and recovery of catalysts after degradation treatment. For example, Yan and co-workers synthesized in situ $NiCo_2O_4$ nanosheets with lily shape on waste biochar (Figures 2.5i and 2.5j), which can efficiently oxidize various phenolic pollutants from water via photocatalysis.[79] Therefore, the introduction of metal structures into these biomass solid wastes provides a new pathway for their applications in the environmental field.

However, the introduction of these metal structures may affect the internal pore structures or other active component compositions in the biochar-based materials, thus having a certain impact on their adsorption/catalytic performances. In addition, the problem of metal dissolution cannot be completely avoided. As mentioned in above section, heteroatom doping of non-metallic elements (for example, nitrogen, sulfur, oxygen, phosphorus, boron) can remarkably change the geometric structures, electronic properties and carrier effects of biochar, thereby improving the adsorption/catalytic performances of biochar after doping. In recent years, the synthesis of biochar-based materials with co-doping metal/non-metal (such as Fe/N, Co/N, Mn/N) has been reported.[10, 11, 32, 80, 81] This metal-nonmetallic co-doping has some special advantages, including: (i)

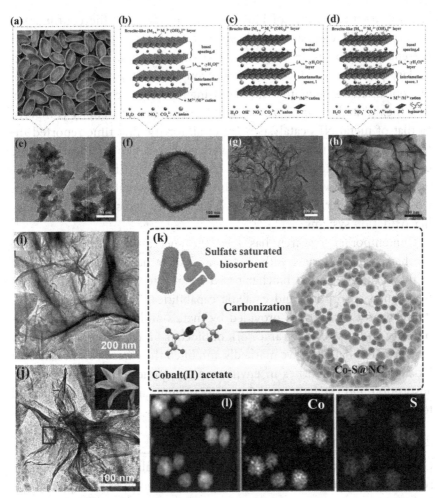

FIGURE 2.5 Schematic diagrams of (a) biochar from pistachio shells, (b) CoNi-LDHs, (c) CoNi-LDHs/biochar, (d) CoNi-LDHs/biochar after use, as well as (e-h) TEM images of above samples [76]. Reprinted with permission from Copyright 2024 Elsevier。 (i) TEM image and (j) HRTEM image of waste biomass anchored with lily-shape NiCo oxides [79] Reprinted with permission from Copyright 2022 Elsevier; (k) Fabrication of Sulphur-doped Co_3O_4 from saturated biosorbent; (f) HRTEM of Sulphur-doped Co_3O_4 and relevant EDS mappings [80] Reprinted with permission from Copyright 2020 Elsevier.

Transition metal/nonmetallic co-doping can greatly enrich the active sites in the modified biochar, and improve their adsorption or catalytic properties towards various pollutants; (ii) Transition metal/non-metal co-doping can effectively improve the surface structure, electron distribution and metal-carrier effect of biochar; (iii) In addition, the coordination of non-metallic elements with metallic elements can effectively achieve the encapsulation of metal ions, greatly avoiding the leaching and secondary pollution of metal ions. For example, Qi et al. reported the Co-N_4 coordinated single-atom catalyst using lignin as the carbon precursor, which exhibited superior oxidation performance and stability compared to that of metal NPs catalysts.[10] Du et al. for the first time reported the fabrication of sulfur-doped Co_3O_4 with sulfate-saturated adsorbent and cobalt salts as precursors (Figures 2.5k and 2.5l).[80] The sulfur could be well coordinated with the Co_3O_4 to improve the geometric property of the Co_3O_4 and further facilitate the catalytic activity and stability in Fenton-like oxidation systems.

Contemporary research has seen the successful integration of the biochar-based materials with metals and their derivatives, culminating in the creation of metal/biochar-based composites distinguished by their heightened adsorption and catalytic capabilities. These composites capitalize on the active sites provided by metals and leverage the inherent porosity and supportive framework of biochar-based materials Such synergistic combinations have markedly amplified the efficacy of these metal/biochar-based composites in environmental applications, notably in the adsorption and catalytic breakdown of contaminants within aquatic and terrestrial environments.

2.5 CONCLUSION

Over the past decade, the development of materials derived from various solid waste–predominantly that based on biomass–and their applications in environmental contexts have advanced considerably. The preparation of these solid waste-based materials includes a series of pretreatments (for example, chemical, physical, biological pretreatments), carbonization processes (for example, traditional pyrolysis and microwave assisted pyrolysis and hydrothermal carbonization), and further chemical modification (for example, acid modification, metal/non-metal modification). Beyond water and wastewater treatment, these solid waste-based materials are being employed in burgeoning environmental sectors, including soil remediation, biosensor development, and new energy resources. Notably, biomass-based materials, often have the advantages of a wide range of raw material sources, cheap prices, and a variety of available preparation methods. However, the high cost of carbon production (energy costs or waste disposal costs) of these processes leads to higher prices for biochar products and less desirable performance than commercial products. In addition, the stability and suitability of biochar for environmental applications may be key factors in determining its practical application. However, previous studies have mostly been limited to laboratory scales, and the potential effects of industrial-scale production on biochar structure, stability, suitability, and environmental impacts (VOCs or PAHs release) are unknown.

This summary briefly elucidates these preparation methods. Carbonization processes are energy-intensive, and most modification techniques, such as acid and metal/non-metal chemical modifications, are inextricably linked to complex chemical reactions. These methods have the characteristics of complex operation process, high cost and easy to reduce harmful waste. In addition, the researchers developed these preparation methods with the goal of using these materials for the adsorption/catalytic removal of one pollutant or a class of pollutants in lab, without extensive application in real-world environments. Therefore, it is imperative to further study the stabilities, cost-effectiveness, and secondary pollution associated with the production of solid waste-based materials. To address these challenges, the exploration of potential solutions and future prospects is essential:

1. Improving the pretreatment efficiency and increasing the pyrolysis process efficiency may be an effective means and strategy to reduce the cost of solid waste-based materials in the future.

2. The development of new treatment methods or the combination of multiple treatment methods is one of the means of further improving the applicability of solid waste-based materials.

3. The stability and usability of solid waste-based materials can be further assessed by the detections of PAHs, VOCs, and heavy metals released during their environmental applications.

4. In the future, the environmental applications of solid waste-based materials should be moved from the labs to real environments, and the applicability of the preparation methods for solid waste-based materials will be strengthened in this way.

5. Biochar is a critical class of solid waste-based materials, and most of the current preparation methods of solid waste-based materials are mainly developed for the productions of various biochar. In the last decade, developing the nano-biochar for environmental applications

is considered to be more promising, because its nano-structural properties give it advantages over traditional biochar in terms of material geometry and application. However, the study on the preparation strategies of various nano-biochar is still in its infancy. Therefore, the preparation method and industrialization process of nano-biochar still need to be further developed.

REFERENCES

1. Raud, M.; Kikas, T.; Sippula, O.; Shurpali, N. J. Potentials and challenges in lignocellulosic biofuel production technology. *Renew. Sustain. Energ. Rev.* 2019, 111, 44–56.

2. Li, Y.; Xing, B.; Ding, Y.; Han, X.; Wang, S. A critical review of the production and advanced utilization of biochar via selective pyrolysis of lignocellulosic biomass. *Bioresour. Technol.* 2020, 312, 123614.

3. Gupta, J.; Kumari, M.; Mishra, A.; Swati; Akram, M.; Thakur, I. S. Agroforestry waste management – A review. *Chemosphere* 2022, 287, 132321.

4. Kim, J. R.; Kan, E. Heterogeneous photocatalytic degradation of sulfamethoxazole in water using a biochar-supported TiO_2 photocatalyst. *J. Environ. Manage.* 2016, 180, 94–101.

5. Che, H.; Wei, G.; Fan, Z.; Zhu, Y.; Zhang, L.; Wei, Z.; Huang, X.; Wei, L. Super facile one-step synthesis of sugarcane bagasse derived N-doped porous biochar for adsorption of ciprofloxacin. *J. Environ. Manage.* 2023, 335, 117566.

6. Tripathi, M.; Sahu, J. N.; Ganesan, P. Effect of process parameters on production of biochar from biomass waste through pyrolysis: A review. *Renew. Sustain. Energ. Rev.* 2016, 55, 467–481.

7. Xiao, R.; Awasthi, M. K.; Li, R.; Park, J.; Pensky, S. M.; Wang, Q.; Wang, J. J.; Zhang, Z. Recent developments in biochar utilization as an additive in organic solid waste composting: A review. *Bioresour. Technol.* 2017, 246, 203–213.

8. Yu, S.; Zhang, W.; Dong, X.; Wang, F.; Yang, W.; Liu, C.; Chen, D. A review on recent advances of biochar from agricultural and forestry waste: Preparation, modification and applications in wastewater treatment. *J. Environ. Chem. Eng.* 2024, 12 (1), 111638.

9. Peng, L.; Duan, X.; Shang, Y.; Gao, B.; Xu, X. Engineered carbon supported single iron atom sites and iron clusters from Fe-rich Enteromorpha for Fenton-like reactions via nonradical pathways. *Appl. Catal. B: Environ.* 2021, 287, 119963.

10. Qi, Y.; Li, J.; Zhang, Y.; Cao, Q.; Si, Y.; Wu, Z.; Akram, M.; Xu, X. Novel lignin-based single atom catalysts as peroxymonosulfate activator for pollutants degradation: Role of single cobalt and electron transfer pathway. *Appl. Catal. B: Environ.* 2021, 286, 119910.
11. Chen, C.; Ma, T. F.; Shang, Y. N.; Gao, B. Y.; Jin, B.; Dan, H. B.; Li, Q.; Yue, Q. Y.; Li, Y. W.; Wang, Y.; Xu, X. In-situ pyrolysis of Enteromorpha as carbocatalyst for catalytic removal of organic contaminants: Considering the intrinsic N/Fe in Enteromorpha and non-radical reaction. *Appl. Catal. B: Environ.* 2019, 250, 382–395.
12. Li, R.; Deng, H.; Zhang, X.; Wang, J. J.; Awasthi, M. K.; Wang, Q.; Xiao, R.; Zhou, B.; Du, J.; Zhang, Z. High-efficiency removal of Pb(II) and humate by a CeO_2–MoS_2 hybrid magnetic biochar. *Bioresour. Technol.* 2019, 273, 335–340.
13. Lyu, H.; Lim, J. Y.; Zhang, Q.; Senadheera, S. S.; Zhang, C.; Huang, Q.; Ok, Y. S. Conversion of organic solid waste into energy and functional materials using biochar catalyst: Bibliometric analysis, research progress, and directions. *Appl. Catal. B: Environ.* 2024, 340, 123223.
14. Luo, M.; Lin, H.; He, Y.; Li, B.; Dong, Y.; Wang, L. Efficient simultaneous removal of cadmium and arsenic in aqueous solution by titanium-modified ultrasonic biochar. *Bioresour. Technol.* 2019, 284, 333–339.
15. Yan, L.; Liu, Y.; Zhang, Y.; Liu, S.; Wang, C.; Chen, W.; Liu, C.; Chen, Z.; Zhang, Y. $ZnCl_2$ modified biochar derived from aerobic granular sludge for developed microporosity and enhanced adsorption to tetracycline. *Bioresour. Technol.* 2020, 297, 122381.
16. Li, J.; Dai, J.; Liu, G.; Zhang, H.; Gao, Z.; Fu, J.; He, Y.; Huang, Y. Biochar from microwave pyrolysis of biomass: A review. *Biomass Bioenerg.* 2016, 94, 228–244.
17. Suresh, A.; Alagusundaram, A.; Kumar, P. S.; Vo, D.-V. N.; Christopher, F. C.; Balaji, B.; Viswanathan, V.; Sankar, S. Microwave pyrolysis of coal, biomass and plastic waste: A review. *Environ. Chem. Lett.* 2021, 19 (5), 3609–3629.
18. Motasemi, F.; Afzal, M. T. A review on the microwave-assisted pyrolysis technique. *Renew. Sustain. Energ. Rev.* 2013, 28, 317–330.
19. Quereshi, S.; Ahmad, E.; Pant, K. K. K.; Dutta, S. Insights into microwave-assisted synthesis of 5-ethoxymethylfurfural and ethyl levulinate using tungsten disulfide as a catalyst. *ACS Sustain. Chem. Eng.* 2020, 8 (4), 1721–1729.
20. Mankar, A. R.; Pandey, A.; Modak, A.; Pant, K. K. Pretreatment of lignocellulosic biomass: A review on recent advances. *Bioresour. Technol.* 2021, 334, 125235.

21. Zakaria, M. R.; Fujimoto, S.; Hirata, S.; Hassan, M. A. Ball milling pretreatment of oil palm biomass for enhancing enzymatic hydrolysis. *Appl. Biochem. Biotechnol.* 2014, 173 (7), 1778–1789.

22. Dora, S.; Bhaskar, T.; Singh, R.; Naik, D. V.; Adhikari, D. K. Effective catalytic conversion of cellulose into high yields of methyl glucosides over sulfonated carbon based catalyst. *Bioresour. Technol.* 2012, 120, 318–321.

23. Gallego-García, M.; Moreno, A. D.; Manzanares, P.; Negro, M. J.; Duque, A. Recent advances on physical technologies for the pretreatment of food waste and lignocellulosic residues. *Bioresour. Technol.* 2023, 369, 128397.

24. Lee, G.; Eui Lee, M.; Kim, S.-S.; Joh, H.-I.; Lee, S. Efficient upcycling of polypropylene-based waste disposable masks into hard carbons for anodes in sodium ion batteries. *J. Ind. Eng. Chem.* 2022, 105, 268–277.

25. Wang, J.; Ma, D.; Lou, Y.; Ma, J.; Xing, D. Optimization of biogas production from straw waste by different pretreatments: Progress, challenges, and prospects. *Sci. Total Environ.* 2023, 905, 166992.

26. Yu, Q.; Liu, R.; Li, K.; Ma, R. A review of crop straw pretreatment methods for biogas production by anaerobic digestion in China. *Renew. Sustain. Energ. Rev.* 2019, 107, 51–58.

27. Xu, P.; Shu, L.; Li, Y.; Zhou, S.; Zhang, G.; Wu, Y.; Yang, Z. Pretreatment and composting technology of agricultural organic waste for sustainable agricultural development. *Heliyon* 2023, 9 (5), e16311.

28. Hashmi, M.; Sun, Q.; Tao, J.; Wells, T.; Shah, A. A.; Labbé, N.; Ragauskas, A. J. Comparison of autohydrolysis and ionic liquid 1-butyl-3-methylimidazolium acetate pretreatment to enhance enzymatic hydrolysis of sugarcane bagasse. *Bioresour. Technol.* 2017, 224, 714–720.

29. Karnaouri, A.; Asimakopoulou, G.; Kalogiannis, K. G.; Lappas, A. A.; Topakas, E. Efficient production of nutraceuticals and lactic acid from lignocellulosic biomass by combining organosolv fractionation with enzymatic/fermentative routes. *Bioresour. Technol.* 2021, 341, 125846.

30. Li, C.; Knierim, B.; Manisseri, C.; Arora, R.; Scheller, H. V.; Auer, M.; Vogel, K. P.; Simmons, B. A.; Singh, S. Comparison of dilute acid and ionic liquid pretreatment of switchgrass: Biomass recalcitrance, delignification and enzymatic saccharification. *Bioresour. Technol.* 2010, 101 (13), 4900–4906.

31. Xiang, Q.; Zhang, H.; Liu, Z.; Zhao, Y.; Tan, H. Engineered structural carbon aerogel based on bacterial Cellulose/Chitosan and graphene Oxide/Graphene for multifunctional piezoresistive sensor. *Chem. Eng. J.* 2024, 480, 147825.

32. Yin, K.; Peng, L.; Chen, D.; Liu, S.; Zhang, Y.; Gao, B.; Fu, K.; Shang, Y.; Xu, X. High-loading of well dispersed single-atom catalysts derived from Fe-rich marine algae for boosting Fenton-like reaction: Role identification of

iron center and catalytic mechanisms. *Appl. Catal. B: Environ.* 2023, 336, 122951.

33. Yu, K. L.; Show, P. L.; Ong, H. C.; Ling, T. C.; Chi-Wei Lan, J.; Chen, W.-H.; Chang, J.-S. Microalgae from wastewater treatment to biochar – Feedstock preparation and conversion technologies. *Energ. Convers. Manage.* 2017, 150, 1–13.

34. Amalina, F.; Razak, A. S. A.; Krishnan, S.; Sulaiman, H.; Zularisam, A. W.; Nasrullah, M. Biochar production techniques utilizing biomass waste-derived materials and environmental applications – A review. *J. Hazard. Mater. Adv.* 2022, 7, 100134.

35. Dhyani, V.; Bhaskar, T. A comprehensive review on the pyrolysis of lignocellulosic biomass. *Renew. Energ.* 2018, 129, 695–716.

36. Thomas, P.; Lai, C. W.; Bin Johan, M. R. Recent developments in biomass-derived carbon as a potential sustainable material for super-capacitor-based energy storage and environmental applications. *J. Anal. Appl. Pyrolysis* 2019, 140, 54–85.

37. Lee, D.-J.; Lu, J.-S.; Chang, J.-S. Pyrolysis synergy of municipal solid waste (MSW): A review. *Bioresour. Technol.* 2020, 318, 123912.

38. Li, J.; Lin, L.; Ju, T.; Meng, F.; Han, S.; Chen, K.; Jiang, J. Microwave-assisted pyrolysis of solid waste for production of high-value liquid oil, syngas, and carbon solids: A review. *Renew. Sustain. Energ. Rev.* 2024, 189, 113979.

39. Beneroso, D.; Bermúdez, J. M.; Arenillas, A.; Menéndez, J. A. Influence of the microwave absorbent and moisture content on the microwave pyrolysis of an organic municipal solid waste. *J. Anal. Appl. Pyrolysis* 2014, 105, 234–240.

40. Lestinsky, P.; Grycova, B.; Pryszcz, A.; Martaus, A.; Matejova, L. Hydrogen production from microwave catalytic pyrolysis of spruce sawdust. *J. Anal. Appl. Pyrolysis* 2017, 124, 175–179.

41. Mushtaq, F.; Mat, R.; Ani, F. N. A review on microwave assisted pyrolysis of coal and biomass for fuel production. *Renew. Sustain. Energ. Rev.* 2014, 39, 555–574.

42. Li, J.; Ju, T.; Lin, L.; Meng, F.; Han, S.; Meng, Y.; Du, Y.; Song, M.; Lan, T.; Jiang, J. Biodrying with the hot-air aeration system for kitchen food waste. *J. Environ. Manage.* 2022, 319, 115656.

43. Zaker, A.; Chen, Z.; Wang, X.; Zhang, Q. Microwave-assisted pyrolysis of sewage sludge: A review. *Fuel Process. Technol.* 2019, 187, 84–104.

44. Das, S.; Lee, S. H.; Kumar, P.; Kim, K.-H.; Lee, S. S.; Bhattacharya, S. S. Solid waste management: Scope and the challenge of sustainability. *J. Clean. Prod.* 2019, 228, 658–678.

45. Eghbal Sarabi, F.; Liu, J.; Stankiewicz, A. I.; Nigar, H. Reverse traveling microwave reactor – Modelling and design considerations. *Chem. Eng. Sci.* 2021, 246, 116862.

46. Ge, S.; Yek, P. N. Y.; Cheng, Y. W.; Xia, C.; Wan Mahari, W. A.; Liew, R. K.; Peng, W.; Yuan, T.-Q.; Tabatabaei, M.; Aghbashlo, M.; Sonne, C.; Lam, S. S. Progress in microwave pyrolysis conversion of agricultural waste to value-added biofuels: A batch to continuous approach. *Renew. Sustain. Energ. Rev.* 2021, 135, 110148.

47. Sturm, G. S. J.; Verweij, M. D.; Stankiewicz, A. I.; Stefanidis, G. D. Microwaves and microreactors: Design challenges and remedies. *Chem. Eng. J.* 2014, 243, 147–158.

48. Chen, Z.; Wei, W.; Ni, B.-J.; Chen, H. Plastic waste derived carbon materials for green energy and sustainable environmental applications. *Environ. Funct. Mater.* 2022, 1, 34–48.

49. Jiang, M.; Wang, X.; Xi, W.; Zhou, H.; Yang, P.; Yao, J.; Jiang, X.; Wu, D. Upcycling plastic waste to carbon materials for electrochemical energy storage and conversion. *Chem. Eng. J.* 2023, 461, 141962.

50. Jie, X.; Li, W.; Slocombe, D.; Gao, Y.; Banerjee, I.; Gonzalez-Cortes, S.; Yao, B.; AlMegren, H.; Alshihri, S.; Dilworth, J.; Thomas, J.; Xiao, T.; Edwards, P. Microwave-initiated catalytic deconstruction of plastic waste into hydrogen and high-value carbons. *Nat. Catal.* 2020, 3 (11), 902–912.

51. Kumar, K.; Kumar, R.; Kaushal, S.; Thakur, N.; Umar, A.; Akbar, S.; Ibrahim, A. A.; Baskoutas, S. Biomass waste-derived carbon materials for sustainable remediation of polluted environment: A comprehensive review. *Chemosphere* 2023, 345, 140419.

52. Nangan, S.; Kanagaraj, K.; Kaarthikeyan, G.; Kumar, A.; Ubaidullah, M.; Pandit, B.; Govindasamy, R.; Natesan, T. Sustainable preparation of luminescent carbon dots from syringe waste and hyaluronic acid for cellular imaging and antimicrobial applications. *Environ. Res.* 2023, 237, 116990.

53. Zhou, X.; Zhu, L.; Dong, W.; Jiang, M. Solving two environmental problems simultaneously: Microporous carbon derived from mixed plastic waste for CO_2 capture. *Chemosphere* 2023, 345, 140546.

54. Sonu; Rani, G. M.; Pathania, D.; Abhimanyu; Umapathi, R.; Rustagi, S.; Huh, Y. S.; Gupta, V. K.; Kaushik, A.; Chaudhary, V. Agro-waste to sustainable energy: A green strategy of converting agricultural waste to nano-enabled energy applications. *Sci. Total Environ.* 2023, 875, 162667.

55. Khedulkar, A. P.; Pandit, B.; Dang, V. D.; Doong, R.-a. Agricultural waste to real worth biochar as a sustainable material for supercapacitor. *Sci. Total Environ.* 2023, 869, 161441.

56. Liu, S.; Han, G.; Zhang, J.; Wang, H.; Huang, X. Ultrafast, scalable and green synthesis of amorphous iron-nickel based durable water oxidation electrode with very high intrinsic activity via potential pulses. *Chem. Eng. J.* 2022, 428, 130688.

57. Wu, J.; Yang, J.; Huang, G.; Xu, C.; Lin, B. Hydrothermal carbonization synthesis of cassava slag biochar with excellent adsorption performance for Rhodamine B. *J. Clean. Prod.* 2020, 251, 119717.

58. Yao, F.; Ye, G.; Peng, W.; Zhao, G.; Wang, X.; Wang, Y.; Zhu, W.; Jiao, Y.; Huang, H.; Ye, D. Preparation of activated biochar with adjustable pore structure by hydrothermal carbonization for efficient adsorption of VOCs and its practical application prospects. *J. Environ. Chem. Eng.* 2023, 11 (2), 109611.

59. Zhou, N.; Chen, H.; Xi, J.; Yao, D.; Zhou, Z.; Tian, Y.; Lu, X. Biochars with excellent Pb(II) adsorption property produced from fresh and dehydrated banana peels via hydrothermal carbonization. *Bioresour. Technol.* 2017, 232, 204–210.

60. Cai, J.; Li, B.; Chen, C.; Wang, J.; Zhao, M.; Zhang, K. Hydrothermal carbonization of tobacco stalk for fuel application. *Bioresour. Technol.* 2016, 220, 305–311.

61. White, R. J.; Yoshizawa, N.; Antonietti, M.; Titirici, M.-M. A sustainable synthesis of nitrogen-doped carbon aerogels. *Green Chem.* 2011, 13 (9), 2428–2434.

62. Liu, S.; Wang, L.; Zheng, C.; Chen, Q.; Feng, M.; Yu, Y. Cost-effective asymmetric supercapacitors based on nickel cobalt oxide nanoarrays and biowaste-derived porous carbon electrodes. *ACS Sustain. Chem. Eng.* 2017, 5 (11), 9903–9913.

63. Zhao, N.; Liu, K.; He, C.; Zhao, D.; Zhu, L.; Zhao, C.; Zhang, W.; Oh, W.-D.; Zhang, W.; Qiu, R. H_3PO_4 activation mediated the iron phase transformation and enhanced the removal of bisphenol A on iron carbide-loaded activated biochar. *Environ. Pollut.* 2022, 300, 118965.

64. Dechapanya, W.; Khamwichit, A. Biosorption of aqueous Pb(II) by H_3PO_4-activated biochar prepared from palm kernel shells (PKS). *Heliyon* 2023, 9 (7), e17250.

65. Kan, Y.; Zhang, R.; Xu, X.; Wei, B.; Shang, Y. Comparative study of raw and HNO_3-modified porous carbon from waste printed circuit boards for sulfadiazine adsorption: Experiment and DFT study. *Chin. Chem. Lett.* 2023, 34 (7), 108272.

66. Liu, S.; Feng, X.; Liu, Z.; Zhao, W.; Li, Y.; Zhang, J. Nitrogen doping in porous biochar from cotton stalk with H_3PO_4 activation for reduction of NO_x with NH_3-SCR at low temperatures: Characteristics and catalytic activity analysis. *Fuel* 2023, 332, 126256.

67. Ma, W.; Fan, J.; Cui, X.; Wang, Y.; Yan, Y.; Meng, Z.; Gao, H.; Lu, R.; Zhou, W. Pyrolyzing spent coffee ground to biochar treated with H_3PO_4 for the efficient removal of 2,4-dichlorophenoxyacetic acid herbicide: Adsorptive behaviors and mechanism. *J. Environ. Chem. Eng.* 2023, 11 (1), 109165.

68. Wang, P.; Cao, J.; Mao, L.; Zhu, L.; Zhang, Y.; Zhang, L.; Jiang, H.; Zheng, Y.; Liu, X. Effect of H_3PO_4-modified biochar on the fate of atrazine and remediation of bacterial community in atrazine-contaminated soil. *Sci. Total Environ.* 2022, 851, 158278.

69. Ouyang, J.; Chen, J.; Chen, W.; Zhou, L.; Cai, D.; Ren, C. H_3PO_4 activated biochars derived from different agricultural biomasses for the removal of ciprofloxacin from aqueous solution. *Particuology* 2023, 75, 217–227.

70. Wang, L.; Wang, T.; Hao, R.; Wang, Y. Synthesis and applications of biomass-derived porous carbon materials in energy utilization and environmental remediation. *Chemosphere* 2023, 339, 139635.

71. Liu, S.; Dong, K.; Guo, F.; Wang, J.; Tang, B.; Kong, L.; Zhao, N.; Hou, Y.; Chang, J.; Li, H. Facile and green synthesis of biomass-derived N, O-doped hierarchical porous carbons for high-performance supercapacitor application. *J. Anal. Appl. Pyrolysis* 2024, 177, 106278.

72. Gong, X.; Xie, J.; Pan, X.; Luo, X. S, N co-doped carbon material functionalized catalytic membrane for efficient peroxymonosulfate activation and continuous refractory pollutants flow-treatment. *Chem. Eng. Sci.* 2023, 282, 119353.

73. Ren, Z.; Xu, X.; Wang, X.; Gao, B.; Yue, Q.; Song, W.; Zhang, L.; Wang, H. FTIR, Raman, and XPS analysis during phosphate, nitrate and Cr(VI) removal by amine cross-linking biosorbent. *J. Colloid Interface Sci.* 2016, 468, 313–323.

74. Wang, X.; Zhang, Q.; Jing, J.; Song, G.; Zhou, M. Biomass derived S, N self-doped catalytic Janus cathode for flow-through metal-free electrochemical advanced oxidation process: Better removal efficiency and lower energy consumption under neutral conditions. *Chem. Eng. J.* 2023, 466, 143283.

75. Xu, X.; Gao, Y.; Gao, B.; Tan, X.; Zhao, Y.-Q.; Yue, Q.; Wang, Y. Characteristics of diethylenetriamine-crosslinked cotton stalk/wheat stalk and their biosorption capacities for phosphate. *J. Hazard. Mater.* 2011, 192 (3), 1690–1696.

76. Saghir, S.; Xiao, Z. Synergistic approach for synthesis of functionalized biochar for efficient adsorption of Lopinavir from polluted water. *Bioresour. Technol.* 2024, 391, 129916.

77. Shang, Y.; Xu, X.; Gao, B.; Yue, Q. Highly selective and efficient removal of fluoride from aqueous solution by ZrLa dual-metal hydroxide anchored bio-sorbents. *J. Clean. Prod.* 2018, 199, 36–46.

78. Farhan, A.; Khalid, A.; Maqsood, N.; Iftekhar, S.; Sharif, H. M. A.; Qi, F.; Sillanpää, M.; Asif, M. B. Progress in layered double hydroxides (LDHs): Synthesis and application in adsorption, catalysis and photoreduction. *Sci. Total Environ.* 2024, 912, 169160.

79. Li, X.; Yang, Z.; Wu, G.; Huang, Y.; Zheng, Z.; Garces, H. F.; Yan, K. Fabrication of ultrathin lily-like $NiCo_2O_4$ nanosheets via mooring NiCo bimetallic oxide on waste biomass-derived carbon for highly efficient removal of phenolic pollutants. *Chem. Eng. J.* 2022, 441, 136066.

80. Du, W.; Zhang, Q.; Shang, Y.; Wang, W.; Li, Q.; Yue, Q.; Gao, B.; Xu, X. Sulfate saturated biosorbent-derived Co-S@NC nanoarchitecture as an efficient catalyst for peroxymonosulfate activation. *Appl. Catal. B: Environ.* 2020, 262, 118302.

81. Yin, K.; Wu, R.; Shang, Y.; Chen, D.; Wu, Z.; Wang, X.; Gao, B.; Xu, X. Microenvironment modulation of cobalt single-atom catalysts for boosting both radical oxidation and electron-transfer process in Fenton-like system. *Appl. Catal. B: Environ.* 2023, 329, 122558.

CHAPTER 3

Application in Solid Waste Disposal

Jinglan Wang, Zhanjun Cheng, Xiaoqiang Cui,
Jian Li, Lan Mu, Beibei Yan, Ning Li,
and Guanyi Chen

3.1 INTRODUCTION

Biochar is a carbon-enriched solid material produced through the pyrolysis of organic substances under anaerobic conditions and at temperatures typically below 900°C. Initially utilized for soil enhancement, biochar serves to sequester carbon, improve nutrient availability, reduce soil compaction, and enhance soil pH. Its modern applications have expanded to encompass energy production, biochemical process optimization, climate change mitigation, and construction materials.[1] The physicochemical attributes of biochar, which include a stable structure, thermal and mechanical robustness, chemical inertness, high surface acidity, and environmental compatibility, are pivotal in determining its diverse applications. These extend to the adsorption of water and air pollutants,[2] the creation of activated carbon,[3] the enhancement of anaerobic digestion,[4] and utilization in the construction industry[5] and agriculture for soil conditioning[6] and compost additives.[7] Notably, biochar's catalytic properties have been harnessed in sustainable processes such as pyrolysis, gasification, and hydrothermal treatments, favoured for their environmental benefits and economic viability. Also, biochar is beneficial in the anaerobic fermentation and composing due to its a large specific surface area, porosity, and abundance of

DOI: 10.1201/9781003535409-3

functional groups. In this context, this chapter comprehensively discusses these varied applications of biochar.

3.2 APPLICATION OF BIOCHAR IN CATALYTIC PYROLYSIS

As shown in Figure 3.1, biochar is a solid residue resulting from the thermo-chemical conversion of lignocellulosic biomass by pyrolysis. Biochar is a rich resource with a unique structure and multiple functions.[8] It contains various elements such as C, N and P, which have value-adding potential and provide a sustainable and environmentally friendly pathway for the study of catalysts or catalyst carriers.[9] The preparation and application of biochar is one of the effective methods of biomass recycling which overcomes the shortcomings of the high cost of traditional activated carbon.

3.2.1 Biochar-Catalysed Biomass Pyrolysis

The introduction of a biochar catalyst decreases the yield of bio-oil but increases the yield of gaseous products. This is because biochar promotes the secondary cracking of pyrolyzed volatiles, leading to several reactions such as deoxygenation and bond breaking, resulting in more gaseous products.[10] There is an increase of hydrogen from 0.82% to 3.74%, carbon dioxide from 21.16% to 32.33%, and carbon monoxide from 16.49% to 23.19% in the gas-phase composition of woodchip pyrolysis with the addition of biochar as compared to the non-biochar pyrolysis.[11] With a biochar catalysed pyrolysis of sawdust, the content of H_2 and CO in the syngas was as high as 20.43 Vol% and 43.03 Vol%, respectively, whereas, during a catalytic pyrolysis of dry biomass, these contents were 27.02 Vol% and 38.34 Vol%, respectively.[12] ThermoGravimetric (TG) analysis showed that the original and recovered biochar catalysts had good thermal stability.

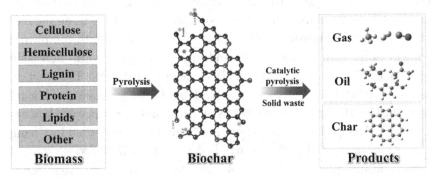

FIGURE 3.1   Catalytic pyrolysis of solid waste by biomass preparation of biochar.

There are two main reaction mechanisms for the increase in H_2 and CO concentrations. One is that H_2O and CO enter the biochar catalyst and undergo a water-gas shift reaction due to the presence of some metals (for example, Fe and Cu), increasing H_2 concentration. The other reason is the dry conversion of methane, where CH_4 decomposes or reacts with CO_2 to produce large amounts of H_2 and C.[10] When activating biochar with KOH, K_2CO_3, $KHCO_3$, or CH_3COOK, the oxygen-containing groups on the surface of the biochar can be increased. With the introduction of activated biochar catalysts, the bio-oil yields decreased and the gaseous product yields increased, especially CO, CO_2, and CH_4. After catalytic pyrolysis, the reactive O-containing groups were all reduced, especially -OH, O-C double bond O, and C-O groups, which had higher catalytic activity for biomass pyrolysis. The activated biochar catalysts act as both catalysts and reactants in the catalytic pyrolysis process.[13] Metal loading also affects the catalytic effect of biochar. the Fe/AC catalyst increased the yield of light gas. A peak yield of 37.09 wt% of the gaseous fraction was achieved with the 1%-Fe/AC catalyst. The presence of iron improved the selectivity of the deoxygenation products. Catalytic reforming of PKS pyrolysis vapours over an Fe/AC catalyst resulted in 75.12 vol% H_2 production.[14]

In the biomass pyrolysis process, the biochar catalyst not only enhances the syngas but also improves the quality of bio-oil. Raw bio-oils from non-catalytic pyrolysis have very high levels of guaiacol, plus amounts of furans, phenols, acids, sugars and other unclassified compounds. Guaiacol and acids are considered undesirable, especially for liquid transportation fuel production purposes. Highly oxygenated compounds such as guaiacol, esters and ethers reduce the calorific value of bio-oil and make it chemically unstable. Organic acids are corrosive and make it difficult to use bio-oils in engines. Upgrading is necessary to obtain a more stable and energy-intensive liquid product.[15] Organic acids were almost undetectable in bio-oil produced from biomass catalysed pyrolysis using biochar catalysts. In the catalytic pyrolysis of raw Douglas fir pellets, the phenol content ranged from 30% to 38% and increased slightly with increasing biochar catalyst loading, while guaiacol decreased slightly. The reaction mechanism for the conversion of guaiacol may be the generation of phenol from the O-CH$_3$ cleavage of methyl by guaiacol in the presence of a biochar catalyst.[12] Four pathways for phenol production were proposed by Zhang et al.[16] Firstly, most phenols are mainly derived from lignin pyrolysis, and some unstable phenols, such as 4-ethyl-phenol and 2-methoxyphenol, may undergo demethylation and demethoxylation reactions catalysed by high temperatures, which explains the decrease of 4-ethyl-phenol and the increase of phenol content at 700°C and 800°C. In addition, ketones from cellulose or hemicellulose pyrolysis undergo oligomerization, decarbonylation, and the Diels-Alder reaction to produce phenols and the release of CO, CO_2, and CH_4, and AAEM in biochar may contribute to decarbonylation reactions. It is hypothesized that a significant reduction in 2,3-dihydrobenzofurans is responsible for phenolic production from the ring opening, while other furans such as furfural rearrangements also contribute to the production of additional phenols. In addition, since high temperatures favour the decarbonylation reaction of ketones, higher phenolic yields are obtained as the temperature is increased below 700°C, which could well explain the results of 1,2-cyclopentanedione and 3-methyl-1,2-cyclopentanedione which are detected only at 500°C.

The N-doped biochar catalyst greatly promoted the generation of phenols (up to 82%), especially the valuable 4-vinylphenol (31% content, 6.65 wt% yield) and 4-ethylphenol (16% content, 3.04 wt% yield). It also promoted the formation of aromatic hydrocarbons while inhibiting the production of O-species and acetic acid and releasing more CO_2 and H_2O.[17] N-doped biochar catalysts promoted the generation of phenolic compounds, which may be attributed to the facts that:[17]

1. The reactive O and N groups in the N-doped biochar catalysts promoted the breaking of β-O-4 linkages and the formation of more phenol intermediates

2. The N-doped biochar catalysts also promoted the reaction of H-donors (for example, water, aldehydes, and alcohols) generated from the decomposition of hemicellulose and cellulose with the phenol intermediates (H-receptors) and the formation of more phenols[18]

3. N-doped biochar catalysts promoted the cleavage of -O-CH_3 in phenol intermediates to produce more monomeric phenols[19]

4. N-doped biochar catalysts also promoted the decomposition of lignin and inhibited the cleavage of hemicellulose and cellulose, which increased phenol content of the bio-oil.[20]

When metallic elements are introduced on the surface of biochar, they provide more active sites on the surface of the biochar, which improves the biomass pyrolysis reaction.[21] Behnam Nejati et al.[22] investigated the catalytic reaction of Fe-loaded biochar on microalgae. The carbon content in the bio-oil increased by 17.14% with the Fe10/CAC catalyst as compared to the non-catalysed experiments. The liquid product should contain more aliphatic or aromatic hydrocarbons. The catalytic temperature was 650°C and the bio-oil yield was 46.23% using 10% Fe/AC as catalyst. The HHV and ER values of the bio-oil were 31.26 and 71.58, respectively.

3.2.2 Biochar-Catalysed Polymers Pyrolysis

Catalytic up-cycling is a promising waste management strategy to improve the recyclability of polymer waste by converting it into high-value-added products. Catalysts play a key role in up-cycling, affecting catalytic capacity, conversion of polymer waste, selective formation, yield and purity of the target product, and energy requirements for up-cycling.[23] The distribution of gaseous products from pyrolysis varies widely, and the presence of a biochar catalyst significantly affects the gas distribution. Non-catalytic pyrolysis of LDPE produced a high concentration of H_2 (57.75 vol%). However, C_3H_8 (49.74 vol%) was the major gas product of catalytic pyrolysis of LDPE, and the H_2 content decreased to 1.40 vol%. Furthermore, the addition of biochar catalyst reduced the CO_2 and C_2H_6 contents from 3.72% and 8.89% to zero, respectively. The high content of hydrogen was obtained because hydrogen may not have enough time to react with the free radicals of hydrocarbons. In the presence of the biochar catalyst, the hydrocarbon and hydrogen radicals may have had more time to react, or the biochar may have catalysed the cracking of the reaction intermediates, yielding propane as the main gaseous product.[24] To facilitate the production of high-quality gas products from plastics, heterogeneous catalytic technology has been developed. The plastics are pyrolyzed in a first-stage reactor and the pyrolyzed volatiles are directed to a second-stage reactor for catalytic steam reforming to produce hydrogen-rich syngas. The hydrogen yield of polyolefin plastics was 0.44 g plastic^{-1} with a biochar catalyst, but only 0.32 g plastic^{-1} with a solid waste charcoal catalyst. At a steam input of 10 g h^{-1} catalyst^{-1} and a catalyst temperature of 1000°C, when biochar was used as a catalyst, High-Density PolyEthylene (HDPE) produced the highest syngas (H_2, CO) yield of 3.83 g plastic^{-1} and the highest syngas (H_2, CO) yield of 2.73 g plastic^{-1} when solid waste charcoal was used as catalyst.[25]

Corn stover and Douglas-fir derived biochar catalysts were used for catalytic pyrolysis of model Low-Density PolyEthylene (LDPE) and actual waste plastics. The corn stover-derived biochar produced a liquid yield of approximately 40 wt% with no wax formation. The liquid product contained approximately 60% C_8-C_{16} aliphatic, 20% monoaromatic, and 20% C_{17}-C_{23} aliphatic hydrocarbons.[26] Chenxi Wang et al.[27] catalysed pyrolysis over a corn stover-derived biochar catalyst and for the first-time enhanced hydrogen and monoaromatic hydrocarbon production. Up to 53 wt% waxes were observed in the pyrolysis of masks (mainly PP) at 550°C in the absence of biochar, whereas at a biochar/PP ratio of 2, about 41 wt% liquid oil was produced and no waxes were formed. Ni metal loaded on lignocellulosic biomass was prepared for the catalytic pyrolysis of plastics. The Ni/lignocellulosic biochar showed a chemoselectivity of 76.2% for gasoline hydrocarbons (C_4-C_{12} hydrocarbons) and significant catalytic cracking of LDPE. The catalytic performance was comparable to that of Ru-based catalysts, which contributed 85.6% to gasoline hydrocarbons. The nickel/lignin-catalysed cracking of LDPE, PolyPropylene (PP), and PolyStyrene (PS) simulated plastic waste showed a maximum selectivity of 87.9% for C_4-C_{12} hydrocarbons and a yield of 78.1 mg/g of Benzene, Toluene, and Xylene (BTX). Unprocessed biochar showed little catalytic efficiency for plastic conversion because of its insignificant acidic sites. However, the functional groups on the surface of lignin biochar act like pincers and facilitate the deposition and dispersion of metals. As a result, the catalytic efficiency of the metal sites was improved.[28]

3.2.3 Biochar-Catalysed Pyrolysis of Other Solid Waste

The biochar catalyst was used for the pyrolysis of tyre waste. The biochar catalyst had a significant influence on the evolution of volatiles and char properties while having less influence on the yields of gas, tar, and char products. Pyrolysis of tyre waste was carried out at 500°C using biochar produced by gasification of poplar wood at 850°C. The biochar catalyst catalysed the cracking of limonene, the main liquid product in tar, resulting in the formation of more propane in gas and more alkanes or olefins in tar.[29]

Pyrolysis of Oily Sludge (OS) is a viable technology that meets the principles of minimization and recycling, but its feasible environmental destination is difficult to determine and meet the corresponding requirements. In the catalytic pyrolysis process, biochar acts as a catalyst to enhance the removal of difficult-to-degrade petroleum hydrocarbons at the expense of liquid product yield. At the same time, biochar acts as an adsorbent to inhibit the release of micromolecular gaseous pollutants (for example, HCN, H_2S, and HCl) and stabilize heavy metals. The pyrolysis reaction of OS is more likely to occur due to the auxiliary effect of biochar and requires lower temperatures to achieve the same effect.[30]

Various waste management methods are available for treating food waste; however, these methods are often challenged by issues such as high costs, generation of toxic by-products, and environmental pollution. Pyrolysis has recently attracted increasing interest as a potential sustainable and environmentally friendly solution to valorise food waste by developing new products.[31] Jong-Min Jung et al. investigated a low-cost, highly efficient, and porous catalytic material (namely, chicken manure biochar), which was made from chicken manure by pyrolysis, for the conversion of waste cooking oil to Fatty Acid Methyl Ester (FAME) via a transesterification reaction (FAME, namely, biodiesel). The biochar contains many inorganic compounds (mainly $CaCO_3$), which can accelerate the catalytic activity during the transesterification of waste cooking oil. Compared to silica, chicken manure biochar lowers the transesterification temperature (350°C), where FAME yields are highest (95%).[32]

3.3 GASIFICATION

3.3.1 Biomass Gasification Enhanced by In-Situ Solid Waste-Based Catalysts

The biomass gasification process can be enhanced by two methods. From the perspective of tar removal, the in-situ catalyst can enhance in-situ tar elimination, thereby increasing the yields of products (especially for syngas) and optimizing product distribution. From the perspective of the gasification process, the in-situ catalyst can facilitate the gasification reaction and thus increase the gasification efficiency. Given the demanding conditions of gasification, in-situ catalysts should exhibit greater resistance to attrition and agglomeration, and be easier to separate and recover than ex-situ catalysts.[33] With these considerations in mind, catalysts for in-situ tar removal fall into two categories: those based on biomass char/ash and those derived from other solid waste, which mainly include minerals, noble metals, and non-noble metal catalysts (Table 3.1).

3.3.1.1 Biomass Char/Ash as In-Situ Catalysts

Biomass char, which is a by-product of biomass gasification, has emerged as a potential carbon-based catalyst for tar removal and promoting gasification due to its highly-developed matrix structure, trace metal content and

TABLE 3.1 Features and performance of typical in-situ catalysts

Catalyst	Features and performance	Ref.
Calcium oxide	The use of 2,3 and 6 wt% in-bed CaO promoted the conversion of Class1, 4 and 5 tars to Class 3 tars to varying degrees Tar yields decreased ranging from 16~35%; Tar dew point decreased ranging from 37~60°C.	47
Ru-based catalyst	Tar yield decreased from 32.1 wt% to 19.3 wt%; Yield of combustible gas reached 84 wt%; A low char yield (3.9 wt%) can be obtained for Ru-based catalyst.	48
Dolomite and olivine	Tar yield was reduced by 92% for dolomite; Tar yield was reduced by 40% for olivine; Olivine exhibited better slag resistance. Tar content achieved 2 g/Nm³ at the exit.	49
Activated carbon	The high specific surface areas and relatively more defects in activated carbon are primary factors for tar upgrading; The heavy tars were decomposed to light tars and gases.	50
Calcium oxide	Tar yield decreased with the increasing temperature, reaching the minimum at 1000°C (1 wt%); Calcium oxide reduced tar yield from 3.8 wt% to 0 wt% at 900°C.	51
Dolomite	Tar yield was decreased by 23.16% and 26.11% when 3% and 6% (w/w) dolomite was added.	52
Ilmenite	Low levels of ilmenite reduced the tar yield by similar to 50 wt%; Lower fluidization velocity gave the highest reduction of the tar yield.	53

low cost. The effects of activated carbon were examined by Jin et al.[34] The results showed that the heavy tars were converted to light tars and gases with the benefit of activated carbon. Zhang et al.[35] researched the effects of biochar catalysts on tar inhibition, showing a maximum tar decomposition efficiency of 91.75% at 800°C. Since carbon-based catalysts suffer from accumulated coke, the synergistic application of carbon-based catalyst with hydrogen catalysis not only effectively promotes the decomposition of tar and coke, but also upgrades the quality of syngas and allows continuous regeneration of carbon-based catalysts.

The ash-based catalyst, which is also known as the residue-based catalyst, has the advantage of low cost, high metal content, and allows material recycling in the gasification system. Common ash-based catalysts include Paper Sludge Ash (PSA), petroleum residues, and ash derived from the pyrolysis and gasification of industrial solid waste and municipal waste.

PSA has been proven to reduce tar yields while increasing the yields of H_2, CH_4 and CO_2, with the yield of H_2 almost doubling.[36] At the same time, PSA shows good catalytic reactivity and stability, which achieves a tar conversion rate of 62.8% at 700°C. As a cheap and widely available catalyst, ash has a positive impact on tar inhibition, biomass conversion and gasification rates. For example, by increasing the coal bottom ash composition from 0.02% to 0.07%, the palm kernel shell conversion rate during steam gasification increased from 69% to 91%,[37] and the gasification rate increased from 28% to 82%. Since alkaline metal oxides and CaO are widely present in ash, tar can be easily reduced. In addition, the presence of Al_2O_3 and Fe_2O_3 helps to resist the carbon deposition and sintering of the in-situ catalyst.[38] The presence of SiO allows the ash to be used as the bed material. This phenomenon could significantly affect downstream processes such as cogeneration or Fischer-Tropsch synthesis. As ash can be continuously generated during gasification, the recirculation of ash will maintain the amount of catalyst, thus stabilizing the product yield and providing a positive impact on the economic viability and sustainability of gasification system.[39]

3.3.1.2 Other Solid Waste-Based Catalysts

Mineral catalysts are usually low-cost and widely available, but suffer from attrition and agglomeration. Dolomite is a widely available catalyst that is frequently used for in-situ tar removal. It is effective for the decomposition of tar at approximately 900°C.[40] The in-situ catalyst shows more contact with volatiles than ex-situ catalysts, which results in easier coke accumulation and agglomeration. Metal loading and calcination can produce more stable and active dolomite catalysts, reaching a high tar removal efficiency of 97% at 750°C.[41] A mineral aluminium silicate catalyst can not only reduce tar yield of approximately 50%, but also significant increase gasification efficiency, resulting the lower heating value of syngas from 8.66 MJ/Nm^3 to 13.38 MJ/Nm^3.[42]

Metal-based catalysts are widely used due to the catalytic active sites and excellent catalytic performance. The noble-metal catalysts, mainly including Rh, Pt, Pd and Ru-based catalysts, show advantages of high thermal/chemical stability and catalytic capacity. For example, Rh-CeO_2-SiO_2 catalysts have shown an extremely stable conversion capacity of H_2, CO and CH_4.[43] The carbon utilization rate can reach 77% over the entire catalytic reaction, while the Ni-CeO_2-SiO_2 catalysts showed a decrease in carbon utilization rate from 70% to 58%. At 823 K, most noble-metal catalysts have shown stronger tar catalytic activity than non-noble-metal catalysts (Rh > Pt > Pd > Ni = Ru). In contrast, non-noble-metal catalysts are slightly inferior in terms of catalytic efficiency and stability, but are easy to synthesize, cheaper and widely available. The char-supported iron catalysts can reduce the tar yield by 84% and 96% at 800°C and 850°C respectively, and the char-supported nickel catalysts showed less apparent degradation even at low temperatures.[44, 45] Shen et al.[46] developed in-situ tar conversion methods in a fluidized-bed gasifier using rice-husk-ash-supported nickel-iron bimetallic catalysts. Under optimum conditions, 93% tar conversion efficiency was achieved using rice-husk-char-supported Ni. And the gasification products were optimized.

3.3.2 Tar Removal with the Application of Solid Waste-Based Materials

3.3.2.1 Absorption of Tar by Solid Waste

Choosing appropriate solid waste as the gasification tar adsorbent has significant advantages and potential. Firstly, solid waste as adsorbent has the advantages of accessibility and low cost. Secondly, solid waste as adsorbent also helps promote waste management and resource recycling. Converting waste into adsorbents not only reduces waste generation and landfilling but also allows for their recycling. Most importantly, solid waste as adsorbent can effectively remove gasification tar and improve the quality of syngas. Adsorbents can capture and adsorb harmful substances in gasification tar, reducing pollutant emissions.[54]

The main solid waste materials used for tar adsorption include activated carbon, biomass char, waste ash, alumina, and silica gel. Activated carbon, with its highly developed pore structure and sustainable sourcing, is suitable for adsorbing various gasification tar components. Biomass char, derived from agricultural and wood waste, exhibits excellent adsorption performance for biomass gasification tar.[55] Waste ash, as an industrial byproduct, is characterized by its low cost and availability, and its mineral components contribute to effective tar adsorption.[56] Alumina demonstrates good tar removal efficiency at high temperatures due to its stability.[56] Silica gel, known for its controllable pore structure and high-temperature resistance, is suitable for selectively adsorbing specific tar components. The advantages of these waste materials in tar adsorption include their sustainability, cost-effectiveness, and selective adsorption of different tar components, depending on gasification conditions, tar composition, and equipment requirements.

The performance study of solid waste as adsorbent for gasification tar mainly involves batch adsorption experiments, isotherm studies, and kinetic analysis methods.[57] In batch adsorption experiments, the solid waste adsorbent is brought into contact with a gas containing gasification tar, and the adsorption capacity and efficiency of the adsorbent are evaluated to assess its ability to remove gasification tar. Different types of solid waste adsorbents can be tested in the experiment, and factors such as the dosage of the adsorbent, temperature, and flow rate can be investigated to determine their effects on adsorption performance. By continuously adjusting the experimental conditions, the optimal adsorbent and operating parameters can be identified to achieve efficient removal of gasification tar.

Isotherm study is an important method to evaluate the selectivity and capacity of solid waste adsorbents for different gas-phase components. By plotting and analysing adsorption isotherms, the selectivity and capacity of solid waste adsorbents for different gas-phase components can be understood. The commonly used isotherm models include the Langmuir model and the Freundlich model, which can be used to describe the adsorption capacity and selectivity of the adsorbent.[58, 59] The formulas of these two models are as follows:

(1) Langmuir model: The Langmuir model assumes that there are uniform and independent adsorption sites on the surface of the adsorbent, and the adsorption process is a monolayer molecular adsorption.

$$q = (K * C)/(1 + K * C)$$

where q is the adsorption capacity (mol/mg), C is the concentration of gasification tar in the gas phase (mol/L), and K is the Langmuir adsorption constant ((mol/mg) * (L/mol)$^{-1}$). By fitting the experimental data, the Langmuir adsorption constant b can be obtained.

(2) Freundlich model: The Freundlich model assumes that there are non-uniform adsorption sites on the surface of the adsorbent, and the adsorption process is a multilayer molecular adsorption.

$$q = k * C^N$$

where q is the adsorption capacity (mol/mg), C is the concentration of gasification tar in the gas phase (mol/L), k is the Freundlich adsorption coefficient ((mol/mg) · (mol/L)$^{-n}$)), and n is the Freundlich adsorption index. By fitting the experimental data, the Freundlich adsorption coefficient k and adsorption index n can be obtained.

These isotherms reflect the adsorption behaviour of the adsorbent for the components of gasification tar under different pressures and temperatures.

By studying the isotherms, the affinity and saturation capacity of the adsorbent for the target components can be determined, thereby evaluating the performance of solid waste as an adsorbent and further optimizing the selection and operating conditions of the adsorbent.

Kinetic analysis is an important method for studying the rate and characteristics of adsorption processes.[60] By monitoring the variation of adsorption capacity with time, the rate constants and reaction mechanisms of the adsorption process can be determined, providing insights into the adsorption kinetic behaviour. Kinetic analysis typically involves fitting experimental data with different models and calculating adsorption rates and equilibrium times through parameter fitting. By studying the kinetics of the adsorption process, a better understanding of the performance and response patterns of the adsorbent can be obtained, leading to optimized design and operation of the adsorption process. Typical kinetic models include first-order kinetics and second-order kinetics.

(1) First-order kinetics model: The first-order kinetics model assumes that the adsorption process follows first-order reaction kinetics, where the adsorption rate is proportional to the concentration of the adsorbate in the solution. The equation for the first-order kinetics model is:

$$\frac{dq}{dt} = k_1 * (C_0 - C)$$

where $\frac{dq}{dt}$ is the rate of change of adsorption capacity per unit time (mg/(mg·min)), C_0 is the initial concentration (mg/L), C is the concentration during the adsorption process (mg/L), and k_1 is the first-order adsorption rate constant (1/min).

(2) Second-order kinetics model: The second-order kinetics model assumes that the adsorption process follows second-order reaction kinetics, where the adsorption rate is proportional to the square of the concentration of the adsorbate in the solution. The equation for the second-order kinetics model is:

$$\frac{dq}{dt} = k_2 * (C_0 - C)^2$$

where $\frac{dq}{dt}$ is the rate of change of adsorption capacity per unit time ((mg/(mg·min)), C_0 is the initial concentration (mg/L), C is the concentration during the adsorption process (mg/L), and k_2 is the second-order adsorption rate constant ($(mg/(mg \cdot min))$).

In addition to experimental methods, theoretical simulation and computational methods can also be employed to study the performance of solid waste as an adsorbent. Based on adsorption theories and computational models, the adsorption capacity, selectivity, and kinetic behaviour of the adsorbent can be predicted. Theoretical simulation methods can assist researchers in gaining a better understanding of the adsorption mechanisms and optimizing the design and preparation methods of the adsorbent.

Although solid waste as a gasification tar adsorbent has many advantages and potential, it also faces challenges. Firstly, the surface properties and pore structure of solid waste greatly influence adsorption effectiveness. Therefore, further research and optimization of solid waste preparation methods are needed to enhance their adsorption performance and stability. Secondly, gasification tar consists of complex and diverse components with varying adsorption selectivity. Hence, in-depth investigations into the selectivity of solid waste adsorbents for different gasification tar components are necessary to optimize the adsorption process and improve removal efficiency. Additionally, the regeneration and recycling of solid waste as adsorbent are crucial considerations. Effective regeneration methods can prolong the lifespan and recycling rate of adsorbents while reducing treatment costs. Thus, research on suitable regeneration techniques for solid waste adsorbents is needed. Future research directions also include improving the stability and anti-pollution capabilities of solid waste adsorbents to address the potential impact of other substances (such as dust and sulfides) on adsorption effectiveness during gasification processes.

3.3.2.2 Cracking/Reforming of Tar by Solid Waste-Based Catalysts

Tar cracking and reforming are both tar thermochemical conversion methods, which represent efficient technology in tar conversion methods at present. Thermal cracking refers to the tar compounds in the gasification gas which could be dissociated in a high temperature environment, where the larger molecules of the compound would be dissociated into smaller molecules of gaseous compounds and other products, through dehydrogenation, dealkylation and some other free radical reactions.[61] Tar cracking processes are generally described as follows: tar is thermally decomposed into solid carbon, gas and reactive free radicals, most of which are aromatic. At low temperatures, these free radicals can be polymerized to form larger molecules, and condense at room temperature to form light components of tar; At higher temperatures, reactive free radicals can undergo further pyrolysis to form gas products and deposit carbon. At higher temperatures, the deposited carbon can undergo steam gasification reactions, increasing the conversion of carbon and gas products.

Catalytic reforming refers to the process in which C-C and C-H in tar molecules can be more easily reformed to produce small molecular organic substances under the action of catalysts, and the generated hydrocarbons and coke can further react with H_2O, CO_2 and other gases under the action of catalysts, resulting in an increase in the total amount of syngas and a reduction in tar precursor components.[62] The addition of a catalyst promotes the reforming reaction from two aspects: on the one hand, the catalyst surface provides a base and a place for intermolecular bonding and reaction; on the other hand, the active site of catalyst involved in chemical reaction reduces the activation energy of intermolecular reaction to a certain extent, making intermolecular reaction easier. Tar cracking requires a higher temperature, and the addition of catalysts can reduce the tar pyrolysis temperature, and the tar conversion rate can be similar to that of 1000–1200°C thermal cracking at 700–900°C.[63] At the same temperature, the product gas obtained by catalytic reforming contains higher H_2 and lower CH_4, C_2H_4, C_2H_6 components than that obtained by thermal cracking.[64]

At present, the catalysts added in the biomass tar cracking process are mainly natural ores and synthetic catalysts, natural ore catalysts are dolomite and olivine, synthetic catalysts are nickel-based catalysts, alkali metal catalysts, new metal catalysts and so on.[65] Among synthetic catalysts, carbon-supported metal catalysts prepared by activated carbon or semi-coke supported by active metals (such as Ni, Co, Fe, and the like) using solid waste as raw materials have attracted much attention in recent years.[66] They usually have the characteristics of highly developed pore structure, abundant surface acid groups and oxygen-containing functional groups, and are cheap and easy to obtain. They can effectively adsorb light tar compounds and achieve biomass tar removal under the action of active metal sites.

A large number of researchers have proved through experiments that carbon-based catalysts prepared from solid waste have good tar conversion performance. Du et al. prepared a Ni/BC catalyst by a one-step impregnation method showing a conversion rate of 98.7% to toluene at a low catalytic temperature (600°C) and S/C of 3, with activity stable for 10 h. From the perspective of catalyst structure, 5% NiBC shows the highest specific surface area (524.6 vol m^2/g) and the smallest metal Ni particle size (4.8 nm), and therefore the highest metal dispersion and conversion frequency (TOF) on the catalyst.[67] Tian et al. synthesized monolithic biochar-supported cobalt-based catalysts with unique long and through mesopores, based on the biological channels of biomass materials through simple impregnation and carbonization, finding that the average toluene reforming rate reached around 97% with H_2 and CO as the main product by using PC@CoNi as the catalyst.[68] In addition, some researchers have introduced microwaves as a heat source for catalytic cracking, further improving the efficiency of tar catalytic cracking, because carbon is a good microwave absorber and can efficiently convert microwave energy into heating. Li et al. investigated toluene cracking and CO_2 reforming under microwave irradiation on biomass derived char, finding that microwave heating had a promoting effect on toluene conversion.[69] Li et al. converted toluene with char-supported nickel-iron and nickel-cerium catalysts in an ex-situ hot gas condition, and proved that the conversion efficiency of the nickel-cerium catalysts could still be higher than 90% after an 8 h successive run.[70,71]

3.4 ANAEROBIC DIGESTION

Anaerobic fermentation mainly refers to the decomposition of organic substances in organic solid waste under specific anaerobic conditions by facultative bacteria and anaerobic bacteria (mainly including fermentative bacteria, acid-producing bacteria, and methanogens), and finally the production of methane and carbon dioxide, accompanied by the production of a large number of value-added chemicals such as volatile fatty acids, lactic acid, ethanol, and so forth.[72] After purification, the methane produced can be used to generate heat and electricity. Metabolites such as organic acids and alcohols produced in the fermentation process can be used as alternative raw materials for industrial chemicals, thereby replacing fossil fuels to a certain extent. The anaerobic fermentation process has two stage, three stage, and four stage theory as shown in Figure 3.2.

Two Stage Theory[73]

The two-stage theory was proposed in 1906. According to the theory, anaerobic fermentation can be divided into two stages or groups of metabolic bacteria, including the acidogenic phase and the methanogenic phase. The two groups of metabolic flora were fermentative bacteria without methanogenesis and methanogenic bacteria.

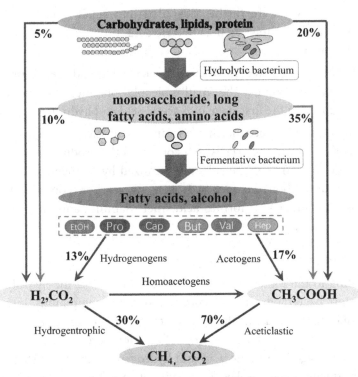

FIGURE 3.2   Stages and main pathways in anaerobic digestion.

1. The microbial flora in the acid-producing phase (Group I bacteria) hydrolyses and ferments complex organic matter to form components such as volatile fatty acids, alcohols, carbon dioxide, hydrogen, and hydrogen sulfide.

2. The microbial flora in the gas production stage converts the fermentation products of the previous stage into methane and carbon dioxide, so it is called methanogenic flora.

Three Stage Theory[74]

1. In the first stage, the hydrolytic fermentation bacteria hydrolyse and convert complex organic matter into small molecular organic matter, in which carbohydrates such as cellulose and starch are hydrolysed to sugars. Proteins and other organic nitrogen substances are hydrolysed to amino acids, which are then deaminated to form organic acids and ammonia. Lipids are hydrolysed to form glycerol and fatty acids;

Organic acids are further degraded into small molecular organic acids and alcohols, such as butyric acid, propionic acid, acetic acid, ethanol, as well as hydrogen and carbon dioxide.

2. In the second stage, the hydrogen-producing and acetate-producing bacteria decompose the small molecular organic acids and alcohols produced in the first stage into acetic acid and hydrogen.

3. In the third stage, the microorganisms involved in the stage were methanogenic bacteria, which converted the products of the second stage to methane. The substrates utilized by methanogenic bacteria include carbon dioxide, hydrogen, one-carbon compounds (such as CO, methanol, formic acid, methylamine, and so forth.) and acetic acid.

Four Stage Theory

According to the four-stage theory,[75] in the first stage, the hydrolytic fermentation bacteria hydrolyse complex organic compounds into small molecular compounds. In the second stage, the hydrogen-producing and acetate-producing bacteria decomposed various organic acids produced in the first stage into acetic acid and hydrogen. In the third stage, acetic acid producing bacteria played a dominant role and converted hydrogen and carbon dioxide to acetic acid. In the fourth stage, microorganisms are methanogens, which convert carbon dioxide, hydrogen, one-carbon compounds and acetic acid to methane.

3.4.1 Biochar

Biochar is derived from plant-derived agroforestry biomass waste such as straw and is obtained by pyrolysis under hypoxia or limited oxygen supply and relatively low temperature (450–700°C).[76] The elemental composition of biochar generally includes C, H, O, as well as N, S, P, K, Ca, Mg, Na, Si, and so forth. Among them, the content of the C element is the highest, generally above 60%, followed by H and O, and mineral elements mainly exist in ash. The C in biochar is mainly aromatic carbon, which exists as irregular stacking of stable aromatic rings. The types of carbon compounds mainly included fatty acids, alcohols, phenols, esters, and components similar to fulvic acid and humic acid, with relatively high content in fresh biochar, low temperature pyrolysis biochar, and livestock manure biochar. In general, the elemental composition and activity of biochar are related to raw materials, carbonization process conditions, and pH.[77-79] Biomass through the pyrolysis carbonization process can form a rich carbon porous structure with developed pores and a high degree of aromatization, which is closely related to the nature of raw materials and the carbonization process. It is generally believed that the aromatic carbon structure and pore size of biochar increase with higher carbonization temperatures.[78-81] However, temperatures exceeding 700°C may damage some of the biochar's microporous structures, and temperatures above 800°C can destabilize the carbon framework of the biochar.

3.4.1.1 Biochar Sources

Organic pollutants (such as polymers and biomass, which account for 44% of total solid waste[82]) are the most polluting but recyclable type of solid waste. The value of this waste is underutilized. Current organic solid waste management practices are unsatisfactory. Traditional disposal methods such as recycling, incineration/combustion, and landfilling are generally inefficient, have low energy recovery rates, and cause secondary pollution. In contrast, catalytic conversion including catalytic pyrolysis, catalytic cracking,[83, 84] and catalytic carbonation, present a highly promising alternative for solid waste treatment with minimal environmental contamination. These technologies provide good resource recovery rates and convert waste into value-added products. Biochar produced from different raw materials may have different effects on the AD process.[85] For example, the total pore size of rice husk biochar was $2.1 cm^3 \cdot g^{-1}$, while that of sludge biochar was $0.17 cm^3 \cdot g^{-1}$, with a difference of 12.35 times.

3.4.1.1.1 Forestry and Agricultural Residues   Agricultural and forestry waste refers to the organic substances discarded in the whole process of agricultural production. This waste encompasses residues from agricultural and forestry production, by-products from livestock and fishery activities, and remains from agricultural processing, along with rural and urban household garbage. It includes materials such as straw, bark, waste wood, bamboo, coffee bean stems, and cotton shells. Agricultural and forestry waste (namely, corn straw, wheat straw, rice straw, and the like) are abundant biological resources of lignocellulosic waste, which plays an indispensable role in reducing the technical cost of anaerobic digestion and energy sustainability.[83, 86] Among agricultural and forestry waste, straw accounts for a large proportion and is a multi-purpose and renewable biological resource,[72] which is characterized by a high crude fibre content (30%–40%)

and lignin. Recent studies have shown that the application of lignocellulosic waste derived biochar as an additive can stabilize the AD process by alleviating the pH drop problem and promoting the attachment and enrichment of microorganisms on the biochar surface.[87, 88] Plant-derived materials have always been the main raw materials for biochar production.

3.4.1.1.2 Food Industrial Waste   Food industrial waste refers to the substances that food processing enterprises intend to discard or have discarded during the processing of target products using animals and plants as raw materials. Including oil palm shell, bagasse, fruit shell, walnut shell, and so forth. Walnut shell biochar promoted the formation and degradation of microbial biofilm of straw, and increased the biogas yield by 87.55%.[88] Residual sludge from food processing plants, sludge from wastewater treatment plants, and sediment sludge from pulp mill discharge can also be used as feedstock for biochar. Residual biogas produced by anaerobic digestion. Conventional biogas disposal techniques, such as landfilling, incineration, and composting, have disadvantages. Many studies have shown that biochar production by pyrolysis is one of the ideal methods for the treatment of FW anaerobic Digesters (FWD).[89] FWD contain organic carbon, which is conducive to the formation of pores after the evaporation of organic matter with low boiling point during pyrolysis. Therefore, FWD-derived Biochar (FWDB) has good adsorption performance in theory.[90]

3.4.1.1.3 Excrements of Livestock   Wang et al.[91] studied the effects of single addition of Cu^{2+}/Zn^{2+} or dehydrated pig manure biochar and their combination on SW AD. Found 50 mg/L Cu^{2+}/Zn^{2+} processing can temporarily suppress methane generated, and added 20 g/L biochar maybe shortening the methane generated time lag and improve methane production to ease this inhibitory effect.

3.4.1.2 Carbonization Method of Biochar

The pretreatment of raw materials can significantly enhance the quality of biochar, particularly in terms of its purity and adsorption capacity. The typical steps involved in pretreatment include drying, grinding, and sieving. Standard pretreatment experimental conditions often entail drying the biomass material at 105°C to remove its inherent free water, followed by crushing and sieving the material.

3.4.1.2.1 Hydrothermal Treatment Method   HydroThermal Carbonization (HTC) is a moderate hydrothermal process wherein biomass and water are mixed in a reactor at specific ratios and subjected to certain temperatures (180–350°C), durations (4–24 hours), and pressures (1400–27600 kPa) to produce solid fuel as the primary product.[92, 93] HTC operates under relatively mild conditions, resulting in stable and non-toxic solid-phase fuel that is more manageable and storable. Its benefits include no restrictions on the moisture content of the feedstock and a primarily dehydration-based reaction mechanism, which leads to high carbon fixation efficiency. The process is gentle, and the exothermic nature of dehydration and decarboxylation reactions provides energy for the process. Hydrothermal carbonization endows the produced materials with rich oxygen-containing and nitrogen-containing functional groups, making them suitable for a broad array of applications.

3.4.1.2.2 Direct Pyrolysis   Traditional heating: Traditional heating is achieved by changing the outside temperature and transferring the heat source from the container to the inside of the sample. However, in the process of heating biomass in the traditional heating method, the surface of the material heats up faster than the internal heat of the material, which can easily cause uneven heating of the sample and lead to a decline in the quality of the product. In addition, excessive heating can also cause excessive carbonization of biomass raw materials and destroy the pore structure of biochar. In addition, traditional heating also has the shortcomings of high energy consumption and low heating efficiency, resulting in a waste of energy.

Microwave heating: Unlike electric heating or combustion heating methods, microwave heating does not use an external temperature field to heat biomass, but converts microwave energy into thermal energy through the agitation of molecules in the electromagnetic field, and the generated heat diffuses from the inside of the raw material to the outside. Microwave heating is mainly used in rapid pyrolysis (heating rate >10~200°C/s, residence time 0.5~10 s, usually <2 s), which can heat biomass efficiently, shorten the start time of pyrolysis reaction, and have high liquid product yield.

3.4.1.2.3 Catalytic Pyrolysis Method   Catalytic carbonization refers to a carbonization method in which the catalyst and biomass raw materials or biochar are mixed by a certain method, and the biomass is thermally cracked in a closed environment with anoxia and high temperature. The core content of catalytic carbonization technology is to add some kind of catalyst to the raw material. The commonly used catalytic carbonization catalysts include sulfuric acid and various ores to promote the carbonization process, so as to obtain the best carbonization material. The combination of catalytic carbonization and subsequent activation is a common preparation process for the preparation of biomass biochar with high specific surface area. Phosphoric acid and potassium-containing compounds are commonly used catalysts in the early preparation of biochar. During the catalytic pyrolysis process, they will promote the pyrolysis reaction, inhibit the formation of tar, and increase the active centre, thus promoting the formation of the lamellar stacking graphite microcrystalline structure with abundant initial pores.

Compared with the traditional carbonization method for biochar preparation, the catalytic carbonization method reduces the complex steps of biochar preparation from biomass. Moreover, the addition of catalysts in the carbonization process can reduce the carbonization temperature and energy consumption. It can also be divided into in situ catalysis and in situ catalysis.

3.4.1.3 Efficiency

3.4.1.3.1 Maintain PH   Under high food/microbe ratio conditions, the addition of rice husk biochar alleviated the problem of sudden pH drop during AD of sweet sorghum, with a 45% reduction in lag phase and a 25% increase in biomethane production.[94] In the batch anaerobic digestion and fermentation system, biochar is alkaline, which can neutralize the pH value to reduce the inhibitory effect of acid on methanogens.[95]

3.4.1.3.2 Promote Microbial Growth   The porous structure of biochar can provide an ideal environment for the growth of microorganisms.[96] Biochar has a high surface area and can promote microbial colonization and enrichment during anaerobic fermentation. In addition, biochar with a large specific surface area and porous structure provides more sites for microbial colonization and facilitates biofilm formation, making it adaptable to multiple microbial populations, including acidogenic, acidogenic, and methanogenic bacteria, when applied in anaerobic fermentation. Besides, the excellent electrical conductivity of biochar enhances the ability of electron transport between anaerobic microbial species. Biochar increases the number of organic molecules available to acid-producing bacteria by promoting cell decomposition and hydrolysis of large organic molecules, facilitates the attachment of electroactive microbial communities in fungi and archaea, facilitates the conversion of macromolecules to dispersed substrates, and reduces the contents of dissolved salts, total nitrogen ammonia and free ammonia, thereby improving the anaerobic fermentation performance.

3.4.1.3.3 Redox Properties   Redox active groups in biochar may contribute to the synchronization of microorganisms.[97] The capacitance of biochar stems from the abundant functional groups (C-O, C-double bond O, COOH, and so forth), which can receive and contribute electrons as electron shuttles.

3.4.1.4 Mechanism of Action

3.4.1.4.1 Strengthen the Electron Transport Mechanism   Biochar can promote interspecific electron transfer between strophes and methanogens.[98,99] Microorganisms that function as DIET during AD include *Geobacter*, *Shewanella*, *Methanosaeta*, and *Methanosarcina*.[100] Due to its electrochemical activity, biochar can promote direct interspecies electron transfer of anaerobic microorganisms and accelerate the efficiency of methane production.[101]

3.4.1.4.2 Provide Attachment Vectors for Microbial Growth   Biochar can provide a large surface area for microorganisms to attach and selectively enrich functional microorganisms, for example, *Methanosarcina* and *Methanosaeta*.[102] The higher specific surface area is conducive to the growth of bacteria decomposing glucose and further promotes the daily methane production in early AD.[103]

3.4.1.4.3 Adsorption of Harmful Substances, VFAs and Ammonia Nitrogen   The inclusion of biochar in the Anaerobic Digestion (AD) of Waste-Activated Sludge (WAS) enhances Volatile Fatty Acid (VFA) conversion and augments methane production by 48%.[104] Moreover, biochar can sequester and adsorb noxious substances emitted from WAS, including heavy metals and ammonia nitrogen (NH^{4+}), thus mitigating their inhibitory effects on microorganisms[105] and reducing ammonia toxicity inhibition.[106,107]

3.4.2 Carbon-Based Composite Materials, Biochar-MOF Materials

He et al. studied the effects of three carbon-based conductive materials (nano graphite, particulate biochar, carbon cloth) on the anaerobic co-digestion of Fat, Oil and Grease (FOG) and Waste Activated Sludge (WAS).[108] Tiwari et al. explored the effect of Granular Activated Carbon (GAC) and Granular BioChar (GBC) addition on enhancing thermophilic anaerobic co-digestion of wheat husk and sewage sludge.[109] The results showed that GAC and GBC had the highest biogas production at a dosage of 20 g/L, which were 263 and 273 mL/g VS, 22% and 27% higher than the control group, respectively.

In the case of industrial ash slag – steel slag – scrap iron, metal trace elements, fly ash slag, coal slag, the microorganisms that produce methane during anaerobic digestion are methanogenic archaea, whose stimulation of growth and metabolism improves anaerobic processes and methane production. Methanogenic archaea are known to have nutritional requirements for specific metals such as iron, nickel, cobalt, copper, and molybdenum. Lack of sufficient nutrients and trace elements can severely limit microbial growth and thus limit the efficiency of the process, and fly ash is a cheap and abundant by-product (residue) of coal burning in thermal power plants. It contains various oxides (SiO_2, Al_2O_3, Fe_2O_3, CaO, MgO, and so forth), which are generated by oxidation at extremely high temperatures and can release metal ions that are beneficial to anaerobic fermentation performance. The solid residue of fly ash is applied to the anaerobic digestion of sludge, and the released alkali metal has a beneficial effect on the anaerobic digester and improves the acid neutralization capacity.[110]

3.4.2.1 Application of Biochar in Anaerobic Fermentation

3.4.2.1.1 Provide Vectors for Microorganisms   Wang et al.[111] increased methane production by 26.7%, 23.0% and 26.4% respectively after adding 3%, 5% and 7% biochar to the anaerobic digestion system of pig manure. Yang et al.[112] found that adding biochar at an optimized dose increased methane production in SW by 25%.

3.4.2.1.2 Reduce the Toxicity of Inhibitory Substances   Biochar has a strong adsorption effect on adsorption inhibitors in anaerobic fermentation broth.[113] This correlates with the presence of free radicals in biochar. These free radicals can transfer electrons and degrade organic pollutants by activating hydrogen peroxide to produce oxidizing substances, thereby degrading environmental pollutants through oxidation reactions.[114] In addition, some researchers have found[115] that free radicals can also act as catalysts. The catalysis of free radicals in charcoal promotes the production of activated substances, thereby stimulating the biotransformation of microorganisms and the detoxification of environmental pollutants. Based on a semi-continuous fermentation reactor, Huang et al.[116] explored the effect of biochar addition in the biogas slurry circulation system on the characteristics of microorganisms and gas production in the reactor. The results showed that biochar addition enhanced the degradation and metabolic pathways of acetic acid and propionic acid, resulting in a reduction of Volatile Fatty Acids (VFAs), total ammonia and chemical oxygen demand concentration by 55%, 41% and 61%, respectively. The buffer system formed by the binding of NH^{4+} to the VFAs of C_2-C_5 is also enhanced, thus improving the stability of the system. The addition of biochar effectively increased the relative abundance of bacteroidetes, chloramphenicol, spirulina, and synergetic phyla, and enhanced three methanogenic metabolic pathways.

Anaerobic digestive systems usually show a decrease in pH or even acidification. In this case, biochar helps buffer the anaerobic digestion system, and the buffering effect of biochar saves the anaerobic digestion from severe acidification and ensures stable methane production. The simultaneous addition of biochar enables high yield CH_4 production compared to conventional buffer chemicals.[117]

3.4.2.1.3 Promoting Direct Interspecific Electron Transfer Interspecific electron transfer between syntrophic bacteria and methanogenic archaea determines overall AD efficiency. However, there are two ways to determine interspecific electron transfer, one is Mediated Interspecific Electron Transfer (MIET), and the other is Direct Interspecific Electron Transfer (DIET). Current studies have shown that DIET has been demonstrated to be more efficient than MIET as it is not limited by the diffusion rate of electron carriers such as hydrogen and format.[118] However, the biochar mediated DIET shows a better effect on improving AD performance than cytochrome/pili.[98] It has been hypothesized that the electron donor and acceptor capacities of biochar are related to the surface chemistry of biochar and may determine the ability of biochar to promote interspecific electron transfer.[119]

3.5 COMPOSTING

As one of the methods for the safe disposal and recycling of organic solid waste, aerobic composting refers to the process where,[120] under the presence of air, aerobic microorganisms and complex organic matter in the composting material undergo exothermic decomposition reactions. This process facilitates the metabolism of complex organic matter into simpler and stable humus,[121, 122].This process achieves the harmlessness and reduction of organic solid waste,[123]. Simultaneously, it generates organic and inorganic "green" products.[124] Composting materials are typically sourced from solid waste generated in industrial, agricultural, and domestic activities, and are processed under natural or controlled conditions. Compared to traditional methods like landfilling and incineration, composting is a biological, chemical, and heterogeneous reaction process, wherein organic matter is mineralized into substances like CO_2, NH_3, and H_2O.[125] The final product of composting is relatively stable, with lower toxicity and pathogenicity. Additionally, it contains compounds similar to "humic substances," distinguishing it from compounds found in natural soils, coal, and peat. These products can be used as organic fertilizers and soil conditioners.[126]

The aerobic composting process is influenced by various factors, including temperature, pH, particle size, moisture content, aeration, and electrical conductivity, among others.[113, 127] Aeration is one of the critical factors affecting the composting process.[128] Fundamentally, composting is an aerobic process in which oxygen (O_2) is consumed, and gaseous water (H_2O) and carbon dioxide (CO_2) are released.[129] Composting efficiency is positively correlated with oxygen concentration, primarily due to the significant impact of oxygen concentration on microbial activity and population dynamics.[130] Too little aeration can lead to anaerobic conditions in composting, while excessive aeration can result in heat loss, preventing the achievement of the optimal thermophilic conditions required for the efficient decomposition of organic matter.[129] In various composting processes, the optimal range of aeration rates can vary due to differences in composting materials. The ideal aeration rate depends on factors such as the composition of the compost feedstock, moisture content, and the specific goals of the composting operation. Different materials may require different oxygen levels to support the most efficient decomposition and microbial activity. It is essential to tailor aeration rates to the specific composting conditions and materials to achieve the best results.[131] Temperature is indeed a crucial parameter in composting. It plays a dual role as both a result and a determinant of composting effect-iveness.[132] Composting itself is an exothermic process, beginning in the mesophilic phase where mesophilic bacteria are the dominant organisms. During this phase, easily degradable soluble compounds are primarily broken down by these mesophilic bacteria. At the same time, the meta-bolic activity and growth of mesophilic bacteria cause a rapid increase in temperature. This temperature rise leads to the replacement of mesophilic bacteria by thermophilic bacteria, transitioning the compost into the high-temperature phase. During this high-temperature phase, thermophilic bacteria are capable of breaking down more complex compounds such as polysaccharides, proteins, and fats. Simultaneously, the high temperatures in the thermophilic phase serve to kill bacteria and viruses present in the raw materials, achieving the sterilization of the compost. This sterilization is an important aspect of the composting process, as it renders the compost safe and free from harmful pathogens, making it suitable for use as a soil conditioner or fertilizer. The final stage of composting is the maturation phase, during which mesophilic bacteria once again become the dominant microorganisms. This stage further stabilizes and solidifies the compost, marking the formation of mature compost. It contributes to the compost's enhanced stability and solidification, indicating that it has reached a fully developed and usable state.[133] The Carbon-to-Nitrogen ratio (C/N ratio) is another indicator of organic matter decomposition in composting. During the composting process, microorganisms decompose organic compounds to obtain energy for metabolism and acquire nutrients (such as N, P, K) to sustain their population.[123] C, N, P, and K are the primary nutrients required by microorganisms involved in composting. Among these, C and N are the most crucial nutrients: C serves as an energy source, while N is used for building cellular structures. Typically, the C/N ratio grad-ually decreases during the composting process.[134] This is because C is consumed during the biological oxidation process and is lost in the form of CO_2.[135] Research suggests that a C/N ratio ranging from 25:1 to 30:1 is most conducive for the composting process.[136] Another important param-eter is the moisture content, which affects factors such as oxygen absorp-tion rate, free air space, microbial activity, and process temperature.[137] Research has shown that the optimal moisture content required for micro-bial activity during the composting process is in the range of 40% to 70% of the compost's weight. When moisture content is too low, it can lead to a decrease in microbial activity and may even result in the development of anaerobic conditions, potentially causing composting to cease. Therefore, changes in moisture content during composting can serve as an indicative parameter of the rate of organic matter decomposition. pH value is also an important parameter in composting.[128, 138] The composting process is influenced by pH, affecting microbial activity. Generally, the pH level in the early stages of composting gradually decreases, while in the later stages, it gradually increases.[139] Increasing the ratio of NH_3/NH_4 leads to an increase in the volatilization rate. According to reports,[140] in aerobic composting, the pH value is slightly higher than in anaerobic fermentation, possibly due to a higher release of potassium elements. Under optimal conditions, after the complete breakdown of organic acids, the pH initially decreases and then rises from acidity to neutrality.[138] Similarly, some scholars have pointed out[138] that the optimal range for compost pH is between 7 and 8.

Aerobic composting is a widely employed technique that can transform organic solid waste into organic fertilizers and other green bioproducts. The use of organic fertilizers instead of chemical fertilizers can reduce the release of toxic chemicals into the environment, directly promoting environmental and human health.[141] From a recycling perspective, it enables the green recycling of mineral nutrients (N, P, and K).[142] The application of compost products significantly enhances agricultural productivity and the content of organic matter in soil, thus contributing to the assurance of global food security. Besides its utilization as a fertilizer, compost is pivotal in various areas such as bioremediation, plant disease management, weed control, pollution mitigation, erosion prevention, landscaping, and wetland restoration.[120]

To enhance the efficiency of nutrient utilization in compost, reduce gas emissions during the composting process, and prevent nitrogen losses during the process, the method of adding exogenous functional materials (expanding agents, additives) to compost has been proposed. This method is widely considered an effective means to increase composting efficiency and quality while minimizing potential environmental impacts.[143] Research indicates that the addition of various types of additives has different effects on composting efficiency and product quality.[144] Among these, the use of organic solid waste as an additive is widely applied in composting due to its wide availability of raw materials, large quantities, low production costs, and convenient preparation. Examples include biochar, vinasse, and soybean residue, among various types of organic solid waste.

3.5.1 Biochar

Biochar refers to a carbon-rich solid product formed through the pyrolysis of organic materials under anaerobic conditions and at temperatures below 900°C. Biochar is used as a soil amendment in agricultural soils, serving to maintain soil fertility and productivity. Due to its recalcitrant nature, biochar leads to an increase in carbon sequestration in soils and a reduction in greenhouse gas emissions.[143] Biochar is made up mainly of aromatic molecules that are not organized in ideally adherent layers.[144] The structure and properties of biochar are closely related to the preparation process. With the increase of pyrolysis temperature, the carbon content in biochar increases while the content of hydrogen and oxygen decreases, resulting in a reduction in the polarity of biochar. In the range of 400 to 700°C, as the pyrolysis temperature increases, the aromaticity and hydrophobicity of biochar increase, along with an increase in specific surface area and pore volume.[145]

3.5.1.1 The Impact of Biochar on Compost Maturity

Based on the properties and uses of compost products, compost can be categorized as follows:

1. Stabilization stage: the point at which microbial activity ceases
2. Maturation stage: when the toxicity of the product decreases
3. Mature and stable stage: the phase with the best composting results in terms of maturation and stability.[146]

In composting processes, certain values or indicators are commonly used to assess the progress and effectiveness of composting, such as temperature, C/N ratio, Dissolved Organic Carbon content (DOC), NH_4^+/NO_3^- ratio, and Germination Index (GI);[146] The ratio of Humic Acid to Fulvic Acid (HA/FA);[147] Oxygen uptake rate and biochemical composition.[148] Mature compost is considered to have been achieved due to the following indicators: environmental temperature,[149] C/N < 21[146] or C/N < 15,[149] $NH_4^+/NO_3^- < 0.16$,[146] GI b > 50%[146] or GI > 110,[149] HA/FA > 1.6[147] or HA/FA > 1.9.[149]

Research indicates that for the maturity assessment of compost amended with biochar, the C/N ratio cannot accurately determine the degree of compost decomposition. This is because biochar itself has a relatively high stability of carbon elements, and even when the compost is already mature, the C/N ratio can still be higher than 21.[146] Through continuous experimental research, the above conclusion is gradually being validated.[147] For example, the impact of biochar prepared from wood (Quercus serrata, a broadleaf tree) at 550°C (10% fresh weight) in previous research on the organic properties during the composting of cow manure or chicken manure.[147] However, some studies have also indicated that during the composting process, the addition of biochar made from apple pomace, straw, and rice bran results in a decrease in the C/N ratio for both the treatments with and without biochar due to the increase in total nitrogen after substrate mineralization or carbon degradation.[147, 150] However, compared to compost without added biochar, the decrease in the C/N ratio due to the addition of biochar is significantly lower. Some scholars also suggest that this phenomenon may be attributed to the presence of recalcitrant carbon (derived from biochar) and reduced substrate mineralization after adding biochar. It should be noted that these results are due to the higher stability of biochar, resulting in no fundamental changes in the effective carbon-to-nitrogen ratio during the composting process. Furthermore, the addition of biochar leads to an increase in the HA/FA ratio during the composting process, enhancing the degree of humification process products. This is due to the adsorption of humic substances on the surface of biochar, accelerating the formation of aromatic polymers.[147] Similarly, adding biochar during the composting process also leads to an increase in the GI value, indicating that the addition of biochar results in the rapid removal of toxic substances from the compost.[151]

3.5.1.2 The Impact of Biochar on Composting Temperature

Temperature is one of the most crucial parameters in composting. Compared to composting without the use of biochar, the temperature in the process increases more rapidly when biochar is used for composting (typically reaching its peak 6–7 days earlier).[152] In some cases, biochar can also extend the thermophilic phase by 1–6 days. This is due to biochar filling the free spaces between composting materials, thereby reducing heat loss during the composting process.[150] In addition, adding biochar improves the aeration environment, increasing the quantity of microorganisms and thereby accelerating the conversion of organic matter. It also raises the composting temperature. Studies have shown a significant correlation between changes in oxygen concentration, especially in cases where aeration is not sufficient, and temperature variations within the compost. When microbial activity is high, temperature rises and oxygen concentration decreases.[153,154]

3.5.1.3 The Impact of Biochar on Other Elements in Compost

In addition to nitrogen, compost also contains other important nutrients such as phosphorus, potassium, calcium, magnesium, sodium, and sulfur. Their content depends on the type of composting materials, but phosphorus and potassium content are often higher than other essential elements. The loss of these nutrients during the composting process reduces the value of compost as a soil fertilizer.[155] Generally, the addition of biochar to compost has a positive impact on its fertility, owing to the presence of the macro-elements mentioned above. Studies have indicated that adding wheat straw biochar (prepared at temperatures of 500–600°C) at rates of 10% or 15% by wet weight leads to an increased quantity of P, K, Ca, and Mg ions in the compost.[150] Compared to compost without added biochar, the compost with biochar (where the initial Na content in the biochar and compost components before the composting process is not significantly different) does not show any difference in the concentration of Na^+ ions (at a 5% biochar content) or even lower concentrations (within the range of 8.7–16.7%). This is a positive phenomenon, as high concentrations of Na^+ in compost can potentially lead to soil structure degradation when applied to the soil.[150]

3.5.1.4 The Impact of Biochar on Heavy Metals in Compost

Depending on the source of the materials, biomass can contain varying amounts of heavy metals. Complexation between heavy metals and organic compounds forms chemical bonds that reduce the solubility of heavy metals, thereby decreasing their bioavailability. Numerous studies have shown that biochar exhibits strong affinity for heavy metals. Biochar primarily reduces the bioavailability and mobility of heavy metals in compost through its influence on the physicochemical properties of compost products, as well as through physical adsorption, electrostatic interactions, ion exchange, precipitation, and complexation with heavy metals present in the compost.[156] The addition of biochar during the composting process also facilitates the immobilization of heavy metals within the compost. Metals do not disappear from the compost, but their mobility significantly decreases when leaching or application to agricultural soils occurs. Consequently, the environmental risk associated with their transport and transformation (leaching from soil, bioaccumulation) is reduced.[157] Research has shown that adding 5% by dry weight of biochar produced from fungal cultivation waste or rice straw to a mixed compost of chicken manure and sawdust (3:2 by volume) results in reduced total concentrations and bioavailability of zinc, copper, and arsenic (exchangeable and reducible fractions), compared to compost without biochar.[158] Similarly, biochar produced within the temperature range of 450 to 500°C exhibits the highest reduction in the bioavailability of Zn and Cu.[159] Biochar exhibits a strong passivating effect on Cu and reduces the activation of Zn. Wood-based biochar shows good passivation effects on heavy metals such as Zn, Cd, and Pb. Straw-based biochar demonstrates more stable passivation effects on heavy metals, and rice straw biochar effectively reduces the bioavailability of heavy metals like Cu, Cr, Ni, Pb, and Zn. The passivation effect on heavy metals is further enhanced when used in conjunction with vermicomposting.[160]

3.5.1.5 The Effect of Biochar on Organic Pollutants in Compost

The strong adsorption capability of biochar for organic pollutants[161] can reduce the accumulation of organic pollutants in organisms, thereby lowering the environmental risks associated with organic pollution. Compost may contain various organic pollutants such as Polycyclic Aromatic Hydrocarbons (PAHs), pesticides, PolyChlorinated Biphenyls (PCBs), Polychlorinated Dibenzo-p-Dioxins and dibenzoFurans (PCDD/F), Linear Alkylbenzene Sulfonates (LAS), and NonylPhenol (NP), amongst others.[162, 163] Under the action of microorganisms, a portion of the pollutants is transformed into metabolites or mineralized, another portion leaches from the composting materials, and a third part remains in the compost product.[163] The research indicates,[164] after 30 days of storing sewage sludge with the addition of biochar, the bioavailable fraction of PAHs in the sludge decreased from 58.0% to 17.4%. Overall, the addition of biochar to composting is expected to result in a decreased affinity of organic pollutants in the compost,[154] reduced environmental pollution risk.

The addition of biochar plays a significant role in the removal of antibiotics from compost. Biochar contains oxygen-containing functional groups such as hydroxyl (-OH), carboxyl (-COOH), and phenolic surface functional groups, which effectively bind various pollutants in the soil, reducing their concentration. Adding pig manure biochar compost results in an average removal rate of 90% for antibiotics, with a higher removal rate of 94% for tetracycline antibiotics but a lower removal rate of 80% for sulfonamide antibiotics, indicating a degree of selectivity.[165] Furthermore, the addition of both biochar and natural zeolite can reduce the abundance of antibiotic resistance genes in sewage sludge compost products to varying degrees. The relative abundance of actinobacteria in the biochar-amended treatment significantly decreased in the later stages of composting. At the species level, composting can effectively remove some pathogenic bacteria. However, the addition of biochar and zeolite reduced the abundance of the majority of pathogenic bacteria, with a few pathogenic bacteria nearly completely eliminated, including most of those closely associated with antibiotics.[166]

3.5.1.6 The Influence of Biochar on Gas Emissions in Composting

The gaseous byproducts generated during composting include carbon dioxide (CO_2), methane (CH_4), nitrous oxide (N_2O), carbon monoxide (CO), and ammonia (NH_3).[166] Most of these are greenhouse gases, which have a significant impact on global climate change,[167] and it may also result in the loss of nutrients. Existing research has shown that the application of biochar in soil leads to a reduction in the emission of CH_4,[168] this can be attributed to the influence of biochar on the aeration, gas diffusion, and improvement of living conditions for methane-oxidizing bacteria in the soil.[168] Research has shown that when a 10% dose of biochar is mixed with cow dung, apple pomace, rice bran, and straw and composted in the temperature range of 40 to 60°C, the ratio of methane-producing bacteria to methane-oxidizing bacteria is lower compared to composting without the addition of biochar. This indicates that the growth conditions for methane-oxidizing bacteria are more favourable, leading to a reduction in CH_4 emissions during the composting process. The study confirmed that adding a mixture of hardwood (80%) and softwood (20%) at a dose of 27% dry weight to composted chicken manure in the temperature range of 50–70°C resulted in a 27% to 32% reduction in CH_4 emissions compared to composting without biochar.[169] Another study revealed that composting a mixture of organic waste and the organic fraction of urban solid waste with 10% biochar resulted in an 81.5% reduction in CH_4 emissions and a 51% reduction in carbon dioxide emissions compared to composting without biochar. This reduction is attributed to the adsorption of carbon dioxide by biochar.[148] It could also be due to the sufficient oxygen content in compost enriched with biochar.[170] At the same time, biochar has the capability to adsorb nitrates and Dissolved Organic Carbon (DOC), promoting denitrification and reducing the substrate available for N_2O production. However, there are also reports suggesting that the addition of biochar can lead to increased CO_2 emissions in compost due to improved aeration of the composting mass, which accelerates the degradation of compost materials and increases CO_2 emissions. These variations may arise from differences in composting materials, the inherent properties of the biochar, and the conditions of the composting process.[171] Overall, the addition of biochar in composting demonstrates significant potential for reducing CO_2 emissions. Its products exhibit high stability and low risk of reversal, offering potential co-benefits for food security and soil fertility.

NH_3 is another greenhouse gas produced during the composting process. The generation of NH_3 has two implications: on one hand, it reduces ammonia emissions, which are an indirect greenhouse gas, and on the other hand, it signifies nitrogen loss, which is a component of fertilizers. The use of biochar as an additive in composting has yielded significantly better results.[146, 172] It is widely recognized that adding biochar to compost is considered one of the most effective methods for reducing nitrogen loss. Research indicates that incorporating biochar at doses of 5% and 20% (dry weight) into compost can result in an average reduction of ammonia emissions by 47%,[158] which is primarily due to the adsorption effect of biochar on NH_3.[146, 173]

3.5.1.7 The Impact of Biochar on Microorganisms in Composting
Firstly, the porosity of biochar potentially provides a favourable habitat for aerobic nitrifying bacteria, enhancing microbial activity and shortening composting duration. Studies indicate that during the thermophilic phase, biochar accelerates organic matter degradation and ammonium formation, while during the maturation phase, biochar expedites nitrification. Simultaneously, as a compost additive, it effectively prevents the formation of aggregates larger than 70 mm, enhancing the physical properties of the mixture and accelerating microbial metabolic efficiency. It is estimated that adding 3% of biochar can reduce composting time by 20%.[143]

Secondly, the addition of biochar provides an ample supply of O_2 to microorganisms, enhancing the abundance of aerobic organic matter-degrading bacteria during the initial stages, accelerating organic matter decomposition, and promoting the succession of microbial communities in the composting process.[174] Furthermore, the improved aerobic environment created by biochar inhibits the activity and abundance of certain anaerobic bacteria, significantly reducing NOx emissions throughout the composting process. The suppression of denitrification by biochar is the primary reason for nitrogen loss reduction. It has been reported that nitrogen losses during the composting process, attributed to NH_3 emissions, are estimated to be around 15%.[174]

3.5.1.8 The Influence of Biochar on Humic Substances in Compost Products

Humus, as one of the final products of composting, not only increases soil fertility but also enhances electron transfer capabilities. It promotes the reduction and degradation of organic pollutants and heavy metals, thus having beneficial environmental effects. In the aerobic composting process, humification reactions mainly occur during the initial heating phase and the high-temperature phase in the middle, involving organic matter decomposition reactions and humus polymerization reactions. Organic materials are initially broken down by microorganisms into small molecular products (such as polyphenols, carboxyl groups, and amino acids), which then undergo biochemical reactions within microbial cells to form intermediates. These intermediates are essential precursors for the later polymerization into humus. Research has shown that adding biochar and zeolite to the composting materials of pig manure and wheat straw can promote the increase in fulvic acid content. This is because zeolite enhances the mineralization process, while biochar promotes the humification process, and the synergistic effect of both improves the quality of humic substances.[175] Furthermore, the addition of hydrothermal charcoal in chicken manure composting enhances the humification process and significantly increases the humic content in the resulting products.[176]

In addition to influencing the composting materials and compost properties, biochar itself undergoes certain changes during the composting process. Due to the prevalent high temperatures (up to 70°C), moisture, and intense microbial activity in composting, biochar may also undergo transformations. The oxidation of biochar and the formation of various functional groups are the primary reactions of biochar during the composting process.[177] Co-composting organic matter with biochar can increase the Cation Exchange Capacity (CEC) of biochar.[178] This increases the negative charge on the biochar's surface. Research has shown that adding biochar prepared from bamboo wood at a temperature of 600°C to sewage sludge compost, the acidic functional groups of biochar, especially the carboxyl groups, may increase in number, which could potentially form complexes with NH^{4+}.[179] As a result, nitrogen (a fundamental component of all fertilizers) is less prone to loss due to leaching and gas emissions during composting, indirectly enhancing the fertilizer value of compost products. Similarly, studies have indicated that when biochar is added to the composting process of chicken manure and pine sawdust, the nitrogen content in biochar increases.[180-183] This is attributed to the adsorption of NH_3, NO^{3-}, and NH^{4+} on the surface of the biochar.[183] Typically, when the organic matrix component in composting materials is higher than that of biochar, the component content on the surface of biochar increases. In contrast, biochar "donates" organic matrix components to the composting materials. Furthermore, during the composting process, the specific surface area of biochar decreases, primarily due to the adsorption sites of Dissolved Organic Carbon (DOC) "blocking" the pores of biochar.

3.5.1.9 The Impact of Co-Adding Biochar with Other Materials on Composting

The co-addition of biochar with other functional materials has a positive impact on composting. Research has shown that the addition of garbage enzymes with biochar to composting enhances the total organic carbon degradation rate (up to 21.9%). The combined effect of these two materials promotes the production of fulvic acid and humic acid-like substances, thereby increasing humification.[184] Furthermore, biochar and $CaCO_3$ have a larger adsorption capacity and pH, which is advantageous for NH^{4+} adsorption and N_2O reductase activity.[185] Another research finding indicates that co-adding biochar with bacillus spores and thermophilic actinomycetes in a 1:1 volume ratio to a mixture of pig manure and corn straw for composting reduces nitrogen losses in pig manure composting, increases the overall nutrient content of compost products, and promotes the maturation of compost.[186] The co-addition of biochar, zeolite, and wood vinegar to pig manure composting can shorten the high-temperature phase of composting and enhance the maturation of the compost.[187]

3.5.2 Phosphogypsum and Calcium Phosphate

Gypsum is the primary byproduct generated during the wet process production of phosphoric acid and phosphate fertilizers (approximately 4.5–5.0 tonnes of gypsum are produced for every 1.0 tonne of phosphoric acid). It is also one of the largest solid waste emissions in the chemical industry. Gypsum primarily consists of calcium sulfate dihydrate and contains small amounts of free acids, phosphate, fluoride, organic matter, and trace elements. During the cooling and maturation phase of composting, the presence of gypsum enhances the decomposition of organic matter. At the end of composting, both gypsum and diatomaceous earth can inhibit the growth and activity of methanogenic bacteria, thereby reducing CH_4 emissions.[188] Research also indicates that the addition of phosphorus-containing chemicals, such as calcium phosphate, in pig manure can lower the pH of the compost pile. Simultaneously, Mg^{2+}, Ca^{2+}, and PO_4^{3-} bind with NH^{4+}-N in the pile, making it less prone to conversion to NH_3, thereby reducing nitrogen loss in the form of NH_3 and significantly increasing the humus content after maturation. In the composting of sheep manure, the addition of calcium phosphate and gypsum can substantially reduce the production of CH_4, NH_3, and H_2S, decrease total carbon, total nitrogen, and total sulfur losses. Furthermore, it has been observed that gypsum also significantly inhibits the generation of N_2O.

3.5.3 Mature Compost

Mature compost has been widely used as an alternative in actual composting processes due to its readily available, porous, microbial, and cost-effective characteristics.[189] Adding mature compost or commercial inoculants accelerates the composting of green waste and enhances compost quality. Mature compost contains abundant microorganisms, and inoculating them at the beginning of the composting process has a significant impact. In the composting of sewage sludge, the addition of mature compost increases the diversity of mesophilic and thermophilic bacteria. In the composting of cow manure, adding mature compost can expedite microbial succession.[190] According to reports, the addition of mature compost during kitchen waste composting reduces total greenhouse gas emissions (including CH_4 and N_2O) by 69.2%. Similarly, substituting 50% of sawdust with recycled materials in aerobic composting of food waste not only reduces composting costs but also enhances the heating and maturation processes of compost. The use of composted material generated from the composting process as a supplement in sludge composting not only introduces a rich microbial population, acting as an inoculant but also regulates the sludge's porosity and moisture content. As a result, the final product yield will significantly increase compared to using all or part of the material in a closed-loop composting system. From the perspective of sludge reduction, reusing matured material is more advantageous.

3.5.4 Soybean Dregs

Soybean dregs are byproducts of soybean processing.[191] The addition of soybean dregs promotes the degradation of organic matter in compost, increasing the degree of organic matter degradation by 16.46–25.04%. In addition, adding soybean dregs enhances the aromatic structure, effectively improving the humification of compost and increasing Humification Rate (HR), Humic Acid Percentage (PHA), Polymerization Degree (DP), and Humification Index (HI). The addition of soybean dregs also increases the organic matter content in compost aggregates, which is related to the fact that soybean dregs are rich in carbohydrates, as previously discovered.[192] Adding 10% of catechol improves the maturity and stability of compost.[193]

3.6 HYDROTHERMAL TREATMENT

3.6.1 Biomass Hydrothermal Treatment Enhanced by Solid Waste-Based Materials

The studies on hydrothermal treatment have primarily focused on the development of optimized catalysts to enhance the yields and quality of liquid/solid products (Table 3.2). Compared to homogeneous catalysts, the utilization of heterogeneous catalysts in biomass hydrothermal treatment has been increasingly employed. This is due to their high activity and ease of recovery from the liquid product, ultimately resulting in a reduction in overall production chain costs.[194, 195] To date, the most studied heterogeneous catalysts are biochar-based catalysts.[195-197] and other solid waste-based catalysts, such as sludge, clay, and mineral material catalysts.[194, 198] Significantly, solid waste-based materials with advantages such as low material cost, large surface area, and stability at high temperatures have been gaining increased attention as a preferred heterogeneous catalyst.

3.6.1.1 Biochar-Based Materials Catalysed Hydrothermal Treatment

Biochar, a by-product produced from the thermochemical conversion of biomass in an oxygen-limited environment, has been widely applied for waste management, soil improvement, environmental remediation, and climate change mitigation.[199] In recent years, biochar was employed as a green catalyst during the hydrothermal treatment process owing to its environmental sustainability, cost-effectiveness, and recyclability. Lu et al. investigated the effect of four kinds of low-cost biochars derived from straw, rice husks, peanut shells, and sawdust on the HydroThermal Liquefaction

TABLE 3.2   Solid waste-based catalysts for hydrothermal treatment

Catalyst	Feedstock	Reaction Conditions	Main results	Reference
Sewage sludge-based activated carbons	Sewage sludge	350°C or 400°C, 0.5 h	Catalyst increased the yield and energy density of bio-oil and favoured the risk decrease of HMs	186
Activated carbon felt	Microalgae and sewage sludge	325°C or 375°C, 4 h	Activated carbon felt catalyst gave the highest H/C and lowest O/C of biocrude	201
Activated carbon felt	Municipal sludge	325°C, 0.5 h	Catalyst and formic acid used together increased yield and quality of biocrude	210
Biochar supported metal catalysts	Alkali lignin	260 to 300°C, 15 min	Maximum bio-oil yield (72.0 wt%) was obtained with bimetallic catalyst	197
Biochar loaded with CoO_x and NiO	*Spirulina platensis*	256 to 320°C, 35 min	Biochar and the catalyst carrier showed the most significant improvement in bio-oil yield and quality	202
Iron sludge	*Metasequoia Leaves*	150 to 250 °C, 1 to 5 h	Iron sludge improve the combustion behaviour of HCs	204
Natural mineral additives, such as trona, dolomite, and borax	Residues of leek, cabbage and cauliflower	300 to 600 °C, 1 h	The catalytic effects of natural mineral salts enhanced both the gasification rates and product gas composition	207

(*continued*)

TABLE 3.2 (Continued)

Catalyst	Feedstock	Reaction Conditions	Main results	Reference
Red clay, red mud, and fly ash	Food waste	300°C, 1 h	Food waste is converted with >40% energy recovery into energy-dense biocrude	[205]
Red mud (RM) based catalysts, reduced red mud (RRM) and red mud supported nickel (Ni/RM)	*Tetraselmis* algae	275°C, 1 h	Catalysts increased the yield of *Tetraselmis* derived biocrude	[206]
Hydrochar, zeolite and magnetite powder	Meso-macro structured algal	300 to 350°C, 5 to 30 min	Superior catalytic activity for the production of gasoline (14%) and biodiesel (40%)	[211]
Zeolite molecular sieve	Penicillin residue	280°C, 3 h	Catalyst enhanced the yield and quality of bio-oil	[208]
Hydrotalcite (HT), colemanite (calcium borate mineral), and HT/KOH	Birch wood sawdust	300°C, 0.5 h	Catalysts promoted formation of phenol derivatives and aliphatic compounds	[212]

(HTL) of microalgae. The results showed that biochar could improve the quality of bio-oil. In particular, the sawdust biochar increased the Higher Heating Value (HHV) of the bio-oil, and straw biochar increased the low-boiling point fraction about 15% compared to bio-oil without catalyst.[200] In addition to the plant biomass-based biochar, sewage sludge-derived biochar was also used as a potential catalyst to enhance the efficiency of hydrothermal treatment. Zhai et al. catalysed the hydrothermal HTL of sewage sludge with sewage sludge-based activated carbons. The alkalinity and high

surface area of carbon-based catalyst resulted in significant immobilization of Heavy Metals (HMs), thereby reducing the potential risk associated
with HMs in the final products at 400 °C.[196] In contrast to CoMo and NiMo
catalysts, cheap activated carbon-based catalysts demonstrated significant
promise in the catalytic HTL of microalgae and sewage sludge to enhance
the quality of biocrude oil.[201] The aforementioned studies showed that
the carbon-based catalysts revealed favourable activity. Noteworthy, the
presence of inorganics imparted better catalytic activity for biochar toward
synthesis of a biochar-supported metal catalyst.[197] Alkali lignin was treated
with HTL, resulting in a maximum bio-oil yield of 72.0 wt% using a bimetallic catalyst Ni-Co/AC under an ethanol solvent.[197] Wang et al. loaded Co,
Ni and their oxides to form catalysts based on hydrochar derived from the
hydrothermal treatment of *spirulina platensis*. As a result, bio-oil yield,
quality and denitrification were moderately positively impacted.[202]

3.6.1.2 Other Solid Waste-Based Catalysed Hydrothermal Treatment

The sludge, with a high moisture content (>80%), is unsuitable for conventional dry thermochemical treatment processes but can be effectively
treated using hydrothermal processes.[203] HydroThermal Carbonization
(HTC) of metasequoia leaves was conducted in the presence of iron sludge,
which acted as a catalyst to enhance the dehydration and decarboxylation
of the leaves. This catalytic effect led to the efficient formation of hydrochar
with higher HHV. Simultaneously, the iron sludge was reduced to magnetite, which possesses the property of easy separation.[204] Industrial waste
and natural mineral materials also have the potential to serve as sources of
catalytic properties. Red clay, red mud, and fly ash were evaluated for the
catalysed HTL of food waste. This study achieved over 40% energy recovery
in the conversion of food waste into high-energy biocrude.[205] Rahman
et al. catalysed algal HTL to produce bio-oil using Red Mud (RM)-based
catalysts. The results indicated that the yield of bio-oil followed the order
of no catalyst < reduced RM < Ni/RM.[206] Yildirir et al. employed natural
mineral additives, such as trona, dolomite, and borax, as catalysts for the
hydrothermal gasification treatment of residues from leek, cabbage, and
cauliflower. The catalytic effects of natural mineral salts enhanced both the
gasification rates and product gas composition.[207] Hong et al. conducted
homogeneous and heterogeneous catalysed HTL using penicillin residues
to produce bio-oil. This suggested that zeolite molecular sieve catalysts
were significantly superior to organic acid and alkaline catalysts.[208]

The biochar derived from biomass exhibited exceptional catalytic properties. It is characterized by its stable structure, excellent thermal and mechanical stability, chemical inertness, higher acid density, and environmentally friendly properties. However, the practical application of biochar was hindered by the presence of numerous corrosive amino acids and aliphatic acids, while zeolites are limited by low hydrothermal structural stability, coke formation, and a narrow pore size distribution.[209] The development of hydrochar/zeolite composite could effectively address these individual drawbacks. Norouzi et al. designed ternary composites made from zeolite, hydrochar and magnetite powder exhibiting excellent catalytic activity for the production of gasoline and biodiesel. The simultaneous utilization of heterogeneous catalysts has garnered increasing attention in academic research.

3.6.2 Hydrothermal Products Enhanced by Solid Waste-Based Materials

3.6.2.1 Bio-Oil Production Enhanced by Solid Waste-Based Materials

The utilization of catalysts in wet biomass HydroThermal Liquefaction (HTLs) has the potential to enhance bio-crude yields and decrease their oxygen, nitrogen, and sulfur concentrations by means of *in situ* upgrading. The use of catalysts is crucial in enhancing the production of desired products and increasing the selectivity towards high value-added platform chemicals. Wang et al. observed that the bio-oil yield initially increased and then decreased within the biochar dosage range of 0–0.60 g. It was suggested that the primary function of biochar is to facilitate the breakdown of chemical bonds in biomacromolecules present in the feedstock, while also providing active sites for protein and lipid hydrolysis, as well as small molecule compound polymerization.[213] The decrease in bio-oil yield at higher catalyst dosages may be attributed to an increase in secondary pyrolysis leading to the formation of small molecule gases. The key factors influencing the catalytic performance of biochar are metal compounds and carbon. Heavy Metals (HMs) can undergo transformation to a more stable state during pyrolysis, while the stability of organic matter, such as carbon, improves with increasing temperature. Catalysts based on sludge biochar were found to be advantageous in enhancing the yield and energy density of bio-oil at 350°C. Furthermore, higher temperatures led to the production of catalysts with a higher total calorific value of bio-oil due to reduced oxygen content or increased hydrocarbon content.[196]

Biochar-derived activated carbon with metal catalysts might impact lignin depolymerization. The bio-oil obtained from the presence of catalysts exhibited similar molecular weights and distributions, slightly larger than those derived without any catalyst.[212] This indicated that the presence of catalyst enhanced specific condensation/polymerization processes of the reaction intermediates in the HTL process. In general, increasing the catalyst dosage within a certain range will enhance the bio-oil yield. Research has shown that as the number of molecular sieves added increases, more active fragments are converted to products, resulting in reduced solid residues and higher bio-oil yields.[214] However, an excessive amount of catalyst leads to a decrease in bio-oil production. This may be attributed to the fact that an excess of catalyst provides more active surfaces for reactants, thereby increasing the frequency of cracking reactions and gas production.[197,208]

3.6.2.2 Hydrochar Production Enhanced by Solid Waste-Based Materials

As aforementioned, iron sludge enhanced the conversion of volatile carbon to fixed carbon.[204] As showed in Figure 3.3, Acidic zeolites and $FeSO_4$ have been demonstrated to enhance coke formation.[215] The presence of iron

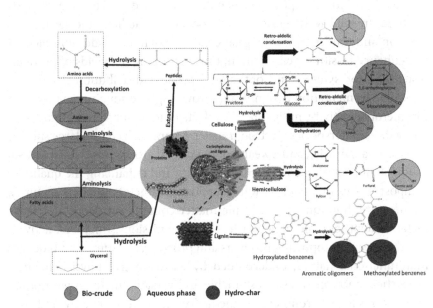

FIGURE 3.3   Reactions involving Hydrothermal process.[217]

oxides in sludge may enhance catalysis by disrupting the hydrogen bonds of the polymer chains. The cellulose and hemicellulose from leaves were hydrolysed into glucose, xylose, and various oligomers. These oligomers were then further hydrolysed into individual monomers, which underwent tandem decomposition to produce furfural derivatives and deeply mineralized intermediates (for example, aldehydes).[216] This subsequently increased the Higher Heating Value (HHV) of the resulting hydrochar. Furthermore, the decomposed furfural derivatives contributed to the transformation of the iron sludge into magnetite for easy separation or catalytic activity. The risk of Cu and Zn in solid products was significantly reduced after catalysing via biochar-based catalysts during HTL of sludge at 350 °C and 400 °C.[196]

3.7 CONCLUSION

This chapter presents an overview of the various applications of biochar in solid waste disposal. These applications include pyrolysis, gasification and tar remover, and hydrothermal treatment, anaerobic digestion, compost additive in solid waste disposal. In solid waste pyrolysis, biochar catalysts can affect the distribution of gaseous, liquid, and solid products.

Biochar catalyst will increase the yield of gaseous products, while decreasing the yield of bio-oil. Meanwhile, the biochar catalysts improve the quality of bio-oil, especially when the nitrogen-doped biochar catalysts are used. In solid waste gasification, biochar can facilitate the gasification reaction and thus increase the gasification efficiency. In addition, biochar can enhance in-situ tar elimination, thus increasing the yields of products (especially for syngas) and optimizing product distribution. Its role in anaerobic fermentation is significant, providing a conducive environment for microbial activity, mitigating the effects of toxic substances, and facilitating direct interspecies electron transfer. Characterized by a large specific surface area, porosity, and a wealth of functional groups, biochar is an ideal candidate for refining composting processes and enhancing the quality of compost. It can influence various compost parameters, such as maturity, temperature, humidity, pH, organic matter content, nitrogen loss, heavy metal content, organic pollutant levels, gas emissions, microbial activity, and the overall composting products and processes. In summary, biochar's multifunctional nature, underpinned by its distinctive physicochemical properties, positions it as a sustainable and versatile material with broad implications for environmental remediation, agriculture, energy, and construction sectors.

REFERENCES

1. Amin, F. R.; Huang, Y.; He, Y.; Zhang, R.; Liu, G.; Chen, C. Biochar applications and modern techniques for characterization. *Clean Technol. Environ.* 2016, 18 (5), 1457–1473.

2. Woolf, D.; Amonette, J. E.; Street-Perrott, F. A.; Lehmann, J.; Joseph, S. Sustainable biochar to mitigate global climate change. *Nat. Commun.* 2010, 1 (1), 56.

3. Kah, M.; Sigmund, G.; Xiao, F.; Hofmann, T. Sorption of ionizable and ionic organic compounds to biochar, activated carbon and other carbonaceous materials. *Water Res.* 2017, 124, 673–692.

4. Pan, J.; Ma, J.; Zhai, L.; Luo, T.; Mei, Z.; Liu, H. Achievements of biochar application for enhanced anaerobic digestion: A review. *Bioresour. Technol.* 2019, 292, 122058.

5. Wang, L.; Chen, L.; Tsang, D. C. W.; Kua, H. W.; Yang, J.; Ok, Y. S.; Ding, S.; Hou, D.; Poon, C. S. The roles of biochar as green admixture for sediment-based construction products. *Cem. Concr. Compos.* 2019, 104, 103348.

6. Hu, Q.; Jung, J.; Chen, D.; Leong, K.; Song, S.; Li, F.; Mohan, B. C.; Yao, Z.; Prabhakar, A. K.; Lin, X; Lim, E.; Zhang, L.; Souradeep, G.; Ok, Y. S.; Kua, H. W.; Li, S. F. Y. L.; Tand, H. T. W.; Dai, Y. J.; Tong, Y. W.; Peng, Y. H.; Joseph, S.; Wang, C. H. Biochar industry to circular economy. *Sci. Total Environ.* 2021, 757, 143820.

7. Wang, D.; Jiang, P.; Zhang, H.; Yuan, W. Biochar production and applications in agro and forestry systems: A review. *Sci. Total Environ.* 2020, 723, 137775.

8. Gholizadeh, M.; Hu, X.; Liu, Q. Progress of using biochar as a catalyst in thermal conversion of biomass. *Rev. Chem. Eng.* 2021, 37 (2), 229–258.

9. Ahmed, A.; Bakar, M. S. A.; Sukri, R. S.; Hussain, M.; Farooq, A.; Moogi, S.; Park, Y. K. Sawdust pyrolysis from the furniture industry in an auger pyrolysis reactor system for biochar and bio-oil production. *Energ. Convers. Manage.* 2020, 226, 113502.

10. Li, P.; Wan, K.; Chen, H.; Zheng, F. J.; Zhang, Z.; Niu, B.; Zhang, Y. Y.; Long, D. H. Value-added products from catalytic pyrolysis of lignocellulosic biomass and waste plastics over biochar-based catalyst: A state-of-the-art review. *Catalysts* 2022, 12 (9), 1067.

11. Jin, W; Singh, K; Zondlo, J. Co-processing of pyrolysis vapors with biochars for ex-situ upgrading. *Renew. Energ.* 2015, 83, 638–645.

12. Ren, S.; Lei, H.; Wang, L.; Bu, Q.; Chen, S.; Wu, J. Hydrocarbon and hydrogen-rich syngas production by biomass catalytic pyrolysis and bio-oil upgrading over biochar catalysts. *RSC Adv.* 2014, 4 (21), 10731–10737.

13. Yang, H.; Chen, Z.; Chen, W.; Chen, Y.; Wang, X.; Chen, H. Role of porous structure and active O-containing groups of activated biochar catalyst during biomass catalytic pyrolysis. *Energy* 2020, 210, 118646.

14. An, Y.; Tahmasebi, A.; Zhao, X.; Matamba, T.; Yu, J. Catalytic reforming of palm kernel shell microwave pyrolysis vapors over iron-loaded activated carbon: Enhanced production of phenol and hydrogen. *Bioresour. Technol.* 2020, 306, 123111.

15. Zhu, L.; Zhang, Y.; Lei, H.; Zhang, X.; Wang, L.; Bu, Q.; Wei, Y. Production of hydrocarbons from biomass-derived biochar assisted microwave catalytic pyrolysis. *Sustain. Energ. Fuels* 2018, 2 (8), 1781–1790.

16. Zhang, S.; Wu, Y.; Wang, Y.; Zhong, M.; Wang, G.; Ban, Y.; Zhao, S.; Hu, H. Q.; Jin, L. J. Facile demineralization of biochar and its catalytic upgrading of bio-oil from fast pyrolysis of bagasse. *Fuel* 2023, 349, 128714.

17. Chen, W.; Fang, Y.; Li, K.; Chen, Z. Q.; Xia, M. W.; Gong, M.; Chen, Y, G.; Yang, H. P.; Tu, X.; Chen, H. P. Bamboo waste catalytic pyrolysis with N-doped biochar catalyst for phenols products. *Appl. Energ.* 2020, 260, 114242.

18. Bu, Q.; Lei, H.; Ren, S.; Wang, L.; Zhang, Q.; Tang, J.; Ruan, R. Production of phenols and biofuels by catalytic microwave pyrolysis of lignocellulosic biomass. *Bioresour. Technol.* 2012, 108, 274–279.

19. Chen, W.; Li, K.; Xia, M.; Yang, H.; Chen, Y.; Chen, X.; Che, Q. F.; Chen, H. P. Catalytic deoxygenation co-pyrolysis of bamboo waste and microalgae with biochar catalyst. *Energy* 2018, 157, 472–482.

20. Zhang, Z. B.; Lu, Q.; Ye, X. N.; Li, W. T.; Hu, B.; Dong, C. Q. Production of phenolic-rich bio-oil from catalytic fast pyrolysis of biomass using magnetic solid base catalyst. *Energ. Convers. Manage.* 2015, 106, 1309–1317.

21. Zainan, N. H.; Srivatsa, S. C.; Li, F.; Bhattacharya, S. Quality of bio-oil from catalytic pyrolysis of microalgae *Chlorella vulgaris*. *Fuel* 2018, 223, 12–19.

22. Nejati, B.; Adami, P.; Bozorg, A.; Tavasoli, A.; Mirzahosseini, A. H. Catalytic pyrolysis and bio-products upgrading derived from *Chlorella vulgaris* over its biochar and activated biochar-supported Fe catalysts. *J. Anal. Appl. Pyrolysis* 2020, 152, 104799.

23. Deng, L.; Guo, W.; Ngo, H. H.; Zhang, X.; Wei, D.; Wei, Q.; Deng, S. Novel catalysts in catalytic upcycling of common polymer waste. *Chem. Eng. J.*2023, 471, 144350.

24. Li, C.; Zhang, C.; Gholizadeh, M.; Hu, X. Different reaction behaviours of light or heavy density polyethylene during the pyrolysis with biochar as the catalyst. *J. Hazard. Mater.* 2020, 399, 123075.

25. Li, Y.; Williams, P. T. Catalytic biochar and refuse-derived char for the steam reforming of waste plastics pyrolysis volatiles for hydrogen-rich syngas. *Ind. Eng. Chem. Res.* 2023, 62 (36), 14335–14348.

26. Wang, C.; Lei, H.; Qian, M.; Huo, E.; Zhao, Y.; Zhang, Q.; Mateo, W.; Lin, X. N.; Kong, X.; Zou, R.; Ruan, R. Application of highly stable biochar catalysts for efficient pyrolysis of plastics: A readily accessible potential solution to a global waste crisis. *ACS Sustain. Energ. Fuels* 2020, 4 (9), 4614–4624.

27. Wang, C.; Zou, R.; Lei, H.; Qian, M.; Lin, X.; Mateo, W.; Wang, L.; Zhong, X. S.; Ruan, R. Biochar-advanced thermocatalytic salvaging of the waste disposable mask with the production of hydrogen and mono-aromatic hydrocarbons. *J. Hazard. Mater.* 2022, 426, 128080.

28. Huang, P.; Zhou, W.; Jin, K.; Wang, Y.; Qian, J.; Liu, L.; Peng, H. Y.; Wu, J. B.; Hu, J. Y.; Wang, M. J.; Wang, W. X.; Luo, T.; Fan, L. One-step carbothermal reduction synthesis of metal-loaded biochar catalyst for in situ catalytic upgrading of plastic pyrolysis products. *ACS Sustain. Energ. Fuels* 2023, 11 (2), 696–707.

29. Chao, L.; Zhang, C.; Zhang, L.; Gholizadeh, M.; Hu, X. Catalytic pyrolysis of tyre waste: Impacts of biochar catalyst on product evolution. *Waste Manage.* 2020, 116, 9–21.

30. Li, J.; Lin, F.; Yu, H.; Tong, X.; Cheng, Z.; Yan, B.; Song, Y. J.; Chen, G. Y.; Hou, L. A.; Crittenden, J. C. Biochar-assisted catalytic pyrolysis of oily sludge to attain harmless disposal and residue utilization for soil reclamation. *Environ. Sci. Technol.* 2023, 57 (17), 7063–7073.

31. Kim, S.; Lee, Y.; Lin, K. Y. A.; Hong, E.; Kwon, E. E.; Lee, J. The valorization of food waste via pyrolysis. *J. Clean. Prod.* 2020, 259, 120816.

32. Jung, J. M.; Oh, J. I.; Baek, K.; Lee, J.; Kwon, E. E. Biodiesel production from waste cooking oil using biochar derived from chicken manure as a porous media and catalyst. *Energ. Convers. Manage.* 2018, 165, 628–633.

33. Sutton, D.; Kelleher, B.; Ross, J. R. H. Review of literature on catalysts for biomass gasification. *Fuel Process. Technol.* 2001, 73, 155–173.

34. Jin, L. J.; Bai, X. Y.; Li, Y.; Dong, C.; Hu, H. Q.; Li, X. In-situ catalytic upgrading of coal pyrolysis tar on carbon-based catalyst in a fixed-bed reactor. *Fuel Process. Technol.* 2016, 147, 41–46.

35. Zhang, L.; Yao, Z. L.; Zhao, L. X.; Li, Z. H.; Yi, W. M.; Kang, K.; Jia, J. X. Synthesis and characterization of different activated biochar catalysts for removal of biomass pyrolysis tar. *Energy* 2021, 232, 120927.

36. Guo, F.; Peng, K.; Li, T.; Zhao, X.; Dong, Y.; Rao, Z. Catalytic cracking of primary tar vapor from biomass over high ash-containing paper sludge ash. *Energ. Fuels* 2018, 32, 12514–12522.

37. Shahbaz, M.; Yusup, S.; Inayat, A.; Patrick, D. O.; Pratama, A. Application of response surface methodology to investigate the effect of different variables on conversion of palm kernel shell in steam gasification using coal bottom ash. *Appl. Energ.* 2016, 184, 1306–1315.

38. Shahbaz, M.; Yusup, S.; Inayat, A.; Patrick, D. O.; Ammar, M. The influence of catalysts in biomass steam gasification and catalytic potential of coal bottom ash in biomass steam gasification: A review. *Renew. Sustain. Energ. Rev.* 2017, 73, 468–476.

39. Kirnbauer, F.; Wilk, V.; Kitzler, H.; Kern, S.; Hofbauer, H. The positive effects of bed material coating on tar reduction in a dual fluidized bed gasifier. *Fuel* 2012, 95, 553–562.

40. Simell, P.; Kurkela, E.; Ståhlberg, P.; Hepola, J. Catalytic hot gas cleaning of gasification gas. *Catal. Today.* 1996, 27, 55–62.
41. Wang, T.; Chang, J.; Lv, P.; Zhu, J. Novel catalyst for cracking of biomass tar. *Energ. Fuels* 2004, 19, 22–27.
42. Chiang, K. Y.; Lu, C. H.; Chien, K. L. The aluminum silicate catalyst effect on efficiency of energy yield in gasification of paper-reject sludge. *Int. J. Hydrogen Energ.* 2013, 38, 15787–15793.
43. Tomishige, K.; Miyazawa, T.; Asadullah, M.; Ito, S.; Kunimori, K. Catalyst performance in reforming of tar derived from biomass over noble metal catalysts. *Green Chem.* 2003, 5, 399–403.
44. Min, Z.; Yimsiri, P.; Asadullah, M.; Zhang, S.; Li, C. Z. Catalytic reforming of tar during gasification. Part II. Char as a catalyst or as a catalyst support for tar reforming. *Fuel* 2011, 90, 2545–2552.
45. Min, Z.; Zhang, S.; Yimsiri, P.; Wang, Y.; Asadullah, M.; Li, C. Catalytic reforming of tar during gasification. Part IV. Changes in the structure of char in the char-supported iron catalyst during reforming. *Fuel* 2013, 106, 858–863.
46. Shen, Y.; Zhao, P.; Shao, Q.; Takahashi, F.; Yoshikawa, K. In situ catalytic conversion of tar using rice husk char/ash supported nickel-iron catalysts for biomass pyrolytic gasification combined with the mixing-simulation in fluidized-bed gasifier. *Appl. Energ.* 2015, 160, 808–819.
47. Jordan, C. A.; Akay, G. Effect of CaO on tar production and dew point depression during gasification of fuel cane bagasse in a novel downdraft gasifier. *Fuel Process. Technol.* 2013, 106, 654–660.
48. Hurley, S.; Li, H.; Xu, C. C. Effects of impregnated metal ions on air/CO$_2$-gasification of woody biomass. *Bioresour. Technol.* 2010, 101, 9301–9307.
49. Sancho, J. A.; Aznar, M. P.; Toledo, J. M. Catalytic air gasification of plastic waste (polypropylene) in fluidized bed. Part I: use of in-gasifier bed additives. *Ind. Eng. Chem. Res.* 2008, 47, 1005–1010.
50. Jin, L.; Bai, X.; Li, Y.; Dong, C.; Hu, H.; Li, X. In-situ catalytic upgrading of coal pyrolysis tar on carbon-based catalyst in a fixed-bed reactor. *Fuel Process. Technol.* 2016, 147, 41–46.
51. Slatter, N. L.; Vichanpol, B.; Natakaranakul, J.; Wattanavichien, K.; Suchamalawong, P.; Hashimoto, K.; Tsubaki, N.; Vitidsant, T.; Charusiri, W. Syngas production for Fischer-Tropsch synthesis from rubber wood pellets and eucalyptus wood chips in a pilot horizontal gasifier with CaO as a tar removal catalyst. *ACS Omega* 2022, 7, 44951–44961.
52. Yu, H. M.; Yang, X. Y.; Jiang, L. H.; Chen, D. Z. Experimental study on co-gasification characteristics of biomass and plastic waste. *Bioresources* 2014, 9, 5615–5626.

53. Larsson, A.; Israelsson, M.; Lind, F.; Seemann, M.; Thunman, H. Using ilmenite to reduce the tar yield in a dual fluidized bed gasification system. *Energ. Fuels* 2014, 28, 2632–2644.

54. Xiao, X.; Chen, B.; Chen, Z.; Zhu, L.; Schnoor, J. L. Insight into multiple and multilevel structures of biochars and their potential environmental applications: A critical review. *Environ. Sci. Technol.* 2018, 52, 5027–5047.

55. Chen, G.; Wang, J.; Yu, F.; Wang, X.; Xiao, H.; Yan, B.; Cui, X. A review on the production of P-enriched hydro/bio-char from solid waste: Transformation of P and applications of hydro/bio-char. *Chemosphere* 2022, 301, 134646.

56. Wu, W.; Yan, B.; Zhong, L.; Zhang, R.; Guo, X.; Cui, X.; Lu, W.; Chen, G. Combustion ash addition promotes the production of K-enriched biochar and K release characteristics. *J. Clean. Prod.* 2021, 311, 127557.

57. Cui, X.; Zhang, J.; Wang, X.; Pan, M.; Lin, Q.; Khan, K. Y.; Yan, B.; Li, T.; He, Z.; Yang, X.; Chen, G. A review on the thermal treatment of heavy metal hyperaccumulator: Fates of heavy metals and generation of products. *J. Hazard. Mater.* 2021, 405, 123832.

58. Tran, H. N.; Lima, E. C.; Juang, R. S.; Bollinger, J. C.; Chao, H. P. Thermodynamic parameters of liquid-phase adsorption process calculated from different equilibrium constants related to adsorption isotherms: A comparison study. *J. Environ. Chem. Eng.* 2021, 9 (6), 106674.

59. Ghaedi, M.; Hajjati, S.; Mahmudi, Z.; Tyagi, I.; Agarwal, S.; Maity, A.; Gupta, V. K. Modeling of competitive ultrasonic assisted removal of the dyes – Methylene blue and Safranin-O using Fe_3O_4 nanoparticles. *Chem. Eng. J.* 2015, 268, 28–37.

60. Fu, Z.; Xue, Y.; Li, J.; Yan, B.; Han, Z.; Chen, G. Steam gasification of yak manure: Kinetic modeling by a sequential and coupling method. *Fuel* 2022, 329,125464.

61. Anis, S.; Zainal, Z. A. Tar reduction in biomass producer gas via mechanical, catalytic and thermal methods: A review. *Renew. Sustain. Energ. Rev.* 2011, 15, 2355–2377.

62. Kuhn, J. N.; Zhao, Z.; Senefeld-Naber, A.; Felix, L. G.; Slimane, R. B.; Choi, C. W.; Ozkan, U. S. Ni-olivine catalysts prepared by thermal impregnation: Structure, steam reforming activity, and stability. *Appl. Catal. A: Gen.* 2008, 341, 43–49.

63. Mun, T. Y.; Kim, J. W.; Kim, J. S. Air gasification of railroad wood ties treated with creosote: Effects of additives and their combination on the removal of tar in a two-stage gasifier. *Fuel* 2012, 102, 326–332.

64. Wang, T. J.; Chang, J.; Wu, C. Z.; Fu, Y.; Chen, Y. The steam reforming of naphthalene over a nickel-dolomite cracking catalyst. *Biomass Bioenerg.* 2005, 28, 508–514.

65. Sutton, D.; Kelleher, B.; Ross, J. R. Review of literature on catalysts of bio-mass gasification. *Fuel Process. Technol.* 2001, 73 (3), 155–173.

66. Guo, F.; Jia, X.; Liang, S.; Zhou, N.; Chen, P.; Ruan, R. Development of biochar-based nanocatalysts for tar cracking/reforming during biomass pyrolysis and gasification. *Bioresour. Technol.* 2020, 298, 122263.

67. Du, Z. Y.; Zhang, Z. H.; Xu, C.; Wang, X. B.; Li, W. Y. Low-temperature steam reforming of toluene and biomass tar over biochar-supported Ni nanoparticles. *ACS Sustain. Chem. Eng.* 2018, 7, 3111–3119.

68. Tian, B.; Mao, S.; Guo, F.; Bai, J.; Shu, R.; Qian, L.; Liu, Q. Monolithic biochar-supported cobalt-based catalysts with high-activity and superior-stability for biomass tar reforming. *Energy* 2022, 242, 122970

69. Li, L.; Song, Z.; Zhao, X.; Ma, C.; Kong, X., Wang, F. Microwave-induced cracking and CO_2 reforming of toluene on biomass derived char. *Chem. Eng. J.* 2016, 284, 1308–1316.

70. Li, J.; Tao, J.; Yan, B.; Cheng, K.; Chen, G.; Hu, J. Microwave reforming with char-supported nickel-cerium catalysts: A potential approach for thor-ough conversion of biomass tar model compound. *Appl. Energ.* 2020, 261, 114375.

71. Ashok, J.; Dewangan, N.; Das, S.; Hongmanorom, P.; Wai, M. H.; Tomishige, K.; Kawi, S. Recent progress in the development of catalysts for steam reforming of biomass tar model reaction. *Fuel Process. Technol.* 2020, 199. 106252.

72. Li, P.; Zhao, H.; Cheng, C.; Hou, T.; Shen, D.; Jiao, Y. A review on anaer-obic co-digestion of sewage sludge with other organic waste for methane production: Mechanism, process, improvement and industrial application. *Biomass Bioenerg.* 2024, 185, 107241.

73. AM Buswell, H. W.; Schlenz, H. E. Studies on two stage sludge diges-tion. 1928.

74. Bryant, M. Microbial methane production – theoretical aspects. *J. Anim. Sci.* 1979.

75. Yu, T.; Deng, Y.; Liu, H.; Yang, C.; Wu, B.; Zeng, G.; Lu, L.; Nishimura, F. Effect of alkaline microwaving pretreatment on anaerobic digestion and biogas production of swine manure. *Sci. Rep.* 2017, 342, 121097.

76. Ahmad, M.; Rajapaksha, A. U.; Lim, J. E.; Zhang, M.; Bolan, N.; Mohan, D.; Vithanage, M.; Lee, S. S.; Ok, Y. S. Biochar as a sorbent for contaminant management in soil and water: A review. *Chemosphere* 2014, 99, 19–33.

77. Janu, R.; Mrlik, V.; Ribitsch, D.; Hofman, J.; Sedláček, P.; Bielská, L.; Soja, G. Biochar surface functional groups as affected by biomass feedstock, biochar composition and pyrolysis temperature. *Carbon Resour. Convers.* 2021, 4, 36–46.

78. Das, S. K.; Ghosh, G. K.; Avasthe, R. K.; Sinha, K. Compositional hetero-geneity of different biochar: Effect of pyrolysis temperature and feedstocks. *J. Environ. Manage.* 2021, 278, 111501.
79. Ippolito, J. A.; Cui, L.; Kammann, C.; Wrage-Mönnig, N.; Estavillo, J. M.; Fuertes-Mendizabal, T.; Cayuela, M. L.; Sigua, G.; Novak, J.; Spokas, K.; Borchard, N. Feedstock choice, pyrolysis temperature and type influ-ence biochar characteristics: A comprehensive meta-data analysis review. *Biochar* 2020, 2(4), 421–438.
80. Yang, C.; Liu, J.; Lu, S. Pyrolysis temperature affects pore characteristics of rice straw and canola stalk biochars and biochar-amended soils. *Geoderma* 2021, 397, 115097.
81. Muzyka, R.; Misztal, E.; Hrabak, J.; Banks, S. W.; Sajdak, M. Various bio-mass pyrolysis conditions influence the porosity and pore size distribution of biochar. *Energy* 2023, 263, 126128.
82. Zhao, Y.; Li, J. Sensor-based technologies in effective solid waste sorting: Successful applications, sensor combination, and future directions. *Environ. Sci. Technol.* 2022, 56 (24), 17531–17544.
83. Faisal, S.; Ebaid, R.; Xiong, M.; Huang, J.; Wang, Q.; El-Hefnawy, M.; Abomohra, A. Maximizing the energy recovery from rice straw through two-step conversion using eggshell-catalytic pyrolysis followed by enhanced anaerobic digestion using calcium-rich biochar. *Sci. Total Environ.* 2023, 858, 159984.
84. Liu, M.; Li, Z.; Qi, X.-e.; Chen, Z.; Ni, H.; Gao, Y.; Liu, X. Improvement of cow manure anaerobic digestion performance by three different crop straw biochars. *Environ. Technol. Inno.* 2023, 31, 103233.
85. Chen, Y.; Zhao, Z.; Zou, H.; Yang, H.; Sun, T.; Li, M.; Chai, H.; Li, L.; Ai, H.; Shi, D.; He, Q.; Gu, L. Digestive performance of sludge with different crop straws in mesophilic anaerobic digestion. *Bioresour. Technol.* 2019, 289, 121595.
86. Kainthola, J.; Kalamdhad, A. S.; Goud, V. V.; Goel, R. Fungal pretreatment and associated kinetics of rice straw hydrolysis to accelerate methane yield from anaerobic digestion. *Bioresour. Technol.* 2019, 286, 121368.
87. Kumar, M.; Dutta, S.; You, S.; Luo, G.; Zhang, S.; Show, P. L.; Sawarkar, A. D.; Singh, L.; Tsang, D. C. W. A critical review on biochar for enhancing biogas production from anaerobic digestion of food waste and sludge. *J. Clean. Prod.* 2021, 305, 127143.
88. Saif, I.; Salama, E. S.; Usman, M.; Lee, D. S.; Malik, K.; Liu, P.; Li, X. Improved digestibility and biogas production from lignocellulosic biomass: Biochar addition and microbial response. *Ind. Crops Prod.* 2021, 171, 113851.

89. Wang, N.; Chen, Q.; Zhang, C.; Dong, Z.; Xu, Q. Improvement in the physicochemical characteristics of biochar derived from solid digestate of food waste with different moisture contents. *Sci. Total Environ.* 2022, 819, 153100.

90. Liu, J.; Huang, S.; Chen, K.; Wang, T.; Mei, M.; Li, J. Preparation of biochar from food waste digestate: Pyrolysis behaviour and product properties. *Bioresour. Technol.* 2020, 302, 122841.

91. Zhang, S.; Ma, X.; Sun, H.; Zhao, P.; Wang, Q.; Wu, C.; Gao, M. Response of semi-continuous anaerobic digestion of food waste to progressively increasing temperature: Methanogen community, correlation analysis, and energy balance. *Ind. Crops Prod.* 2023, 192, 116066.

92. Heidari, M.; Dutta, A.; Acharya, B.; Mahmud, S. A review of the current knowledge and challenges of hydrothermal carbonization for biomass conversion. *J. Energ. Inst.* 2019, 92 (6), 1779–1799.

93. Wang, T.; Zhai, Y.; Zhu, Y.; Li, C.; Zeng, G. A review of the hydrothermal carbonization of biomass waste for hydrochar formation: Process conditions, fundamentals, and physicochemical properties. *Renew. Sustain. Energ. Rev.* 2018, 90, 223–247.

94. Ma, H.; Hu, Y.; Kobayashi, T.; Xu, K. Q. The role of rice husk biochar addition in anaerobic digestion for sweet sorghum under high loading condition. *Biotechnol. Rep.* 2020, 27, e00515.

95. Fagbohungbe, M. O.; Herbert, B. M. J.; Hurst, L.; Ibeto, C. N.; Li, H.; Usmani, S. Q.; Semple, K. T. The challenges of anaerobic digestion and the role of biochar in optimizing anaerobic digestion. *Waste Manage.* 2017, 61, 236–249.

96. Pan, J.; Ma, J.; Zhai, L.; Liu, H. Enhanced methane production and syntrophic connection between microorganisms during semi-continuous anaerobic digestion of chicken manure by adding biochar. *J. Clean. Prod.* 2019, 240, 118178.

97. de Quadros, T. C. F.; Mangerino Sicchieri, I.; Fernandes, F.; Kiyomi Kuroda, E. Selection of additive materials for anaerobic co-digestion of fruit and vegetable waste and layer chicken manure. *Bioresour. Technol.* 2022, 361, 127659.

98. Sun, Z.; Feng, L.; Li, Y.; Han, Y.; Zhou, H.; Pan, J. The role of electrochemical properties of biochar to promote methane production in anaerobic digestion. *J. Clean. Prod.* 2022, 362, 132296.

99. Chiappero, M.; Norouzi, O.; Hu, M.; Demichelis, F.; Berruti, F.; Di Maria, F.; Mašek, O.; Fiore, S. Review of biochar role as additive in anaerobic digestion processes. *Renew. Sustain. Energ. Rev.* 2020, 131, 110037.

100. De Vrieze, J.; Devooght, A .; Walraedt, D.; Boon, N. Enrichment of methanosaetaceae on carbon felt and biochar during anaerobic digestion of a potassium-rich molasses stream. *Appl. Microbiol. Biotechnol.* 2016, 100 (11), 5177–5187.

101. Sunyoto, N. M. S.; Sugiarto, Y.; Zhu, M.; Zhang, D. Transient performance during start-up of a two-phase anaerobic digestion process demonstration unit treating carbohydrate-rich waste with biochar addition. *Int. J. Hydrogen Energ.* 2019, 44 (28), 14341–14350.

102. Lü, F.; Luo, C.; Shao, L.; He, P. Biochar alleviates combined stress of ammonium and acids by firstly enriching methanosaeta and then methanosarcina. *Water Res.* 2016, 90, 34–43.

103. Qin, Y.; Yin, X.; Xu, X.; Yan, X.; Bi, F.; Wu, W. Specific surface area and electron donating capacity determine biochar's role in methane production during anaerobic digestion. *Bioresour. Technol.* 2020, 303, 122919.

104. Duan, S.; He, J.; Xin, X.; Li, L.; Zou, X.; Zhong, Y.; Zhang, J.; Cui, X. Characteristics of digested sludge-derived biochar for promoting methane production during anaerobic digestion of waste activated sludge. *Bioresour. Technol.* 2023, 384, 129245.

105. Peng, Y.; Li, L.; Dong, Q.; Yang, P.; Liu, H.; Ye, W.; Wu, D.; Peng, X. Evaluation of digestate-derived biochar to alleviate ammonia inhibition during long-term anaerobic digestion of food waste. *Chemosphere* 2023, 311, 137150.

106. Cai, Y.; Zhu, M.; Meng, X.; Zhou, J. L.; Zhang, H.; Shen, X. The role of biochar on alleviating ammonia toxicity in anaerobic digestion of nitrogen-rich waste: A review. *Bioresour. Technol.* 2022, 351, 126924.

107. Wang, S.; Zhang, H.; Huang, H.; Xiao, R.; Li, R.; Zhang, Z. Influence of temperature and residence time on characteristics of biochars derived from agricultural residues: A comprehensive evaluation. *Process. Saf. Environ.* 2020, 139, 218–229.

108. He, X.; Guo, Z.; Lu, J.; Zhang, P. Carbon-based conductive materials accelerated methane production in anaerobic digestion of waste fat, oil and grease. *Bioresour. Technol.* 2021, 329, 124871.

109. Tiwari, S. B.; Dubey, M.; Ahmed, B.; Gahlot, P.; Khan, A. A.; Rajpal, A.; Kazmi, A. A.; Tyagi, V. K. Carbon-based conductive materials facilitated anaerobic co-digestion of agro waste under thermophilic conditions. *Waste Manage.* 2021, 124, 17–25.

110. Montalvo, S.; Cahn, I.; Borja, R.; Huilinir, C.; Guerrero, L. Use of solid residue from thermal power plant (fly ash) for enhancing sewage sludge anaerobic digestion: Influence of fly ash particle size. *Bioresour. Technol.* 2017, 244, 416–422.

111. Wang, J.; Hao, X.; Liu, Z.; Guo, Z.; Zhu, L.; Xiong, B.; Jiang, D.; Shen, L.; Li, M.; Kang, B.; Tang, G.; Bai, L. Biochar improves heavy metal passivation during wet anaerobic digestion of pig manure. *Environ. Sci. Pollut. Res.* 2020, 28 (1), 635–644.

112. Yang, S.; Chen, Z.; Wen, Q. Impacts of biochar on anaerobic digestion of swine manure: Methanogenesis and antibiotic resistance genes dissemination. *Bioresour. Technol.* 2021, 324, 124679.

113. Zhang, K.; Sun, P.; Faye, M. C. A. S.; Zhang, Y. Characterization of biochar derived from rice husks and its potential in chlorobenzene degradation. *Carbon* 2018, 130, 730–740.

114. Li, H.; Pan, B.; Liao, S.; Zhang, D.; Xing, B. Formation of environmentally persistent free radicals as the mechanism for reduced catechol degradation on hematite-silica surface under UV irradiation. *Environ. Pollut.* 2014, 188, 153–158.

115. Oh, S. Y.; Son, J. G.; Hur, S. H.; Chung, J. S.; Chiu, P. C. Black carbon-mediated reduction of 2,4-dinitrotoluene by dithiothreitol. *J. Environ. Qual.* 2013, 42 (3), 815–821.

116. Huang, X.; Miao, X.; Chu, X.; Luo, L.; Zhang, H.; Sun, Y. Enhancement effect of biochar addition on anaerobic co-digestion of pig manure and corn straw under biogas slurry circulation. *Bioresour. Technol.* 2023, 372, 128654.

117. Chen, L.; Fang, W.; Liang, J.; Nabi, M.; Cai, Y.; Wang, Q.; Zhang, P.; Zhang, G. Biochar application in anaerobic digestion: Performances, mechanisms, environmental assessment and circular economy. *Resour. Conserv. Recycl.* 2023, 188, 106720.

118. Qi, Q.; Sun, C.; Zhang, J.; He, Y.; Wah Tong, Y. Internal enhancement mechanism of biochar with graphene structure in anaerobic digestion: The bioavailability of trace elements and potential direct interspecies electron transfer. *Chem. Eng. J.* 2021, 406, 126833.

119. Lü, C.; Shen, Y.; Li, C.; Zhu, N.; Yuan, H. Redox-active biochar and conductive graphite stimulate methanogenic metabolism in anaerobic digestion of waste-activated sludge: Beyond direct interspecies electron transfer. *ACS Sustain. Chem. Eng.* 2020, 8 (33), 12626–12636.

120. Ayilara, M.; Olanrewaju, O.; Babalola, O.; Odeyemi, O. Waste management through composting: Challenges and potentials. *Sustainability – basel.* 2020, 12 (11), 4456.

121. Bernal, M. P.; Alburquerque, J. A.; Moral, R. Composting of animal manures and chemical criteria for compost maturity assessment. A review. *Bioresour. Technol.* 2009, 100 (22), 5444–5453.

122. Akyol, C.; Ince, O.; Ince, B. Crop-based composting of lignocellulosic digestates: Focus on bacterial and fungal diversity. *Bioresour. Technol.* 2019, 288, 121549.

123. Onwosi, C. O.; Igbokwe, V. C.; Odimba, J. N.; Eke, I. E.; Nwankwoala, M. O.; Iroh, I. N.; Ezeogu, L. I. Composting technology in waste stabilization: On the methods, challenges and future prospects. *J. Environ. Manage.* 2017, 190, 140–157.

124. Toledo, M.; Siles, J. A.; Gutierrez, M. C.; Martin, M. A. Monitoring of the composting process of different agroindustrial waste: Influence of the operational variables on the odorous impact. *Waste. Manage.* 2018, 76, 266–274.

125. Das, M.; Uppal, H. S.; Singh, R.; Beri, S.; Mohan, K. S.; Gupta, V. C.; Adholeya, A. Co-composting of physic nut (*Jatropha curcas*) deoiled cake with rice straw and different animal dung. *Bioresour. Technol.* 2011, 102 (11), 6541–6546.

126. Cai, Q. Y.; Mo, C. H.; Wu, Q. T.; Zeng, Q. Y.; Katsoyiannis, A. Concentration and speciation of heavy metals in six different sewage sludge-composts. *J. Hazard. Mater.* 2007, 147 (3), 1063–1072.

127. Li, Z.; Lu, H.; Ren, L.; He, L. Experimental and modeling approaches for food waste composting: A review. *Chemosphere* 2013, 93 (7), 1247–1257.

128. Chen, R.; Wang, Y.; Wang, W.; Wei, S.; Jing, Z.; Lin, X. N_2O emissions and nitrogen transformation during windrow composting of dairy manure. *J. Environ. Manage.* 2015, 160, 121–127.

129. Awasthi, M. K.; Pandey, A. K.; Khan, J.; Bundela, P. S.; Wong, J. W.; Selvam, A. Evaluation of thermophilic fungal consortium for organic municipal solid waste composting. *Bioresour. Technol.* 2014, 168, 214–221.

130. Nakasaki, K.; Tran le, T. H.; Idemoto, Y.; Abe, M.; Rollon, A. P. Comparison of organic matter degradation and microbial community during thermophilic composting of two different types of anaerobic sludge. *Bioresour. Technol.* 2009, 100 (2), 676–682.

131. Gao, M.; Li, B.; Yu, A.; Liang, F.; Yang, L.; Sun, Y. The effect of aeration rate on forced-aeration composting of chicken manure and sawdust. *Bioresour. Technol.* 2010, 101 (6), 1899–1903.

132. Agnew, J. M.; Leonard, J. J. The physical properties of compost. *Compost Sci. Util.* 2003, 11 (3), 238–264.

133. Bhatia, A.; Madan, S.; Sahoo, J.; Ali, M.; Pathania, R.; Kazmi, A. A. Diversity of bacterial isolates during full scale rotary drum composting. *Waste. Manage.* 2013, 33 (7), 1595–1601.

134. Yang, F.; Li, G.; Shi, H.; Wang, Y. Effects of phosphogypsum and superphosphate on compost maturity and gaseous emissions during kitchen waste composting. *Waste. Manage.* 2015, 36, 70–76.

135. Lazcano, C.; Gomez-Brandon, M.; Dominguez, J. Comparison of the effectiveness of composting and vermicomposting for the biological stabilization of cattle manure. *Chemosphere* 2008, 72 (7), 1013–1019.

136. Huang, G. F.; Wong, J. W.; Wu, Q. T.; Nagar, B. B. Effect of C/N on composting of pig manure with sawdust. *Waste. Manage.* 2004, 24 (8), 805–813.

137. Petric, I.; Helic, A.; Avdic, E. A. Evolution of process parameters and determination of kinetics for co-composting of organic fraction of municipal solid waste with poultry manure. *Bioresour. Technol.* 2012, 117, 107–116.

138. Chan, M. T.; Selvam, A.; Wong, J. W. Reducing nitrogen loss and salinity during 'struvite' food waste composting by zeolite amendment. *Bioresour. Technol.* 2016, 200, 838–844.

139. Turan, N. G. The effects of natural zeolite on salinity level of poultry litter compost. *Bioresour. Technol.* 2008, 99 (7), 2097–2101.

140. Kalemelawa, F.; Nishihara, E.; Endo, T.; Ahmad, Z.; Yeasmin, R.; Tenywa, M. M.; Yamamoto, S. An evaluation of aerobic and anaerobic composting of banana peels treated with different inoculums for soil nutrient replenishment. *Bioresour. Technol.* 2012, 126, 375–382.

141. Proietti, P.; Calisti, R.; Gigliotti, G.; Nasini, L.; Regni, L.; Marchini, A. Composting optimization: Integrating cost analysis with the physical-chemical properties of materials to be composted. *J. Clean. Prod.* 2016, 137, 1086–1099.

142. Wang, Y.; Ai, P.; Cao, H.; Liu, Z. Prediction of moisture variation during composting process: A comparison of mathematical models. *Bioresour. Technol.* 2015, 193, 200–205.

143. Sanchez-Garcia, M.; Alburquerque, J. A.; Sanchez-Monedero, M. A.; Roig, A.; Cayuela, M. L. Biochar accelerates organic matter degradation and enhances N mineralisation during composting of poultry manure without a relevant impact on gas emissions. *Bioresour. Technol.* 2015, 192, 272–279.

144. Godlewska, P.; Schmidt, H. P.; Ok, Y. S.; Oleszczuk, P. Biochar for composting improvement and contaminants reduction. A review. *Bioresour. Technol.* 2017, 246, 193–202.

145. Lehmann, J.; Joseph, S. Biochar for environmental management: Science and technology. *Forest. Policy Econ.* 2009, 11, 535–536.

146. Khan, N.; Clark, I.; Sanchez-Monedero, M. A.; Shea, S.; Meier, S.; Bolan, N. Maturity indices in co-composting of chicken manure and sawdust with biochar. *Bioresour. Technol.* 2014, 168, 245–251.

147. Jindo, K.; Sonoki, T.; Matsumoto, K.; Canellas, L.; Roig, A.; Sanchez-Monedero, M. A. Influence of biochar addition on the humic substances of composting manures. *Waste. Manage.* 2016, 49, 545–552.

148. Vandecasteele, B.; Sinicco, T.; D'Hose, T.; Vanden Nest, T.; Mondini, C. Biochar amendment before or after composting affects compost quality

and N losses, but not P plant uptake. *J. Environ. Manage.* 2016, 168, 200–209.

149. Raj, D.; Antil, R. S. Evaluation of maturity and stability parameters of composts prepared from agro-industrial waste. *Bioresour. Technol.* 2011, 102 (3), 2868–2873.

150. Zhang, J.; Chen, G.; Sun, H.; Zhou, S.; Zou, G. Straw biochar hastens organic matter degradation and produces nutrient-rich compost. *Bioresour. Technol.* 2016, 200, 876–883.

151. Chen, Y. X.; Huang, X. D.; Han, Z. Y.; Huang, X.; Hu, B.; Shi, D. Z.; Wu, W. X. Effects of bamboo charcoal and bamboo vinegar on nitrogen conservation and heavy metals immobility during pig manure composting. *Chemosphere* 2010, 78 (9), 1177–1181.

152. Lopez-Cano, I.; Roig, A.; Cayuela, M. L.; Alburquerque, J. A.; Sanchez-Monedero, M. A. Biochar improves N cycling during composting of olive mill waste and sheep manure. *Waste. Manage.* 2016, 49, 553–559.

153. Czekala, W.; Malinska, K.; Caceres, R.; Janczak, D.; Dach, J.; Lewicki, A. Co-composting of poultry manure mixtures amended with biochar: The effect of biochar on temperature and C-CO$_2$ emission. *Bioresour. Technol.* 2016, 200, 921–927.

154. Zhang, J.; Lu, F.; Shao, L.; He, P. The use of biochar-amended composting to improve the humification and degradation of sewage sludge. *Bioresour. Technol.* 2014, 168, 252–258.

155. Zhou, Y.; Kurade, M. B.; Sirohi, R.; Zhang, Z.; Sindhu, R.; Binod, P.; Jeon, B. H.; Syed, A.; Verma, M.; Awasthi, M. K. Biochar as functional amendment for antibiotic resistant microbial community survival during hen manure composting. *Bioresour. Technol.* 2023, 385, 129393.

156. Arnau, A. S.; Lamon, F.; Dekker, H.; Lorin, A.; Giacomazzi, M.; Decorte, M. EBA Activity Report, 2021.

157. Ignatowicz, K. The impact of sewage sludge treatment on the content of selected heavy metals and their fractions. *Environ. Res.* 2017, 156, 19–22.

158. Cui, E.; Wu, Y.; Zuo, Y.; Chen, H. Effect of different biochars on antibiotic resistance genes and bacterial community during chicken manure composting. *Bioresour. Technol.* 2016, 203, 11–17.

159. Awasthi, M. K.; Wang, Q.; Huang, H.; Li, R.; Shen, F.; Lahori, A. H.; Wang, P.; Guo, D.; Guo, Z.; Jiang, S.; Zhang, Z. Effect of biochar amendment on greenhouse gas emission and bio-availability of heavy metals during sewage sludge co-composting. *J. Clean. Prod.* 2016, 135, 829–835.

160. Liao, Y. Q.; Jaehac, K.; Yuan, T. G. Effect of sewage sludge derived biochar addition on methane production and microbial community structure during anaerobic digestion of food waste. *Chin. J. Environ. Eng.* 2020, 14 (2), 523–534.

161. Semple, K. T.; Riding, M. J.; McAllister, L. E.; Sopena-Vazquez, F.; Bending, G. D. Impact of black carbon on the bioaccessibility of organic contaminants in soil. *J. Hazard. Mater.* 2013, 261, 808–816.

162. Houot, S.; Verge-Leviel, C.; Poitrenaud, M. Potential mineralization of various organic pollutants during composting. *Pedosphere* 2012, 22 (4), 536–543.

163. Lashermes, G.; Barriuso, E.; Houot, S. Dissipation pathways of organic pollutants during the composting of organic waste. *Chemosphere* 2012, 87 (2), 137–143.

164. Oleszczuk, P.; Zielinska, A.; Cornelissen, G. Stabilization of sewage sludge by different biochars towards reducing freely dissolved polycyclic aromatic hydrocarbons (PAHs) content. *Bioresour. Technol.* 2014, 156, 139–145.

165. Zhang, X.; Jiao, P.; Wang, Y.; Wu, P.; Li, Y.; Ma, L. Enhancing methane production in anaerobic co-digestion of sewage sludge and food waste by regulating organic loading rate. *Bioresour. Technol.* 2022, 363, 127988.

166. Yellezuome, D.; Zhu, X.; Liu, X.; Liu, X.; Liu, R.; Wang, Z.; Li, Y.; Sun, C.; Abd-Alla, M. H.; Rasmey, A. H. M. Integration of two-stage anaerobic digestion process with in situ biogas upgrading. *Bioresour. Technol.* 2023, 369, 128475.

167. Yasmin, N.; Jamuda, M.; Panda, A. K.; Samal, K.; Nayak, J. K. Emission of greenhouse gases (GHGs) during composting and vermicomposting: Measurement, mitigation, and perspectives. *Energ. Nexus* 2022, 7, 100092.

168. Sonoki, T.; Furukawa, T.; Jindo, K.; Suto, K.; Aoyama, M.; Sanchez-Monedero, M. A. Influence of biochar addition on methane metabolism during thermophilic phase of composting. *J. Basic Microbiol.* 2013, 53 (7), 617–621.

169. Chowdhury, M. A.; de Neergaard, A.; Jensen, L. S. Potential of aeration flow rate and bio-char addition to reduce greenhouse gas and ammonia emissions during manure composting. *Chemosphere* 2014, 97, 16–25.

170. Wang, C.; Lu, H., Dong, D.; Deng, H.; Strong, P. J.; Wang, H.; Wu, W. Insight into the effects of biochar on manure composting: Evidence supporting the relationship between N_2O emission and denitrifying community. *Environ. Sci. Technol.* 2013, 47, 7341–7349.

171. Steiner, C.; Das, K. C.; Melear, N.; Lakly, D. Reducing nitrogen loss during poultry litter composting using biochar. *J. Environ. Qual.* 2010, 39 (4), 1236–1242.

172. Chen, H.; Awasthi, S. K.; Liu, T.; Duan, Y.; Ren, X.; Zhang, Z.; Pandey, A.; Awasthi, M. K. Effects of microbial culture and chicken manure biochar on compost maturity and greenhouse gas emissions during chicken manure composting. *J. Hazard. Mater.* 2020, 389, 121908.

173. Liu, T.; Kumar Awasthi, M.; Verma, S.; Qin, S.; Awasthi, S. K.; Liu, H.; Zhou, Y.; Zhang, Z. Evaluation of cornstalk as bulking agent on greenhouse gases emission and bacterial community during further composting. *Bioresour. Technol.* 2021, 340, 125713.

174. Wang, S. P.; Wang, L.; Sun, Z. Y.; Wang, S. T.; Shen, C. H.; Tang, Y. Q.; Kida, K. Biochar addition reduces nitrogen loss and accelerates composting process by affecting the core microbial community during distilled grain waste composting. *Bioresour. Technol.* 2021, 337, 125492.

175. Huang, X.; He, Y.; Zhang, Y.; Lu, X.; Xie, L. Independent and combined effects of biochar and microbial agents on physicochemical parameters and microbial community succession during food waste composting. *Bioresour. Technol.* 2022, 366, 128023.

176. Shan, G.; Li, W.; Liu, J.; Zhu, L.; Hu, X.; Yang, W.; Tan, W.; Xi, B. Nitrogen loss, nitrogen functional genes, and humification as affected by hydrochar addition during chicken manure composting. *Bioresour. Technol.* 2023, 369, 128512.

177. Wiedner, K.; Fischer, D.; Walther, S.; Criscuoli, I.; Favilli, F.; Nelle, O. Glaser, B. Acceleration of biochar surface oxidation during composting? *J. Agric. Food. Chem.* 2015, 63 (15), 3830–3837.

178. Borchard, N.; Prost, K.; Kautz, T.; Moeller, A.; Siemens, J. Sorption of copper (II) and sulphate to different biochars before and after composting with farmyard manure. *Eur. J. Soil Sci.* 2012, 63 (3), 399–409.

179. Hua, L.; Wu, W.; Liu, Y.; McBride, M. B.; Chen, Y. Reduction of nitrogen loss and Cu and Zn mobility during sludge composting with bamboo charcoal amendment. *Environ. Sci. Pollut. Res. Int.* 2009, 16 (1), 1–9.

180. Ottani, F.; Parenti, M.; Santunione, G.; Moscatelli, G.; Kahn, R.; Pedrazzi, S.; Allesina, G. Effects of different gasification biochar grain size on greenhouse gases and ammonia emissions in municipal aerated composting processes. *J. Environ. Manage.* 2023, 331, 117257.

181. Janczak, D.; Malinska, K.; Czekala, W.; Caceres, R.; Lewicki, A.; Dach, J. Biochar to reduce ammonia emissions in gaseous and liquid phase during composting of poultry manure with wheat straw. *Waste. Manage.* 2017, 66, 36–45.

182. Hestrin, R.; Enders, A.; Lehmann, J. Ammonia volatilization from composting with oxidized biochar. *J. Environ. Qual.* 2020, 49 (6), 1690–1702.

183. Chen, W.; Liao, X.; Wu, Y.; Liang, J. B.; Mi, J.; Huang, J.; Zhang, H.; Wu, Y.; Qiao, Z.; Li, X.; Wang, Y. Effects of different types of biochar on methane and ammonia mitigation during layer manure composting. *Waste Manage.* 2017, 61, 506–515.

184. Jiang, J.; Wang, Y.; Yu, D.; Hou, R.; Ma, X.; Liu, J.; Cao, Z.; Cheng, K.; Yan, G.; Zhang, C.; Li, Y. Combined addition of biochar and garbage enzyme improving the humification and succession of fungal community during sewage sludge composting. *Bioresour. Technol.* 2022, 346, 126344.

185. Yang, X.; Duan, P.; Cao, Y.; Wang, K.; Li, D. Mechanisms of mitigating nitrous oxide emission during composting by biochar and calcium carbonate addition. *Bioresour. Technol.* 2023, 388, 129772.

186. Hossain, M. R.; Khalekuzzaman, M.; Kabir, S. B.; Islam, M. B.; Bari, Q. H. Enhancing faecal sludge derived biocrude quality and productivity using peat biomass through co-hydrothermal liquefaction. *J. Clean. Prod.* 2022, 335, 130371.

187. Wang, Q.; Awasthi, M. K.; Ren, X.; Zhao, J.; Li, R.; Wang, Z.; Wang, M.; Chen, H.; Zhang, Z. Combining biochar, zeolite and wood vinegar for composting of pig manure: The effect on greenhouse gas emission and nitrogen conservation. *Waste Manage.* 2018, 74, 221–230.

188. Lei, L.; Gu, J.; Wang, X.; Song, Z.; Yu, J.; Guo, H.; Xie, J.; Wang, J.; Sun, W. Effects and microbial mechanisms of phosphogypsum and medical stone on organic matter degradation and methane emissions during swine manure composting. *J. Environ. Manage.* 2022, 315, 115139.

189. Yang, W.; Zhang, L. Addition of mature compost improves the composting of green waste. *Bioresour. Technol.* 2022, 350, 126927.

190. Wang, Y.; Tang, Y.; Yuan, Z. Improving food waste composting efficiency with mature compost addition. *Bioresour. Technol.* 2022, 349, 126830.

191. Ruan, C.; Ai, K.; Lu, L. Biomass-derived carbon materials for high-performance supercapacitor electrodes. *RSC Adv.* 2014, 4 (58), 100488.

192. Ortiz-Cornejo, N. L.; Romero-Salas, E. A.; Navarro-Noya, Y. E.; González-Zúñiga, J. C.; Ramirez-Villanueva, D. A.; Vásquez-Murrieta, M. S.; Verhulst, N.; Govaerts, B.; Dendooven, L.; Luna-Guido, M. Incorporation of bean plant residue in soil with different agricultural practices and its effect on the soil bacteria. *Appl. Soil Ecol.* 2017, 119, 417–427.

193. Yang, Y.; Du, W.; Ren, X.; Cui, Z.; Zhou, W.; Lv, J. Effect of bean dregs amendment on the organic matter degradation, humification, maturity and stability of pig manure composting. *Appl. Soil Ecol.* 2020, 708, 134623.

194. Wu, Y.; Wang, H.; Li, H.; Han, X.; Zhang, M.; Sun, Y.; Fan, X.; Tu, R.; Zeng, Y.; Xu, C. C.; Xu, X. Applications of catalysts in thermochemical conversion of biomass (pyrolysis, hydrothermal liquefaction and gasification): A critical review. *Renew. Energ.* 2022, 196, 462–481.

195. Cheng, F.; Li, X. Preparation and application of biochar-based catalysts for biofuel production. *Catalysts* 2018, 8 (9), 346.

196. Zhai, Y. B.; Chen, H. M.; Xu, B. B.; Xiang, B. B.; Chen, Z.; Li, C. T.; Zeng, G. M. Influence of sewage sludge-based activated carbon and temperature on the liquefaction of sewage sludge: Yield and composition of bio-oil, immobilization and risk assessment of heavy metals. *Bioresour. Technol.* 2014, 159, 72–79.

197. Biswas, B.; Kumar, A.; Kaur, R.; Krishna, B. B.; Bhaskar, T. Catalytic hydrothermal liquefaction of alkali lignin over activated bio-char supported bimetallic catalyst. *Bioresour. Technol.* 2021, 337, 125439.

198. Ma, Q. L.; Wang, K.; Sudibyo, H.; Tester, J. W.; Huang, G. Q.; Han, L. J.; Goldfarb, J. L. Production of upgraded biocrude from hydrothermal liquefaction using clays as catalysts. *Energ. Convers. Manage.* 2021, 247, 114764.

199. Cui, X.; Hao, H.; He, Z.; Stoffella, P. J.; Yang, X. Pyrolysis of wetland biomass waste: Potential for carbon sequestration and water remediation. *J. Environ. Manage.* 2016, 173, 95–104.

200. Lu, J. W.; Zhang, Z. Z.; Zhang, L. L.; Fan, G. F.; Wu, Y. L.; Yang, M. D. Catalytic hydrothermal liquefaction of microalgae over different biochars. *Catal. Commun.* 2021, 149, 106236.

201. Prestigiacomo, C.; Costa, P.; Pinto, F.; Schiavo, B.; Siragusa, A.; Scialdone, O.; Galia, A. Sewage sludge as cheap alternative to microalgae as feedstock of catalytic hydrothermal liquefaction processes. *J. Supercrit. Fluid.* 2019, 143, 251–258.

202. Wang, B.; He, Z. X.; Zhang, B.; Duan, Y. B. Study on hydrothermal liquefaction of using biochar based catalysts to produce bio-oil. *Energy* 2021, 230, 120733.

203. Zhang, L. H.; Xu, C. B.; Champagne, P. Energy recovery from secondary pulp/paper-mill sludge and sewage sludge with supercritical water treatment. *Bioresour. Technol.* 2010, 101 (8), 2713–2721.

204. Gu, L.; Li, B. L.; Wen, H. F.; Zhang, X.; Wang, L.; Ye, J. F. Co-hydrothermal treatment of fallen leaves with iron sludge to prepare magnetic iron product and solid fuel. *Bioresour. Technol.* 2018, 257, 229–237.

205. Cheng, F.; Tompsett, G. A.; Murphy, C. M.; Maag, A. R.; Carabillo, N.; Bailey, M.; Hemingway, J. J.; Romo, C. I.; Paulsen, A. D.; Yelvington, P. E.; Timko, M. T. Synergistic effects of inexpensive mixed metal oxides for catalytic hydrothermal liquefaction of food waste. *ACS Sustain. Chem. Eng.* 2020, 8 (17), 6877–6886.

206. Rahman, T.; Jahromi, H.; Roy, P.; Adhikari, S.; Feyzbar-Khalkhali-Nejad, F.; Oh, T. S.; Wang, Q. C.; Higgins, B. T. Influence of red mud catalyst and reaction atmosphere on hydrothermal liquefaction of algae. *Energies* 2023, 16 (1), 491.

207. Yildirir, E.; Cengiz, N.; Saglam, M.; Yüksel, M.; Ballice, L. Valorisation of vegetable market waste to gas fuel via catalytic hydrothermal processing. *J. Energ. Inst.* 2020, 93 (6), 2344–2354.

208. Hong, C.; Wang, Z. Q.; Si, Y. X.; Xing, Y.; Yang, J.; Feng, L. H.; Wang, Y. J.; Hu, J. S.; Li, Z. X.; Li, Y. F. Catalytic hydrothermal liquefaction of penicillin residue for the production of bio-oil over different homogeneous/heterogeneous catalysts. *Catalysts.* 2021, 11 (7), 849.

209. Norouzi, O.; Heidari, M.; Martinez, M. M.; Dutta, A. New insights for the future design of composites composed of hydrochar and zeolite for developing advanced biofuels from cranberry pomace. *Energies.* 2020, 13 (24), 6600.

210. Prestigiacomo, C.; Proietto, F.; Laudicina, V. A.; Siragusa, A.; Scialdone, O.; Galia, A. Catalytic hydrothermal liquefaction of municipal sludge assisted by formic acid for the production of next-generation fuels. *Energy* 2021, 232, 121086.

211. Norouzi, O.; Heidari, M.; Di Maria, F.; Dutta, A. Design of a ternary 3D composite from hydrochar, zeolite and magnetite powder for direct conversion of biomass to gasoline. *Chem. Eng. J.* 2021, 410, 128323.

212. Nazari, L.; Yuan, Z. S.; Souzanchi, S.; Ray, M. B.; Xu, C. B. Hydrothermal liquefaction of woody biomass in hot-compressed water: Catalyst screening and comprehensive characterization of bio-crude oils. *Fuel* 2015, 162, 74–83.

213. Wang, B.; He, Z.; Zhang, B.; Duan, Y. Study on hydrothermal liquefaction of spirulina platensis using biochar based catalysts to produce bio-oil. *Energ.* 2021, 230, 120733.

214. Liao, W. T.; Wang, X.; Li, L.; Fan, D.; Wang, Z. Y.; Chen, Y. Q.; Li, Y.; Xie, X. A. Catalytic alcoholysis of lignin with HY and ZSM-5 zeolite catalysts. *Energ. Fuels* 2020, 34 (1), 599–606.

215. Wang, Y.; Wang, H.; Lin, H. F.; Zheng, Y.; Zhao, J. S.; Pelletier, A.; Li, K. C. Effects of solvents and catalysts in liquefaction of pinewood sawdust for the production of bio-oils. *Biomass Bioenerg.* 2013, 59, 158–167.

216. Jain, A.; Balasubramanian, R.; Srinivasan, M. P. Hydrothermal conversion of biomass waste to activated carbon with high porosity: A review. *Chem. Eng. J.* 2016, 283, 789–805.

217. Lachos-Perez, D.; César Torres-Mayanga, P.; Abaide, E. R.; Zabot, G. L.; De Castilhos, F. Hydrothermal carbonization and liquefaction: Differences, progress, challenges, and opportunities. *Bioresour. Technol.* 2022, 343, 126084.

CHAPTER 4

Application in Water Treatment

Ning Li, Wenjie Gao, Lan Liang,

Yangli Cui, Zhanjun Cheng, Beibei Yan,

Eslam Salama, Mona Ossman, and Guanyi Chen

4.1 INTRODUCTION

With the continuous development of the economy and society, manufacturing, animal husbandry, printing and dyeing, and new industries continue to grow and develop. As society progresses, we are also confronted with serious environmental issues, particularly water pollution, which has become a matter of global concern. In addition to traditional pollutants, the increasing contamination of waterways by pharmaceuticals, plastics, pesticides, and other emerging pollutants has compounded the severity of the problem in recent years.[1] Water treatment processes often require materials such as membranes, resins and carbon. Among these materials, the production of carbon-based materials from solid waste as raw materials is gradually gaining attention from the scientific community. Solid waste-based materials are cheap and easy to obtain, have excellent 3D porous properties, and hold a broad application prospect in the field of water treatment.[2] In addition, the oxygen-containing functional groups on the surface of solid waste-based materials can form hydrogen bonds with organic pollutants containing benzene rings and ionic, covalent and coordination bonds with inorganic heavy metal ions to promote efficient pollutant removal. The solid waste-based materials can also be processed and

DOI: 10.1201/9781003535409-4

reused.[3] In addition, to compensate for the shortcomings of solid waste-based materials, researchers have used various modification methods to change the surface properties of carbon materials to improve their water treatment capacity, and currently the solid waste-based carbon materials have been applied in a variety of fields, such as adsorption, advanced oxidation, and so on.[4]

4.2 ADSORPTION

Water is a vital resource for life, and the availability of fresh water on Earth is limited. It is projected that by 2025, half of the world's population will reside in areas experiencing water scarcity.[5] Water is crucial for human consumption, as well as industrial and agricultural activities. The pollution of water resources stems from both synthetic and natural chemicals released from various anthropogenic and natural sources, including the geological composition of aquifers. Numerous technologies for water treatment and purification have been extensively studied, yet the design and implementation of cost-effective methods remains challenging.

Several water treatment technologies offer unique advantages and disadvantages.[6] Notably, adsorption has been recognized as a viable water purification technology. Adsorption is a process that involves a fluid (in this case, water) and a solid phase (adsorbent). In the fluid phase, one or more dissolved contaminants (adsorbate) are present. These contaminants are transferred from the liquid phase to the surface of the adsorbent, thus purifying the water.[7] Adsorption is currently employed for water treatment due to its low cost, high efficiency, ease of operation, ease of implementation, the potential to use various solids as adsorbents, and the ability to recover both the adsorbent and the adsorbate.[8-10] It is important to highlight that adsorption is particularly effective as a polishing operation when contaminant concentrations range from ng/L to mg/L.

4.2.1 Adsorbent Materials

Developing an adsorbent with all the aforementioned characteristics is indeed challenging. Numerous studies have focused on developing a variety of adsorbents for water treatment. Carbon-based materials such as chars, biochar, activated carbons, coals, and nanomaterials have been utilized for water treatment via adsorption.[11, 12] Another category of adsorbents includes chitin and chitosan-derived materials.[13-15] There is also significant interest in biosorbents and agro-industrial waste.[16-18] Inorganic materials, including zeolites, layered double hydroxides, and geopolymers,[19, 20] have also been employed for water treatment. Additionally, Metal-Organic Frameworks (MOFs) have been designed to adsorb contaminants from water.[21] Silica-based materials are also used as adsorbents. Generally, many suitable adsorbent materials for water treatment have been well-developed and characterized.[22, 23] Nevertheless, continuous innovation in developing new adsorbent materials is essential to achieve the desired characteristics and improve the efficiency of adsorption processes for water treatment.

Regeneration of adsorbents can be achieved through extraction/ leaching, pH changes, thermal desorption, or reaction/degradation.[24] These methods have proven efficient for adsorbent regeneration, but most are not well-suited for adsorbate recovery. Each regeneration method has its advantages and limitations, and its application depends on the specific adsorption system. Significantly, regeneration is a crucial aspect.

4.2.2 Adsorption Operation Mode

The mode of adsorption operation is critical as it directly impacts water treatment costs, the volume of treated water, the physical space required for treatment plant equipment, and the time needed for decontamination. Batch adsorption (batch adsorbers) is the primary mode used for water treatment reported in the literature. In this mode, the adsorbent is added to a tank containing contaminated water. The mixture is stirred until equilibrium or near-equilibrium capacity is reached. Then, solid-liquid separation is performed through decantation, filtration, or centrifugation. At the end of the process, a solid phase (adsorbent loaded with contaminant) and a liquid phase (purified water) are obtained.[25]

The fixed-bed operation mode is mainly used for large-scale treatment applications, though it is less documented in the literature.[26] In this case, the adsorbent is packed in a column through which contaminated water is pumped. The process continues until the column is saturated, at which point the inlet and outlet contaminant concentrations become equal. During the experiment, the contaminant concentration at the column outlet is monitored. This data allows for the construction of breakthrough curves, which are crucial for scaling up the adsorption process.[27, 28]

4.2.3 Application Field

4.2.3.1 Heavy Metals Adsorption

Since the industrial revolution, heavy metals have become a major source of pollution in water. The heavy metals tend to induce cancer and genetic mutation, so heavy metal removal have received extensive attention from researchers. In recent years, low-cost adsorbents from industry, marine, sludge and agriculture sources have been widely used for the cost-effective removal of heavy metal as Ni, Cr, Zn, As, Hg and Cu and so on.[29-34] Solid waste was usually the great choice for adsorbents due to high loading capacity, rapid metal adsorption capacity and high selectivity. Adsorbents prepared from various kinds of solid waste have different abilities to remove heavy metals. For example, agricultural waste such as nut shells/stones, hulls/husks/seed shells, agroforestry peels and others are superior in removing various heavy metals. Studies have shown that nut shells/stones preferentially adsorb Pb and Cd, hulls/husks/seed shells favour Cu and Pb adsorption. Pb and Cd were favoured by shells. Cu and Pb were favoured by hulls/husks/seed coats. Cd and Ni were better adsorbed by chemically modified agroforestry shells. Among them, agroforestry peels have the lowest preparation cost, which is about 0.55$ to prepare one kilogram.[33]

4.2.3.2 Dye Adsorption

Dyes are released globally in excess of 280,000 tonnes per year, posing a huge challenge to environmental safety. Dyes are usually classified as cationic dyes (hydrochloric acid or zinc chloride complexes), anionic dyes (azo, nitro or anthraquinone dyes) and non-ionic dyes.[35] The aromatic structure of dyes determines their toxicity and predisposes to numerous problems like dizziness and development in humans.[36] Thus, it is crucial to seek efficient dye removal processes. Various adsorbents from solid waste have been applied to dye adsorption.[30] In the case of polysaccharides,[37] their main components include chitin and cellulose. However, the dye's adsorption is limited owing to its rigid molecular structure. Various chemical modification techniques including functional group substitution, esterification and oxidation have been developed to further improve performance. Adsorbents from chitosan were mainly used for anionic dyes adsorption owing to their cationic property. The amine-based structures are protonated to act as an adsorption site for negatively charged pollutants.

Besides that, hydroxyl groups of chitosan could be used as chelating sites for dye molecules. Gkika et al measured the cost of modification of chitosan. The results indicated the lowest cost of graphite complexed with magnetic chitosan (6.16€).[38] Compared to commercially available activated carbon (129.3 mg/g), chitosan-based adsorbent showed higher adsorption capacity for methyl orange (175.45 mg/g),[39] suggesting the high potential of chitosan-based adsorbent for dye removal. In addition, agricultural waste-based adsorbents exhibit superior performance in adsorbing cationic dyes compared with anionic dyes. At solution pH greater than the zero-charge point (pH$_{pzc}$) of the adsorbent, the adsorbent exhibits negative charge. The existence of groups including OH$^-$ and COO$^-$ was conducive to cationic dye adsorption. In contrast, the anionic dyes adsorption was favoured when pH < pH$_{pzc}$.[30]

4.2.3.3 Other Organics Adsorption

4.2.3.3.1 Phenol Adsorption   The release of phenol into the environment tends to destroy the ecosystem due to the features such as high solubility, persistence, strong toxicity and low biodegradability. Biochar-based adsorbents prepared from a variety of solid waste materials such as agriculture and forestry were used for phenol removal. Franco et al.[40] prepared activated carbon from Ceiba speciosa by ZnCl$_2$ activation pyrolysis. The maximum capacity under the Langmuir model range of 156.7 ~ 145 mg/g with pH = 7 and 0.83 g/L adsorbent. The adsorption of phenol in this process was spontaneously exothermic and results indicated the application potentials of the material in phenol adsorption. However, the adsorption capacity of conventional biochar was insufficient due to limited specific surface area, inadequate oxygenated groups and inferior porosity. Therefore, further modification was required to improve the properties of biochar. In particular, KOH, metal salts, oxidants and N doping are the main modification methods.[41] Moreover, solid waste could be served as carriers for improving the adsorption performance. Abdel-Gawwad et al.[42] prepared nano-zero valent iron particles using red brick waste as a metal precursor and subsequently embedded in activated carbon as adsorbent for removing phenol in wastewater. The performance of adsorbing phenol was reduced from initial 93.21% to 82.43% after five cycles, which showed a high degree of stability.

4.2.3.4 Emerging Contaminant Adsorption

Recently, emerging contaminants have attracted widespread attention due to biotoxicity and bioaccumulation. They have also been frequently detected in surface water and groundwater as a result of widespread sources. Wastewater treatment plants, as an important source of discharge of emerging contaminants, have been treated poorly by traditional biotechnology. Prepared adsorbents from different agricultural waste were used for BisPhenol A (BPA) adsorption. Research showed that argan shell-based adsorbent exhibited the best performance (1408 mg/g) for BPA adsorption.[43] Modification by activation, polymerization and grafting could improve the adsorbent performance. Numerous studies have reported the mechanisms of BPA removal, including π-π, hydrophobic and hydrogen bonding interactions. The Langmuir/Freundlich model was more applicable for the above process.[44] However, the selection of suitable eluents was crucial for resolving BPA in adsorbents. Moreover, the current experiments for BPA were mainly in the primary stage. The further application of agricultural waste as adsorbents was difficult to promote.

4.2.4 Adsorption Mechanism

The adsorption process involved various mechanisms, including van der Waals forces, electrostatic interactions, π-π interactions, hydrogen bonding and other hydrophobic interactions.[45, 46] These mechanisms can be classified as physical adsorption and chemical adsorption.

4.2.4.1 Physical Adsorption

4.2.4.1.1 Electrostatic Interaction Electrostatic interaction mainly indicates the phenomenon of electrostatic attraction between adsorbent and pollutant with different charges respectively. The pH significantly affects the surface charge potential, thus affecting the adsorption capacity. PFAS is usually negatively charged in water.[47] The critical sites of positively charged adsorbents include -NH-, -OH, -COOH, M-OH, and the like. These functional groups undergo protonation to positively charged ions under low pH conditions, which make it easier to adsorb PFAS for its removal. In contrast, with increasing pH in the solution, the functional groups are deprotonated and gradually change from positively charged to negatively charged. The static attraction between the adsorbent and PFAS is weakened or even repelled.[48, 49] Moreover, the PFAS molecules adsorbed on the adsorbent surface also repel the PFAS anions that are freely present in the solution, thus reducing the adsorption capacity.[50] Besides, negatively charged Natural Organic Matter (NOM) tends to adsorb on the surface of positively charged adsorbents. It would compete for adsorption sites with the target pollutant. And the pH may change the adsorbed materials' charge, thereby affecting the electrostatic interactions.[51]

4.2.4.1.2 Hydrophilic Interaction Hydrophobic interactions are primarily defined as the repulsion of nonpolar long-chain hydrophobic molecules and high affinity forces on hydrophobic surfaces. Typically, negatively charged adsorbents favour PFOS anion adsorption (for example, minerals), indicating that hydrophobic interactions can override electrostatic repulsion.[52] PFAS with long C-F chains could be self-aggregated by hydrophobic interactions, thus forming multilayered aggregate structures on the surface or in the pores of the adsorbent.[52] The maximum adsorption capacity of PFAS as predicted by the Langmuir isotherm model in different experiments was much lower than the experimental data, which further proved that PFAS formed multilayer PFAS structures during adsorption.[53] In addition, clay-based adsorbents have high hydrophobicity. Their combination with metal oxides could enhance hydrophobicity and enhance the adsorption capacity.[54]

4.2.4.2 Chemical Adsorption

4.2.4.2.1 Ion Exchange   Ion exchange refers to the process of ion exchange between an adsorbate in solution and the surface of an adsorbent. The process was usually reversible. The removal rate of pollutants by the adsorbent depended mainly on the ion exchange capacity. Magnetic ion exchange resins have been used to adsorb emerging contaminants such as tetracycline, sulfamethoxazole and amoxicillin from water. It was shown that the adsorption process was mainly an anion-exchange mechanism, which consisted of electron or ion exchange between the magnetic resin and the active site of the antibiotic.[55]

4.2.4.2.2 π-Interactions   π-interactions are typically observed between electron-rich molecules and electron-deficient molecules. They are weak interactions that occur between unsaturated bonds on the aromatic rings of adsorbents and adsorbates. They mainly include cation-π-interactions, anion-π-interactions, π-π-interactions, and π donor-π acceptor interactions. Metal ions could interact with oppositely charged aromatic portions present in organic matter.[56, 57] While π-π-interactions are related to the interaction between lone electron pairs and π-systems in the aromatic ring.[58] Olusegun et al.[59] analysed FTIR of magnetic hematite before and after tetracycline adsorption, suggesting that π-π-interactions were the main mechanism.

4.2.4.2.3 Hydrogen Bonding   Hydrogen bonding between adsorbate and adsorbent surfaces is another mechanism for pollutant removal. Latimer and Rodebush first proposed the interaction in 1920.[60] Hydrogen bonding belongs to dipole-dipole attraction, and which is an interaction between hydrogen atoms (H) and electronegative atoms. It is considerably weaker than covalent or ionic bonds, but more powerful than van der Waals interactions.[61] Hydrogen bonding is important in the process of adsorption between adsorbents and adsorbates containing -NH_2, -COOH, -OH and C= O.[62] FTIR analysis could be used for detecting hydrogen bonding through comparison of changes on adsorbents before and after adsorption.[63]

4.3 ADVANCED OXIDATION PROCESSES

4.3.1 Fenton-Like Oxidation

Persulfate-based advanced oxidation processes (PS-AOPs) are types of Fenton-like oxidation. They use persulfate instead of H_2O_2 as the oxidizing agent.[64] Compared to conventional Fenton-like oxidation, PS-AOPs can be carried out over a wide pH range and do not produce large amounts of iron-containing sludge.[65] Materials prepared from solid waste have been widely used as catalysts in PS-AOPs due to their low cost, recyclability, and environmentally friendliness. Solid waste is prepared as an ideal catalyst by pyrolysis, hydrothermal carbonization and other methods, and is then used to activate persulphates.[66]

There are many kinds of solid waste, and the element composition is different, leading to different properties of catalyst.[67, 68] Table 4.1 lists common types of solid waste used to prepare catalysts and their advantages and disadvantages. Notably, classification is based on the nature of the class, and there may be differences for a particular substance. However, it can still provide reference for the preparation of solid waste-based catalysts. The solid waste base is therefore divided into three main categories based on its natural nature: plant sources (agricultural straw, forest trees), animal sources (animal remains, dung) and other sources (sludge, industrial residues).[69]

TABLE 4.1   Advantages and disadvantages of different solid waste-based catalysts

Feedstock	Advantages	Disadvantages
Plant source	High silicon content High specific surface area Abundant surface functional groups	Low nitrogen doping Low metal sites
Animal source	Autogenous nitrogen doping High content of mineral elements High specific surface area (animal remains)	High ash content Low specific surface area (dung)
Other sources	Autogenous nitrogen doping Metal (such as Fe and Al) loading	High ash content Low specific surface area

The different types of solid waste sources have different characteristics. Compared to plant sources, animal sources have a high ash content due to the high content of natural minerals in their bones and manure. Sludge and industrial residues also contain high levels of ash during the production process.[70] The high ash content has an inhibiting effect on the performance of the catalyst, as the presence of ash tends to cover the active sites on the catalyst's surface. The high ash content brings both high metal and mineral elements, so that animal sources have more mineral elements, whilst other sources (sludge, industrial residues) have a higher metal loading.[71] Plant sources contain a high amount of Si-C bonds in the catalyst due to the accumulation of silicon during growth. Plant sources generally have a higher specific surface area, and the specific surface area of other sources is usually lower due to ash. For animal sources, the residual animal sources have a high specific surface area due to the inherent pore structure of the bones, while the dung animal sources have a much lower specific surface area.[72] The Surface Functional Group (SFG) is the most important factor affecting catalyst performance. Plant-derived catalysts tend to be richer in surface functional groups than other types. No significant difference is found in the SFGs content between animal-derived and sludge-derived catalysts. Plant-derived and sludge-derived catalysts mainly contain microporous and mesoporous structures, while animal-derived catalysts are dominated by mesoporous structures.[73]

It can be seen that the properties of catalysts prepared from different sources of solid waste vary considerably. Therefore, practical needs should be considered when selecting solid waste materials.

4.3.1.1 Performance and Influencing Factors

Solid waste-based materials have shown excellent performance for persulfate activation and pollutant degradation. The whole process is influenced by a number of factors, in addition to the catalyst itself, including the type of oxidant, reaction conditions and water conditions.

4.3.1.1.1 Catalyst   The different sources of solid waste directly determine the fundamental properties of a catalyst and influence the degradation efficiency.

Solid waste-based materials have abundant oxygen-containing functional groups, large specific surface area and excellent electrical conductivity, which can mediate the electron transfer from organics to PeroxyMonoSulfate (PMS). Solid waste is commonly heat-treated, and calcination temperature is important. A moderate increase in temperature can change the structure and crystal of the catalyst. The structure of the catalyst did not change significantly when the pyrolysis temperature was below 300°C. As the pyrolysis temperature rises above 700°C, the graphitic, conductivity and porosity of biochar can be significantly increased, which is favourable to the activation of persulfate.

4.3.1.1.2 Oxidant   Differences in the type of oxidant lead to differences in the active species produced.

The molecular structure of PMS is asymmetric, so the catalyst activates PMS faster than PeroxyDiSulfate (PDS).[74] In addition, when PDS is in contact with transition metals, $SO_4{}^{\bullet-}$ can be formed.[75] In the meantime, the non-polar bond (peroxide bond) of PMS is broken by the catalytic reaction to produce $SO_4{}^{\bullet-}$ and $\cdot OH$.[76] Moderate addition of PMS can have a positive effect on the degradation reaction. The oxidation capacity of $SO_5{}^{\bullet-}$($E^0 = 1.1$ V) produced by the competition between target pollutant and PMS is lower than that of $\cdot OH$ and $SO_4{}^{\bullet-}$ when PMS is in excess.[77,78] Moreover, in excess of PMS, $\cdot OH$ can react with HSO_5 and thus affect the degradation of pollutants.[79] Besides, $SO_4{}^{\bullet-}$ radicals can even quench themselves and reduce the number of active species (Eqs. 1-3).

$$SO_4^{\bullet-} + HSO_5^- \rightarrow SO_5^{\bullet-} + H^+ \tag{1}$$

$$HSO_5^- + H^+ \rightarrow SO_4^{2-} + 3H^+ \tag{2}$$

$$\cdot OH + HSO_5^- \rightarrow SO_5^{\bullet-} + H_2O \tag{3}$$

4.3.1.1.3 Reaction Conditions   The differences in the nature of the components in the system make the reaction conditions critical to the degradation process. As the reaction temperature increases the degradation efficiency is increased. This phenomenon may be related to the increase in available energy to overcome the reaction energy barrier and to the improved production of oxidative species by increasing the reaction temperature continuously.[80] The initial pH of the solution generally has a large effect on AOPs. Solution pH can change the surface charge of the catalyst, as well as affecting the ionic form of PMS and the contaminant molecules in solution. When the solution is strongly acidic (pH = 1) or strongly basic (pH = 11), the degradation efficiency decreases significantly. This phenomenon is related to the presence of H^+, which quenches the $SO_4{}^{\bullet-}$ and $\cdot OH$

radicals under strong acidic conditions (pH = 1), thus reducing the oxidation capacity of the system (Eqs. 4-5).[81] Under strongly alkaline conditions, OH^- can consume reactive substances (HSO_5^-, $SO_4^{•-}$ and $•OH$), reducing the amount of reactive species and decreasing the degradation efficiency (Eqs. 6-8).[82]

$$•OH + H^+ + e^- \rightarrow H_2 \tag{4}$$

$$SO_4^{•-} + H^+ + e^- \rightarrow HSO_4^{2-} \tag{5}$$

$$HSO_5^- + OH^- \rightarrow SO_5^{2-} + H_2O \tag{6}$$

$$SO_5^{2-} + •OH \rightarrow SO_5^{•-} + OH^- \tag{7}$$

$$SO_4^{•-} + OH^- \rightarrow •OH + SO_4^{2-} \tag{8}$$

4.3.1.1.4 Water Quality Conditions   Water quality conditions may affect the degradation process positively or negatively. Organic and inorganic substances in water can also have a significant effect on pollutant degradation. The $H_2PO_4^-$, CI^-, NO_3^-, CO_3^{2-} and HCO_3^- anions quench the active substances and affect pollutant degradation. $H_2PO_4^-$ depletes the active species and significantly weakens the degradation efficiency (Eq. 9). The CI^- can react with $SO_4^{•-}$ to form $CI^•$ and $CI_2^{•-}$ with weaker oxidation capacity (Eqs. 10 and 11), while CI^- can also react directly with PMS (Eq. 12), causing a significant decrease in degradation efficiency. The addition of NO_3^- causes the conversion of $SO_4^{•-}$ and $•OH$ to the less reactive species $NO_3^•$ (Eqs. 13 and 14). In addition, the CO_3^{2-} and HCO_3^- react with $•OH$ and $SO_4^{•-}$ to form other less active species (Eqs. 15-18). The asymmetric structure of PMS is vulnerable to be attacked by the nucleophilic HCO_3^-, depleting the PMS concentration (Eq. 19). Moreover, the natural organic matter in the water also competes for PMS, thus reducing the effectiveness of pollutant treatment.

$$SO_4^{•-} + H_2PO_4^{2-} \rightarrow SO_4^{2-} + H_2PO_4^{•-} \tag{9}$$

$$SO_4^{•-} + CI^- \rightarrow SO_4^{2-} + CI^• \tag{10}$$

$$CI^• + CI^- \rightarrow CI_2^{•-} \tag{11}$$

$$2CI^- + HSO_5^- + H^+ \rightarrow SO_4^{2-} + CI_2 + H_2O \tag{12}$$

$$NO_3^- + SO_4^{•-} \rightarrow SO_4^{2-} + NO_3^• \tag{13}$$

$$NO_3^- + {}^{\bullet}OH \rightarrow OH^- + NO_3^{\bullet} \tag{14}$$

$$SO_4^{\bullet-} + CO_3^{2-} \rightarrow SO_4^{2-} + CO_3^{\bullet-} \tag{15}$$

$${}^{\bullet}OH + CO_3^{2-} \rightarrow OH^- + CO_3^{\bullet-} \tag{16}$$

$$SO_4^{\bullet-} + HCO_3^- \rightarrow SO_4^{2-} + HCO_3^{\bullet} \tag{17}$$

$${}^{\bullet}OH + HCO_3^- \rightarrow CO_3^{\bullet-} + H_2O \tag{18}$$

$$HSO_5^- + HCO_3^- \rightarrow HCO_4^- + HSO_4^- \tag{19}$$

4.3.1.2 Application Field

PS-AOPs have shown excellent performance in the degradation of a wide range of pollutants.

Organic pollutants (for example, dyes, antibiotics, phenols, amines and endocrine disruptors) in water have attracted great attention worldwide. Among them, the Advanced Oxidation Process (AOP) is a highly competitive water treatment technology that mineralizes organic pollutants in water by producing highly oxidizing Reactive Oxygen Species (ROS). As shown in Table 4.2, the solid waste-based PS system has been investigated extensively.

The toxicity of dyes to aquatic organisms and impediment of natural photosynthesis have affected the healthy circulation of ecosystems. Woody biochar can activate PDS effectively and eliminate AO7 quickly in a wide pH range (3.0–10.0) (99.3% removal rate within 14 min). In addition, Wang et al. used sludge-derived biochar as catalyst. The removal rate of AO7 was 89.1% and 99.1% at pH 9.08 and 2.13, respectively.

As an emerging contaminant, antibiotics are persistent and accumulative, which is difficult to degrade. Antibiotics in natural water can induce the growth of drug-resistant bacteria or drug-resistant genes, posing a potential threat to the ecological environment and human health. Therefore, antibiotics cannot be ignored in the aquatic environment. Antibiotics and endocrine disruptors are emerging contaminants. Previous studies as shown in Table 4.2, defined a removal index that represented the removal efficiency of antibiotic per unit concentration of catalyst (g/L), oxidant (mM) and antibiotic (mg/L) as well as per unit time (min). The removal index of 2,4-dichlorophenol with swine bone activating PDS was 0.0020%, while the index was as high as 0.0126% with shrimp shell. Obviously, shrimp shell activated PDS exhibited better degradation performance for 2,4-dichlorophenol than swine bone.

TABLE 4.2 Performance of persulfate activation by solid waste-based materials

Types	Solid waste	Persulfate type	Persulfate dosage (mM)	Catalyst dosage (g/L)	Pollutants	Catalytic performance	[a] Removal index (%)	Ref.
Plant-derived	Poplar sawdust	PDS	10.00	0.5	Acid orange	99.3% after 14 min	0.0709	83
	Rice straw	PDS	0.47	0.6	Aniline	94.1% within 80 min	0.4170	85
	Corn stalk	PDS	0.47	0.6	Norfloxacin	94.21% removal in 300 min	0.1113	86
	Spirulina residue	PDS	6.00	0.5	Sulfamethoxazole	100% removal in 45 min	0.0370	87
Animal-derived	Pine needle	PMS	8.00	1.0	1,4-dioxane	84.2% in 240 min	0.0025	84
	Shrimp shell	PDS	2.60	0.5	2,4-dichlorophenol	over 98% in 60 min	0.0126	88
	Swine bone	PDS	10.40	0.2	2,4-dichlorophenol	almost 100% in 120 min	0.0020	89
Other feedstocks	Chicken manure	PS	10.00	0.8	p-Nitrophenol	21.9% in 240 min	0.0011	90
	Sludge	PDS	0.93	0.5	Acid orange	99.1% in 60 min	0.1690	91
	Oxidation ditch sludge	PMS	0.89	0.2	Bisphenol A	80% within 30 min	1.4980	92
	Secondary sludge	PMS	0.80	1.0	Triclosan	99.2% within 240 min	0.0525	93

[a] We defined a removal index that represented the removal efficiency of contaminants per unit concentration of catalyst (g/L), oxidant (mM) and contaminants (mg/L) as well as per unit time (min)

It has been shown that using spirulina residue as raw material, the removal rate of sulfamethoxazole (SMX) reached 100% within 45 min. Notably, PDS binds tightly to biochar to form surface-reactive complexes, followed by an attack SMX adsorbed on biochar by an electron transfer mechanism. In the corn stalk biochar/PS system, the Norfloxacin removal rate increased significantly, reaching 94.21% after 300 min, indicating a synergistic effect between corn stalk biochar and PS. In addition, straw biochar can effectively activate PS to degrade aniline. The results showed that the degradation rate of aniline could reach 94.1% within 80 min. Besides, plant-derived solid waste (for example, corn stalk and spirulina residue) also exhibited different catalytic activities in degrading antibiotics when applied to activate PDS, and the corresponding removal indexes of norfloxacin was 0.1113% and sulfamethoxazole was 0.0370%.

Obviously, the persulfate advanced oxidation system has wide applicability and has good development prospects in the field of environmental remediation. However, no studies have attended to the relationship between the types of solid waste-based catalysts and the selectivity of pollutant degradation. It is suggested that the relationship between the characteristics of solid waste-based catalysts and the type of pollutants can be further clarified to promote the application of solid waste-based materials in PS activation.

4.3.1.3 Degradation Mechanism

The degradation mechanism of PS-AOPs consists of radical and non-radical pathways. The activation of PS by the catalyst is mainly influenced by its own properties, such as surface functional groups, and the degree of defects. The specific information of each pathway is discussed below.

Radical pathways can be induced in PS activation by solid waste-based catalysts. $SO_4^{\cdot-}$, $\cdot OH$ and $O_2^{\cdot-}$ are common radicals as Reactive Oxygen Species (ROS). For metal-based solid waste, the metal sites (for example, Fe, Co, Cu) on the catalyst surface can be used as electron donors for the redox reactions (Figure 4.1). PS can be activated to generate $SO_4^{\cdot-}$ and $\cdot OH$ with the elongated rupture of the peroxide bond (O-O) after receiving electrons. Besides, PS can further reduce high-valence metals to low-valence metals, along with the production of $SO_5^{\cdot-}$ with a low redox potential. For non-metal solid waste, carbon matrix promotes electron transfer and PS decomposition to form free radicals (Figure 4.1). The sp^2 hybrid carbon and electron-rich groups, including C-O and C=O, may serve as

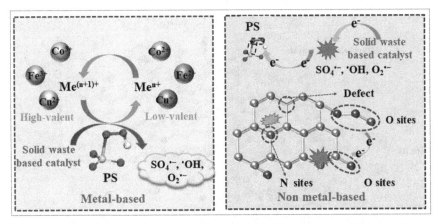

FIGURE 4.1 Free radical pathway during PS activation by solid waste-based catalyst.

interface sites for PMS activation. The accelerated – electron transfer and abundant structural defects of the highly graphitized carbon lattice further promote the migration of unpaired electrons, breaking the O-O bond of the PMS and generating $SO_4^{•-}$ and $•OH$. Moreover, the N-substance attached to C can increase the charge density of positive C atoms, promoting PS activation by weakening the O-O bond, resulting in the formation of $SO_4^{•-}$ and $O_2^{•-}$. Furthermore, defects can also lead to the breaking of O-H bonds in PS to form $SO_5^{•-}$.

Non-radical pathways are another mechanism for degrading organic pollutants. Particularly, a non-radical pathway is a surface reaction, unlike radical processes that occur in aqueous solutions. Compared with a radical pathway, a non-radical pathway can avoid the competitive quenching effect between radicals and effectively improve the utilization of PS. Besides, the non-radical pathway is considered to be resistant to the impacts of inorganic ions and natural organic matter in complex water, which has a wide pH window. Furthermore, non-radical species exhibit selectivity for specific organic substrates. Organics with electron-donating groups are more inclined to be degraded by non-radical pathways. However, non-radical pathways are limited due to relatively weak oxidative capacity. Currently, non-radical pathways dominated by singlet oxygen ($^1O_2$) and direct electron transfer in solid waste-based carbon catalyst-activated PS systems have received intensive attention.

Solid waste-based catalysts act as electron mediators in the direct electron transfer process (Figure 4.2, Path I). PS adsorbs to the sites of the

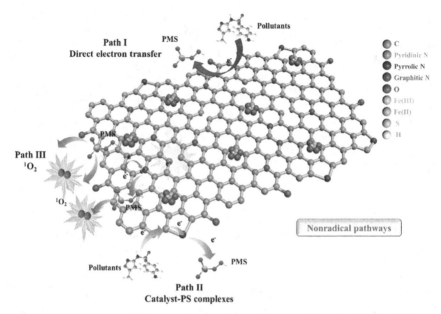

FIGURE 4.2 Proposed three pathways for solid waste-based carbon catalysts-activated PMS.

solid waste-based catalysts. The pollutants are adsorbed on the solid waste-based catalysts through π-π reactions with the carbon or oxygen functional groups of the solid waste-based catalysts. Electrons are rapidly transferred from the organic (electron donor) to the PS (electron acceptor). The chemical interaction in this pathway does not necessarily exist between the PS and the solid waste-based catalyst. Moreover, the chemical structure of the PS is not altered. Hierarchical porous biochar from shrimp shell (PSS-bio)/PDS system was proved to degrade 2,4-dichlorophenol via direct double electron transfer.

Catalyst-PS complexes are formed by strong electrostatic bonding between the positively charged carbon and PS (Figure 4.2, Path II). N-doped bio (NBC) could change the electron distribution of pristine biochar by edge nitration. Hence, a large number of active sites were generated, facilitating the interaction between NBC and PDS to form surface bonding compounds. PDS was reduced on the electron-rich N dopant, while the oxidation of the contaminant occurred around the adjacent electron-deficient C atom.

Unlike the electron transfer mechanism, the $^1O_2$ is produced via the breakage of the peroxide bond in PS (Figure 4.2, Path III). Moreover, the $^1O_2$

has high reactivity to electron-rich organics. The PMS could be activated by C=O on the surface of solid waste-based catalysts to yield $^1O_2$ (Eqs. (20)-(23)). Additionally, graphite N, defective, pyridinic N and pyrrolic N have also been verified to activate PS to generate $^1O_2$.

$$(20)$$

$$(21)$$

$$(22)$$

$$(23)$$

4.3.2 Ozone Catalytic Oxidation

Ozone stands out as a potent oxidizing agent, garnering significant attention for its role in enhancing water and sewage treatment processes. In this process, homogeneous and heterogeneous catalysis is used mostly. Within the water and wastewater sectors, ozone finds application across various domains, including disinfection, management of taste and odour, decolorization, and the oxidation of both organic and inorganic compounds.

Enter Heterogeneous Catalytic Ozonation (HCO), an advanced oxidation method that leverages ozone in conjunction with solid catalysts to instigate the formation of hydroxyl radicals within a solution. This approach has demonstrated promising outcomes in effectively removing diverse classes of contaminants.[94] Over time, a spectrum of heterogeneous catalysts has been explored within the catalytic ozonation framework.[95] These catalysts can be broadly categorized into three main types, as illustrated in Figure 4.3, metal oxides, supported catalysts, and carbon-based materials. Metal oxides primarily encompass transition metal elements such as titanium

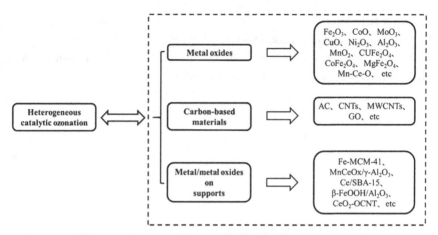

FIGURE 4.3   Heterogeneous catalysts applied in catalytic ozonation process.

(Ti), manganese (Mn), cobalt (Co), iron (Fe), zinc (Zn), and cerium (Ce).[94] Additionally, recent decades have witnessed a surge in research focusing on carbon-based materials for catalysing ozone oxidation. Varieties such as activated carbon, carbon nanotubes, and graphene have displayed notable catalytic prowess within ozone systems. Furthermore, alternative catalysts derived from solid waste, including geopolymer, red mud, and biochar, have been employed to augment the breakdown of ozone molecules into hydroxyl radicals.

Geopolymers (GPMs) derive from the activation of alkalis in metakaolin, various fly ashes, blast furnace slags, or other industrial solid waste.[96] The robustness of the geopolymer catalysts during the ozonation reaction has been demonstrated and verified.[97] Geopolymers with a little magnetite have a positive synergic effect on catalytic ozonation[98] and manganese sand ore,[99] zeolite and ceramics have all demonstrated their catalytic effectiveness on activating ozone oxidation.

Biochar offers a versatile solution for various environmental challenges due to its customizable porosity and surface properties. Specifically, it has been found effective in removing water pollutants, recovering nutrients, managing urban runoff, addressing industrial pollution, and contributing to carbon neutrality through carbon sequestration.[100, 101] Bone Charcoal (BC) is a commonly used biochar variant for water pollutant removal due to its cost-effectiveness and scalability in production. BC particles typically comprise elements such as P, Ca, C, O, Si, Al, Na, and Mg. These elements play crucial roles in water purification processes. Additionally, there's evidence suggesting that these chemical constituents can facilitate the conversion of ozone into OH radicals or other reactive species, enhancing the pollutant removal efficiency of BC. This dual functionality of BC as both an adsorbent and a catalyst makes it a promising candidate for water treatment applications.[102, 103] In addition to BC, biochar from agricultural waste materials may be witnessed as a noticeably effective catalyst. Moussavi and Khosravi[104] prepared biochar from pistachio hull waste and suggested that the utilization of pistachio hull biochar as a catalyst demonstrates promising efficacy in the decolorization and mineralization of reactive azo dyes through catalytic ozonation. This indicates that pistachio hull biochar possesses catalytic properties that enhance the degradation of azo dyes when combined with ozone.

4.3.2.1 Performance and Influencing Factors
Solid waste-based materials have shown excellent performance in ozone-catalysed oxidation and pollutant degradation. The whole process is influenced by multiple factors, including species of catalyst, the properties of organic matter, the initial concentration of pollutants, temperature and the pH value of the solution.[105, 106]

4.3.2.1.1 Catalyst   Catalytic oxidation in heterogeneous systems predominantly takes place at the solid-liquid interface adjacent to the catalyst surface. The surface characteristics of the catalyst are pivotal in governing the dynamics of heterogeneous catalytic reactions. The interplay between the active medium or pollutants and the catalyst surface significantly influences the production of active species and the breakdown of contaminants.[107]

The quantity of catalyst plays a crucial role in determining the efficacy of the treatment process. Typically, there's a positive correlation between catalyst dosage and treatment effectiveness. However, beyond a certain threshold, increasing the catalyst dosage doesn't significantly enhance the treatment efficacy. Conversely, too much catalyst will increase the operation cost.

4.3.2.1.2 Organic Matter   Natural water sources contain organic compounds like humic acid and xanthic acid. These compounds not only compete with target pollutants but can also act as initiators for ozone decomposition. Therefore, when applying catalytic ozonation in natural water treatment, it is crucial to analyse the water composition and optimize experimental conditions accordingly. Additionally, natural waters contain various inorganic anions such as Cl^-, HCO_3^-, CO_3^{2-}, SO_4

$^{2-}$ and PO_4^{3-}. These anions have a strong affinity for the basic sites on the catalyst, leading to rapid occupation of catalytic reaction sites and subsequently reducing catalytic efficiency.

4.3.2.1.3 Pollutants   The initial pollutant concentration plays a significant role in catalytic ozonation system. Higher pollutant concentrations typically result in decreased degradation efficiency. This is attributed to the increased demand for oxidants and the competition between pollutants and ozone molecules for catalytic reaction sites on the catalyst surface, thereby diminishing the catalyst's ability to facilitate ozone decomposition. Moreover, elevated pollutant concentrations lead to the generation of more acidic by-products, consequently lowering the solution's pH and inhibiting ozone's autolytic decomposition.

4.3.2.1.4 Temperature   Temperature also has an effect. Changes in temperature can either facilitate or impede the reaction kinetics. While higher temperatures generally accelerate chemical reaction rates, excessively high temperatures can reduce ozone's half-life and solubility in aqueous solutions.

4.3.2.1.5 PH   The pH of the solution is a critical factor as well. It profoundly affects the reaction mechanism and kinetics of ozone in aqueous environments. When the pH of the aqueous solution is low, ozone molecules react selectively only with specific electrophilic, pro-primary and dipolar addition groups on compound surfaces. Conversely, under basic conditions, ozone can decompose to form highly oxidizing -OH. Additionally, pH significantly influences catalyst surface properties. Under acidic conditions, organic matter removal primarily relies on catalyst adsorption, whereas under neutral or alkaline conditions, carbon-catalysed ozone oxidation becomes pivotal in organic matter removal from solutions.

4.3.3 Applications in Wastewater Treatment

Solid waste-based catalysts are proven effective in ozone oxidation, and can be applied in various kinds of wastewater and pollutant degradation, such as antibiotic drugs, dyes, natural humic acid, and the like. Della Rocca D G, et al.[108] investigated whether the combination of geopolymers containing metakaolin and varying proportions of magnetic mining waste show promise in effectively degrading the antibiotic drug, trimethoprim in water when ozone is present. Nguyen L H, et al.[109] suggested that the successful modification and application of an iron-containing waste natural mineral as a heterogeneous catalyst in the O_3/Fenton process for degrading ofloxacin from wastewater highlights the potential of utilizing waste materials for wastewater treatment.

The presence of azo groups in aromatic azo dyes does indeed contribute to their complexity and resistance to biodegradation. However, the removal of reactive dyes is a hot topic in the fields of biochar-based catalysts and advanced oxidation processes.[110] Also, in the catalytic ozonation process system, bone-char and pistachio hull biochar were successfully demonstrated to remove reactive red 198 dye and black 5 dye respectively.[111] As for other kinds of organic pollutants, red mud and graphene-based materials are expected to be applied in complex practical wastewater treatment. Wang Y, et al.[112] reutilized a spent lithium ion battery and obtained graphene-based materials. They prepared the reduced graphene oxide and exhibited excellent catalytic activity for ozone-catalysed removal of organic pollutants. The study by Ryu et al.[113] on HCl-treated RM (HRM) as an efficient catalyst for toluene removal using ozone at room temperature highlights the potential for advanced catalysts in environmental remediation.

4.3.3.1 Mechanism Summary

The complex nature of the reaction mechanisms in HCO arises from the involvement of gas, solid and liquid phases. Although several typical mechanisms have been established for HCO, their applicability is often limited to specific systems. Mechanistic disagreements and deficiencies have hindered the development of novel active catalysts. Catalytic oxidation of HCO can be broadly categorized into radical-based and non-radical oxidation processes.

Radical oxidation, involving ·OH and/or O_2·$^-$ generated from O_3 dissociation on the active site of catalyst, is recognized as the primary pathway for the degradation of aqueous organics.[114, 115] The catalytic active sites serve as specific locations where ozone molecules dissociate, yielding radicals on the catalyst surface. In metal-based catalyst systems for HCO, surface hydroxyl groups (-OH) can directly interact with O_3 to produce radicals, particularly when functioning as catalytic active sites. The efficacy of this reaction process is heavily influenced by the pH value of the solution. Additionally, Lewis acid sites, characterized by empty orbitals capable of accepting electron pairs, are widely acknowledged as another class of catalytically active sites in metal-based HCO catalysts. Hydroxyl groups on Lewis acid sites and chemisorbed water can interact with O_3 to produce active species. Metal components within Lewis acid sites act as catalytically active sites in conjunction with surface hydroxyl groups, facilitating the decomposition of O_3 and the formation of radicals. Moreover, structural defects, other defects on the surface of crystal such as edges, steps, and kinks have also been observed to serve as catalytically active sites for radical generation in HCO.

Surface functional groups, structural defects, heteroatom doping sites and electron-rich regions have emerged as potential catalytically active sites for HCO on carbonaceous materials.[115] Beyond the presence of surface functional groups and dopant atoms, the role of surface electrons on carbonaceous catalysts is pivotal in activating ozone. Ozone molecules are electrophilic and nucleophilic due to their special resonance structure, so surface electrons have the potential to promote ozone decomposition. Heterogeneous electron distribution on carbonaceous catalysts, particularly in electron-rich regions, serve as catalytic hotspots for organic adsorption and oxidative activation. Organics can donate electrons to electron-poor regions, while oxidizers acquire electrons from electron-rich areas, promoting surface reactions. Surface functional groups, heteroatom dopants, and defective sites contribute to adjusting electron redistribution, thereby forming electron-rich regions that facilitate electrophilic interactions with O_3 and promote the formation of radicals.

Non-radical oxidation processes in HCO rely on surface complexes, surface-adsorbed activated oxygen species or singlet oxygen ($^1O_2$) oxidation.[116, 117] In non-radical oxidation, the degradation of organic matter usually occurred in the vicinity of the catalyst surface, either through direct electron transfer processes or via active non-radical species resulting from ozone dissociation. However, the efficiency of non-radical oxidation is highly sensitive to the adsorption behaviour of ozone and organics on the catalyst surface, as non-radical degradation stems from their surface interactions. The oxidation during non-radical oxidation is milder compared to free radical oxidation and favours the destruction of electron-rich organic matter.

The adsorption behaviour of O_3 molecules on the active site leads to two distinct non-radical oxidation pathways: (a) adsorption of O_3 and/or organic compounds onto the catalyst surface, followed by the degradation of organics through surface complexes; and (b) indirect degradation of

$^1O_2$ facilitated by surface-adsorbed reactive oxygen species or catalytically dissociated O_3 on the active site (Figure 4.4).[115]

The adsorption of both O_3 and organic matter on the surface of a catalyst plays a critical role in non-radical oxidation processes. In certain scenarios, the interaction between adsorbed O_3 and organic molecules can lead to the formation of surface complex structures. These complexes facilitate non-radical oxidation through intermolecular electron transfer mechanisms, without the dissociation of O_3 molecules. The potential oxidation pathways for surface complex reactions in HCO can be categorized as follows: (1) organic molecules chemisorbed onto the catalyst surface directly react with free O_3; (2) O_3 molecules chemically adsorb onto the catalyst surface, forming surface-O_3 complexes that subsequently degrade organic matter; (3) both organic matter and O_3 chemisorb onto the catalyst surface, where the organic compounds undergo degradation via intramolecular electron transfer processes.

The catalytic dissociation of O_3 molecules at catalyst surface active sites may lead to the generation of surface-adsorbed reactive oxygen species. These species may entice immediate oxidative reactions with environmental organic contaminants or further evolution to additional reactive oxygen species because of the high oxidation potential. Thus, when it comes to surface-adsorbed reactive oxygen species, non-radical and radical-based oxidation pathways usually coexist and the presence of multiple oxidation pathways needs to be carefully investigated. Surface atomic oxygen (*O)

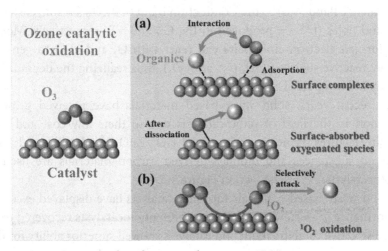

FIGURE 4.4   Non-radical oxidation mechanisms in HCO.

has been identified as the major surface adsorbed substance that destroys organic matter. Its notably high oxidation potential (2.43 V) enables it to facilitate total mineralization processes.[118] *O either attacks the adsorbed organic matter directly or reacts with surrounding water or O_3, leading to the generation of ·OH radicals.

The significant role of singlet oxygen ($^1O_2$) as a non-radical reactive oxygen species in HCO for organic degradation is well recognized. With an oxidation potential of 0.81 V (E_0 ($^1O_2/O_2$·⁻), NHE), $^1O_2$ exhibits mild reactivity, rendering it a highly selective oxidant. $^1O_2$ predominantly targets unsaturated organic compounds through the electrophilic addition and electron abstraction. Additionally, the production of $^1O_2$ often coincides with the generation of other reactive oxygen species, as its formation depends on electron transfer or a chemical reaction involving previously formed reactive oxygen species, and thus promises a multiple oxidation pathway (Eqs. 24-26).

$$^*O_2 \rightarrow 1O_2 \tag{24}$$

$$O_2^{·-} + HO_2^· + H^+ \rightarrow 1O_2 + H_2O_2 \tag{25}$$

$$O_2^{·-} + ·OH \rightarrow 1O_2 + OH- \tag{26}$$

4.3.4 Photocatalytic Oxidation

Photocatalytic oxidation process utilizes UltraViolet (UV) or visible light to irradiate the photocatalysts, causing the promotion of electrons from the Valence Band (VB) to the Conduction Band (CB). As a result, electrons (e⁻) and holes (h⁺) are produced in the CB and VB, respectively. Mostly, the formed electron–hole pairs can react with O_2 and H_2O to generate highly reactive species (·OH, O_2·⁻ and $^1O_2$), thus realizing the degradation of organic compounds.

In recent years, solid waste-based materials have received growing attention in the field of photocatalysis due to their low cost and eco-friendly features. Typically, waste-based TiO_2, ZnO, metal sulfide, calcium-inspired substances and biomass-derived carbon materials are used as photocatalysts,[119, 120] as shown in Figure 4.5.

Solid waste-based materials for photocatalysis have displayed excellent performance in wastewater treatment. TiO_2 photocatalysts recovered from waste selective catalytic reduction catalysts showed superior ability for dye effluent purification. Under visible light irradiation, the regenerated TiO_2

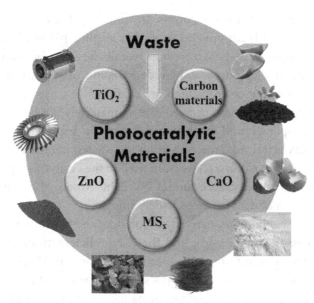

FIGURE 4.5   Types of solid waste-based photocatalytic materials.

(1.3 g/L) achieved 66.9% and 99.0% photodegradation of 10 ppm rhodamine B (RhB) and methylene blue (MB) within 180 min, respectively.[121] Notably, the waste-derived TiO_2 contained Fe elements, thus exhibiting better photocatalytic performance than P25 TiO_2. Additionally, a ZnO/graphite composite created from Zn dust and PET bottle mixed waste realized complete degradation of organic dyes (20 ppm MB and Malachite Green (MG)) after 60–70 min under UV irradiation, with 1.0 g/L of the catalyst added.[122] Interestingly, wasted pig bristles were employed to synthesize Cu_2S carbon composite photocatalysts.[123] Pig bristles-derived Cu_2S catalysts (0.1 g/L) were explored for the degradation of methyl red (50 ppm) under Light Emitting Diode (LED) irradiation, especially the activity of Cu_2S under different light (white, blue, green and red visible light). As a result, the best methyl red removal with ~40% in 180 min was achieved under the white light. Likewise, eggshell driven CaO nano-photocatalysts (1.0 g/L) almost completely degraded MB and toluidine blue dyes (10 ppm) within 15 min under sunlight.[124] And pH 9 was determined to be the optimal pH, with the acidic condition being detrimental to the photocatalytic removal of dyes. Furthermore, lemon pulp waste, a type of biomass, was reported to produce Green Carbon Dots (GCDs) for photocatalytic MB degradation.[125] The complete removal of MB (10 ppm) was observed within 30 min under UV–visible light irradiation using GCD-loaded TiO_2

nanofibres (1.0 g/L) as photocatalysts. In summary, the catalyst dosage of 1 g/L, dye level of 10–20 ppm and neutral to alkaline pH levels are usually suitable for the photocatalytic reaction. Overall, the waste-based substances exhibited outstanding photocatalytic degradation of various organic dye wastewater under UV or visible light irradiation.

The structure of biomass-derived carbon materials can be designed. At present, the application of biochar in the catalytic degradation of organic pollutants has attracted great attention in wastewater treatment. Solid waste-based carbon catalysts have been used in the catalytic degradation of dyes, antibiotics and pesticides under UV light. Among them, photocatalytic degradation of dyes has been widely studied.

Multi-coloured biomass-derived carbon dots and hydrothermal carbon were prepared from green tea, and the Z-type heterostructure of C_3N_4/BiOBr was introduced to promote electron transport and band regulation in the structure efficiently. The results showed that the catalysts exhibited good photocatalytic degradation performance for rhodamine B (RhB). The photocatalytic degradation rate of RhB reached 98.48% within 15 min.[126] Furthermore, Li et al. selected animal waste feathers as raw materials to prepare high-efficiency adsorption-photocatalytic materials, which were also applied in the oxidative degradation of dyes and drugs (for example, tetracycline, amoxicillin and levofloxacin). TiO_2 is an effective photocatalyst, and the efficiency of photocatalytic degradation of organic contaminants can be improved effectively by loading TiO_2 onto sludge biochar. The TINCs were prepared by blending titanium-containing crosslinker with hydrolysed N-embedded chicken feathers. A multilayer graphene oxide structure similar to that of TiO_2-graphene oxide material was observed on the TINCs. Correspondingly, the degradation rate of RhB was 90.91% with TINCs under visible light in 240 min. Simultaneously, the Total Organic Carbon (TOC) of the wastewater decreased by 56.26%.[127] In addition, Fe_3O_4@SiO_2/TiO_2/g-C_3N_4@biochar composites (magnetic biochar-semiconductor heterojunction) were successfully prepared for photocatalytic oxidation by a simple sol-gel and calcination method. The photocatalytic system removed 91% tetracycline and 79% hexavalent chromium from wastewater simultaneously.[128] Similarly, Jamil et al. prepared TiO_2/sludge biochar catalysts by the sol-gel method for photocatalytic degradation of tame lite dyes. Compared with the oxidation efficiency of TiO_2 alone (less than 20%), TiO_2/sludge has a higher photocatalytic oxidation efficiency (more than 90%), which proved that TiO_2 has a synergistic effect with sludge biochar.

The improved photocatalytic efficiency might be attributed to the porous structure of sludge biochar, which provided more active sites. Additionally, the carbon structure and metals (for example., Al, Mg) in sludge biochar reduced the visible light response band gap of TiO_2 and accelerated the separation of electron-hole pairs.[129] The $TiO_2/Fe/Fe_3C$ hybrid sludge biochar composites was synthesized by sludge pyrolysis and a chitosan titanium leaching method, which showed adsorption-catalytic properties for the degradation of Methylene Blue (MB). The MB was adsorbed to the surface of sludge biochar by π-π and hydrogen bonds. Meanwhile, quaternary nitrogen, Fe_3C and Fe^{2+} catalyse the formation of $\cdot OH$ from H_2O_2, and Ti^{3+} promoted the generation of $O_2^{\cdot-}$ from O_2.[130] The inorganic components in sludge biochar helped to reduce the photocatalytic excitation energy by combining with the metal components. Chen et al. synthesized hydrothermal carbonized sludge-based materials and constructed a visible light/O_2 activation system. HydroThermal Carbonized Sludge (HTC-S) could promote the degradation of various organic pollutants in combination with oxalate. The results showed that the iron in HTC-S could chelate oxalic acid to activate O_2 to form H_2O_2 under visible light and promote the activation of H_2O_2 to produce $\cdot OH$. The degradation rate of contaminants in the HTC-S/oxalic acid system was almost 5~20 times higher than that of iron oxide/oxalic acid system. At present, the application of sludge biochar catalysts in photocatalysis was mainly limited to the loading of catalysts on the surface of sludge biochar or the synthesis of composite materials with photocatalysts. However, the synergistic effect of photocatalysts and sludge biochar composites on the improvement of catalytic efficiency has not been studied, and further improving the photocatalytic efficiency is also the direction of future development.

Photocatalytic treatment of dye-polluted wastewater has attracted much attention in recent years. Azo dyes commonly applied in the textile industry have high solubility, stability, and are difficult to remove or degrade. Most azo dyes are basically non-biodegradable under aerobic conditions but have photochemical activity. The mechanism of photocatalytic degradation of dyes can be divided into three categories: charge-injected allergenic dyes, oxidation/reduction indirect degradation and direct photolysis.

According to the dye sensitization mechanism, an electron is excited from from the valence band to the conduction band of the semiconductor metal oxide when the dye absorbs photons with an energy equal to or greater than the band gap of the material, thus forming electron pairs and positive holes. Light-induced positive holes and electrons interact with dye molecules, producing an excited dye. The unstable excitation dye is then converted into anionic (Dye⁻) or cationic (Dye⁺) radicals. Finally, these anionic and cationic radicals decompose into degradation products spontaneously.

The e⁻ and h⁺ are the main active species on the photocatalyst's surface, which can react with dissolved O_2 and H_2O to form other active species (namely, $O_2^{·-}$ and $·OH$). For example, $O_2^{·-}$ and $·OH$ released by photocatalytic materials played a major role in the degradation of RhB. Similarly, toxic tetracycline was removed in the $Fe_3O_4@SiO_2/TiO_2/g-C_3N_4@biochar$ photocatalysis by $O_2^{·-}$ and $·OH$, with a removal rate of 91.88% within 3 h. The superior performance can be attributed to the high specific surface area, enhanced light response, and the presence of magnetic nanoparticles and biochar to accelerate charge transfer, and the Z-heterojunction to enhance the spatial separation of photogenerated carriers.

In addition, the photocatalytic degradation of antibiotic pollutants has also received extensive attention. The combination of biochar and n-n heterojunction can accelerate charge separation and transport. $Ag_3BiO_3/ZnO/BC$ is used for photocatalytic degradation of levofloxacin (LFX) antibiotics.[131] The n-n heterojunction formed inside $Ag_3BiO_3/ZnO/BC$ and the introduction of biochar can effectively promote the separation and transfer of photogenerated electrons and hole pairs, and indirectly reduce recombination. Similarly, the biochar can produce a graphene-like structure at high temperatures, which can interact with the triazine ring of $g-C_3N_4$ π-π, resulting in an electron delocalization effect to reduce the recombination rate, which can be applied to the visible light efficient activation of PMS to degrade tetracycline hydrochloride.[132]

4.3.5 Electrocatalytic Oxidation

Catalytic oxidation technology is considered a highly efficient water treatment technology, belonging to the redox reaction category, by directly or indirectly generating strong oxidizing substances to interact with organic pollutants, partially transforming them or eventually completely mineralizing them into CO_2 and H_2O.[133-135] The Electro-Fenton (EF) process is considered a promising advanced oxidation technology for the removal of recalcitrant organic pollutants from wastewater.

The main principle of Electrocatalytic Oxidation (EO) is that, in the presence of an electric field, large amounts of highly active $·OH$ are generated with the help of an electric field, which undergoes electron transfer reactions such as addition, substitution, electron transfer, and bond breaking with the organic compounds, and the difficult-to-degrade macromolecule organic matter in wastewater is degraded into small molecules until it is completely mineralized into CO_2, water, and inorganic ions.

During the Electrochemical Fluorination (EF) process, hydrogen peroxide (H_2O_2) can be generated in situ at the cathode by a two-electron oxygen reduction reaction (2e-ORR) and subsequently reacted with ferrous ions (Fe^{2+}) to generate powerful hydroxyl radicals ($\cdot OH$) via the Fenton reaction. The standard redox potential of the hydroxyl radical is up to 2.8 volts (for a Standard Hydrogen Electrode (SHE)), which exhibits a strong oxidizing ability and can rapidly and non-selectively degrade a variety of organic pollutants.

Carbon-based materials currently employed in bioelectrical Fenton systems include the Activated Carbon Fibre (ACF), carbon felts, Carbon NanoTubes (CNTs), and graphite rods. In addition to the modification of existing Carbonaceous Material (CM) cathodes, there has been a growing interest in the use of recycled waste materials for the fabrication of carbon-based electrodes, with the objective of further reducing the cost of the bioelectric Fenton process. It is well established that biomass is a plentiful and ubiquitous resource globally. With the gradual depletion of traditional fossil fuels, biomass is receiving increasing attention as a potential energy and carbon source. Derived biomass char is emerging as a promising carrier for catalysts, offering advantages such as low cost, environmental protection, and a large specific surface area. The porous structure of biochar is rich in N, S and other active functional groups and heteroatoms, which can provide abundant active sites for oxygen reduction, shorten the diffusion path of oxygen reduction, and effectively improve the efficiency of the EF reaction while lowering the cost of materials. Therefore, pollutants such as antibiotics, perfluorinated compounds and dyes have been degraded by solid waste-based epoxy systems (EO-SWM).[136,137]

4.3.5.1 Advantages and Disadvantages of Solid Waste-Based Materials in EO Systems

Carbonaceous materials are considered to be effective cathode carriers due to their strong in-situ H_2O_2 generation capability and the compound's high Fe^{2+} regeneration rate and wide pH applicability, which indicate promising prospects for the removal of new pollutants from wastewater. However, despite these advantages of carbon cathodes, there are still some limitations in practical applications. For instance, the complexity and diversity of actual wastewater constituents, coupled with the diverse composition of the substance of interest, may severely hinder the removal of target contaminants. In particular, the selectivity of ·OH oxidation for these trace pollutants is very low, and it is usually necessary to generate excessive amounts of ·OH or input excessive amounts of energy to achieve effective removal of the target pollutants. This not only increases the treatment cost, but also may generate other toxic by-products or cause secondary pollution due to catalyst leakage. Therefore, subsequent studies should focus on improving the selective oxidizing ability of the cathode for the target pollutants and paying attention to the potential environmental risks associated with the pollutant degradation intermediates.[136, 137]

It is important to note that while carbon cathodes are inexpensive, catalytic, and widely adaptable, cost-benefit calculations for the large-scale preparation of such cathodes, as well as assessments of the long-term stability of such cathodes in a wide range of states of practice, are still lacking. Consequently, the large-scale use of carbon electrodes is still challenging, and further studies and tests are necessary to fully understand their strength, environmental impact, and energy consumption in different systems. In the meantime, it is recommended that the joint application of electro-Fenton technology with biotechnology or membrane separation technology be actively pursued in order to expand the application scope of carbon cathodes. In conclusion, although there are numerous issues associated with the use of carbonaceous materials as the cathode of the electro-Fenton process, it still has the potential for application and development in water pollution control, which is worthy of extensive attention.

4.3.5.2 Influence Factors in an EO-SWM System for Wastewater Treatment

4.3.5.2.1 Effect of Electrolytic Voltage The driving force of the electro-Fenton reaction is the applied voltage. As the applied voltage increases, so does the rate of organic matter removal. However, when the applied voltage exceeds a certain value, there will be a large amount of electrical energy consumed in the side reaction. Therefore, do not apply excessive voltage, generally take 5–25 V is appropriate.[138]

4.3.5.2.2 Influence of Solution Impedance (Conductivity) Because of the poor electrical conductivity of organic wastewater solutions, it is advisable to add an appropriate supporting electrolyte when making the electrolysis work. Although many soluble inorganic salts can be found as electrolytes, chlorides and nitrates will generate chlorine gas with irritating odour or various toxic and harmful nitride gases on the anode during electrolysis, and thus should not be used. Among the various inorganic solutions, the specific conductance of KOH solution is larger. However, what will be carried out in this electrolysis is the O reduction to generate H_2O, so the reaction must be carried out in an acidic or neutral solution, so Na_2SO_4 solution is generally selected as the supporting electrolyte. Although the specific resistance of Na_2SO_4 solution is much larger than that of KOH solution, with the increase of concentration, the conductivity of Na_2SO_4 solution increases rapidly, at the same time, because it does not take part in the electrolysis reaction, so it can always keep the solution conductivity stable.[138]

4.3.5.2.3 Effect of Solution PH   The acidity or alkalinity (pH) of a solution has a great influence on the cathodic reduction of oxygen. In an alkaline banyan solution, oxygen will undergo a four-electron reduction reaction; it is in an acidic solution that oxygen undergoes a two-electron reduction reaction to produce hydrogen peroxide. Therefore, the acidity or alkalinity of the solution determines the current efficiency of hydrogen peroxide generation, which in turn affects the subsequent formation of hydroxyl radicals and the degradation reaction of organic matter. With a solution pH greater than 4, Fe^{2+} is easily oxidized to form $Fe(OH)_3$ precipitate, a Fenton reaction cannot be carried out smoothly, and hence the initial solution pH should not be greater than 3.50.[138]

4.3.5.2.4 Effect of Current Density   In theory, the increase in electric current flow is limited by the unit area of the electrode in the electrolytic cell, as the electrode's surface area remains fixed. As the current increases, so does electrode polarization, which leads to side reactions at both the anode and cathode. For example, at higher anode current densities, iron dissolution is accompanied by oxygen formation. Similarly, increased cathode current density results in hydrogen gas formation. This hydrogen generation raises the pH of the solution, which is detrimental to hydrogen peroxide production and radical reactions.[138]

4.3.5.2.5 Effect of Aeration Rate   The electro-Fenton reaction has two major stages, mass transfer and reaction, and the mass transfer process of the electro-Fenton reaction is mainly influenced by the air flow rate. In the absence of air supply, only dissolved oxygen in solution and oxygen produced by electrolysis can be used in the electro-Fenton reaction, and thereby, it cannot effectively constitute a Fenton reaction, resulting in unsatisfactory removal results. Under the condition of air supply, air on the one hand can play the role of mixing, strengthen the mass transfer process in the reactor; on the other hand, it can replenish the oxygen that is constantly consumed in the reaction process. When the air flow rate is large enough to reach a certain value into the reaction control process, the effect of aeration on the removal of organic matter is reduced.[138]

4.3.5.3 Application of the EO-SWM System

At present, electrocatalytic oxidation systems with solid waste-based materials have been used effectively for the treatment of phenolic wastewater, dye wastewater and pharmaceutical wastewater.

4.3.5.3.1 Phenolic Compounds Phenolic compounds are common pollutants in industrial wastewater with recalcitrance and toxicity. According to reports, phenol, 4-nitrophenol, ibuprofen and bisphenol A could be removed by the EO-SWM system with a degradation efficiency of more than 88%.[139-142] In addition, the EO-SWM system has been employed in the treatment of actual coking wastewater. Coking wastewater is a typical phenol-containing wastewater with high toxicity and carcinogenicity, containing phenols, cyanide, aromatic hydrocarbons and other insoluble substances.[143] Zhang et al. prepared Ti-Sn-Ce/bamboo BioChar (BC) and used it as a particle electrode in 3-Dimensional Electrochemical Reaction systems (3DERs) for the effective treatment of coking wastewater.[144] The removal rate of Chemical Oxygen Demand (COD) and Dissolved Organic Carbon (DOC) reached 92.91% and 74.66% at a current density of 30 mA after 150 min. Besides, UV_{254} was reduced to 1.22 cm^{-1}, indicating that the biochemistry of the coking wastewater was improved greatly.

4.3.5.3.2 Pharmaceutical Wastewater Currently, antibiotics have been applied widely in the medical industry and animal husbandry. Besides, the antibiotics were detected frequently in sewage treatment plants, groundwater and surface water due to the discharge of a large amount of pharmaceutical wastewater. Antibiotics could accumulate in the environment owing to high persistence and non-biodegradability. Therefore, it is important to remove antibiotics. An EOB system has been employed in the degradation of sulfadiazine (SDZ),[145] sulfathiazole (STZ),[136] sulfamethoxazole (SMX),[146] norfloxacin (NOR).[147] As a result, the degradation rate could reach more than 87%. In addition, the Total Organic Carbon (TOC) removal rates of NOR and SDZ were 87.05% and 86.9% respectively after electrocatalytic oxidation treatment. Thus, the EO-SWM system can degrade effectively antibiotics.

The excellent performance of the EO-SWM system was presented in treating real pharmaceutical wastewater. Pharmaceutical wastewater has been treated by a Fe/N/biochar-catalysed electro-Fenton process. Consequently, COD and TOC decreased from 740 mg/L and 559 mg/L to 200 mg/L and 211 mg/L, respectively. Moreover, more than 93% of PO_4^{3-}-P was also removed in pharmaceutical wastewater. Therefore, the EO-SWM has system has shown great potential in the treatment of practical pharmaceutical wastewater.

4.3.5.3.3 Dyes   Dyes exist in the textile industry as well as in printing and dyeing wastewater. Common types of dyes in wastewater include Orange II, crystal violet, methylene blue, methyl orange, and Rhodamine B, and the like. Particularly, the EO-SWM system has been employed for dye removal with a degradation efficiency of over 90%.[148, 149] Real textile wastewater can also be treated based on waste biomass-derived biochar cathodes in an EF system.[150] The removal efficiencies of dyes and TOC were 96.47% and 25.67% after 120 min respectively in the EF system with a CO_2-700–1 h catalytic biochar cathode. In addition, almost all the pollutants in textile wastewater were transformed into small molecule pollutants after 5 h under the optimal process conditions within the EF system. Thus, the EO-SWM system can treat dye wastewater efficiently and stably.

4.3.5.4 Mechanism in an EO-SWM System for Wastewater Treatment
The degradation of pollutants is dominated by the free radical pathway in the EO-SWM system. Additionally, the degradation process changed according to the function of solid waste-based materials in the EO process.

The mechanism of pollutant degradation was divided into two types when solid waste-based materials were involved directly in organics degradation within the EO-SWM system (Figure 4.6a). (i) Free radical pathways were dominated by Persistent Free Radicals (PFR). Oxygen-containing functional groups on the surface of solid waste-based materials

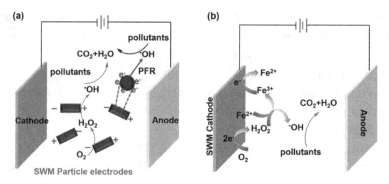

FIGURE 4.6   The mechanisms of pollutant degradation in the EO-SWM system.

can promote the production of persistent free radicals, which act as electron donors for $\cdot$OH generation.[151] (ii) There is the formation of microelectrodes for redox reactions. Solid waste-based materials have been used as particle electrodes for 3DERs.[152] Solid waste-based particles were polarized and formed microelectrodes in the presence of an electric field, triggering electrocatalytic oxidation reactions. H_2O_2 was generated through a reduction reaction of dissolved oxygen at the cathode of the microelectrode. Then, $\cdot$OH was produced at the anode of the microelectrode to degrade the pollutant.

Conventional EF technology generated H_2O_2 via electrochemical reactions. Subsequently, $\cdot$OH was formed through the reaction of H_2O_2 and Fe^{2+}. The oxygen-containing functional groups promoted the 2e-ORR reaction to generate H_2O_2 with the introduction of solid waste-based materials to the EF system as a cathode material (Figure 5.6b). Fe^{2+} activated H_2O_2 to produce $\cdot$OH.

4.4 CONCLUSION

Over the past decade, solid waste-based materials have demonstrated remarkable potential in the field of water treatment. Utilizing low-cost solid waste as a substrate for water pollution control aligns with the waste-to-wealth. While extensive research has been conducted on adsorption and advanced oxidation processes using these materials, several aspects require further improvement for future advancement: (1) Synergistic and efficient multi-pollutant removal: Current studies predominantly focus on the removal of single pollutants. Future research should aim to understand the mechanisms behind the elimination of various pollutants to facilitate the combined treatment of multiple contaminants. (2) Recyclability of solid waste-based materials: With long-term use, the efficacy of solid waste-based materials in water treatment diminishes. Enhancing the longevity of these materials and achieving their recyclability is an imperative challenge to address. (3) Development of large-scale integrated equipment: Present investigations are confined to lab-scale experiments. To escalate the application of solid waste-based materials to an industrial scale, it is crucial to develop large-scale, stable equipment to increase water treatment efficiency.

REFERENCES

1. Niu, J.; Kasuga, I.; Kurisu, F.; Furumai, H. Effects of backwashing on granular activated carbon with ammonium removal potential in a full-scale drinking water purification plant. *Water* 2018, 10, 1830.

2. Indrawirawan, S.; Sun, H.; Duan, X.; Wang, S. Nanocarbons in different structural dimensions (0-3D) for phenol adsorption and metal-free catalytic oxidation. *Appl. Catal. B: Environ.* 2015, 179, 352–362.

3. Zhou, X. New notion of biochar: A review on the mechanism of biochar applications in advanced oxidation processes. *Chem. Eng. J.* 2021, 416, 129027.

4. Liu, C.; Wang, H.; Karim, A. M.; Sun, J.; Wang, Y. Catalytic fast pyrolysis of lignocellulosic biomass. *Chem. Soc. Rev.* 2014, 43, 7594–7623.

5. Bonilla-Petriciolet, A.; Mendoza-Castillo, I.; Luiz Dotto, G.; Duran-Valle, J. Adsorption in water treatment. Reference Module in Chemistry, Molecular Sciences and Chemical Engineering, Elsevier, 2019.

6. Ali, I.; Asim, M.; Khan, A. Low cost adsorbents for the removal of organic pollutants from wastewater. *J. Environ. Manage.* 2010, 113, 170–183.

7. Bonilla-Petriciolet, A.; Mendoza-Castillo, I.; Reynel-Ávila, E. Adsorption processes for water treatment and purification. *J. Mol. Liq.* 2019, 306, 453–462.

8. Cherkasov, N. Liquid-phase adsorption: Common problems and how we could do better. *J. Mol. Liq.* 2020, 301, 112378.

9. Li, W.; Mu, B.; Yang, Y. Feasibility of industrial-scale treatment of dye wastewater via bio-adsorption technology. *Bioresour. Technol.* 2019, 277, 157–170.

10. Choy, H.; Mckay, G. Sorption of metal ions from aqueous solution using bone char. *Environ. Int.* 2005, 31 (6), 845–854.

11. Diraki, A.; Mackey, R.; Mckay, G.; Abdala, A. Removal of emulsified and dissolved diesel oil from high salinity wastewater by adsorption onto graphene oxide. *J. Environ. Chem. Eng.* 2019, 7 (3), 103106.

12. Streit, M.; Côrtes, N.; Druzian, P.; Godinho, M.; Collazzo, C.; Perondi, D.; Dotto, L. Development of high quality activated carbon from biological sludge and its application for dyes removal from aqueous solutions. *Sci. Total. Environ.* 2019, 660, 277–287.

13. Crini, G.; Badot, M. Application of chitosan, a natural aminopolysaccharide, for dye removal from aqueous solutions by adsorption processes using batch studies: A review of recent literature. *Prog. Polym. Sci.* 2008, 33 (4), 399–447.

14. Dotto, L.; Vieira, G.; Pinto, L. A. A. Kinetics and mechanism of tartrazine adsorption onto chitin and chitosan. *Ind. Eng. Chem. Res.* 2012, 51 (19), 6862–6868.

15. Kwok, M.; Koong, F.; Ansari, T.; Mckay, G. Adsorption/desorption of arsenite and arsenate on chitosan and nanochitosan. *Environ. Sci. Pollut. Res.* 2018, 25 (15), 14734–14742.

16. Escudero, B.; Quintas, Y.; Wuilloud, G.; Dotto, L. Recent advances on elemental biosorption. *Environ. Chem. Lett.* 2019, 17 (1), 409–427.

17. TorresPérez, J.; SoriaSerna, A.; Solache-Ríos, M.; Mckay, G. One step carbonization/activation process for carbonaceous material preparation from pecan shells for tartrazine removal and regeneration after saturation. *Adsorpt. Sci. Technol.* 2015, 33 (10), 895–913.

18. Brião, V.; Jahn, L.; Foletto, L.; Dotto, L. Adsorption of crystal violet dye onto a mesoporous ZSM-5 zeolite synthetized using chitin as template. *J. Colloid Interface Sci.* 2017, 508, 313–322.

19. Zubair, M.; Daud, M.; Mckay, G.; Shehzad, F.; Harthi, A. Recent progress in layered double hydroxides (LDH)-containing hybrids as adsorbents for water remediation. *Appl. Clay Sci.* 2017, 143, 279–292.

20. Haldar, D.; Duarah, P.; Purkait, K. MOFs for the treatment of arsenic, fluoride and iron contaminated drinking water: A review. *Chemosphere* 2020, 251, 126388.

21. Fan, T.; Li, J.; Guo, M.; Fan, L.; Zhang, S.; Su, Z.; Wang, J. Removal of arsenic from ammoniacal etching waste liquor by 3-(2-aminoethylamino) propyltrimethoxysilane functionalized silica gel sorbent. *Desalination Water Treat.* 2012, 50 (1–3), 51–58.

22. Fan, T.; Su, J.; Fan, L.; Guo, M.; Wang, J.; Gao, S.; Sun, T. Sol–gel derived organic–inorganic hybrid sorbent for removal of Pb^{2+}, Cd^{2+} and Cu^{2+} from aqueous solution. *J. Sol-Gel Sci. Technol.* 2012, 64 (2), 418–426.

23. Salvador, F.; Martin, N.; Sanchez, R.; Sanchezo, J.; Izquierdo, C. Regeneration of carbonaceous adsorbents. Part I: Thermal regeneration. *Micropor. Mesopor. Mater.* 2015, 202, 259–276.

24. Salvador, F.; MartinSanchez, N.; SanchezHernandez, R.; SanchezMontero, J.; Izquierdo, C. Regeneration of carbonaceous adsorbents. Part II: Chemical,

microbiological and vacuum regeneration. *Micropor. Mesopor. Mater.* 2015, 202, 277–296.

25. Tóth, J. *Adsorption: Theory, Modeling and Analysis.* Marcel Dekker. 2002.

26. Worch, E. Fixed-bed adsorption in drinking water treatment: A critical review on models and parameter estimation. *J. Water Supply Res. Technol.* 2008, 57 (3), 171–183.

27. Brian, W.; Misako, N. Beyond first-line epidermal growth factor receptor (EGFR)-tyrosine kinase inhibitors (TKI): Approach to resistance. *Adv. Oncol.* 2024, 4 (1), 63–72.

28. Xu, Z.; Cai, J.; Pan, B. Mathematically modeling fixed-bed adsorption in aqueous systems. *J. Zhejiang Univ. A* 2013, 14 (3), 155–176.

29. Soliman, N. K.; Moustafa, A. F. Industrial solid waste for heavy metals adsorption features and challenges: A review. *J. Mater. Res. Technol.* 2020, 9 (5), 10235–10253.

30. Aigbe, U. O.; Ukhurebor, K. E.; Onyancha, R. B.; Osibote, O. A.; Darmokoesoemo, H.; Kusuma, H. S. Fly ash-based adsorbent for adsorption of heavy metals and dyes from aqueous solution: A review. *J. Mater. Res. Technol.* 2021, 14, 2751–2774.

31. Neolaka, Y. A. B.; Riwu, A. A. P.; Aigbe, U. O.; Ukhurebor, K. E.; Onyancha, R. B.; Darmokoesoemo, H.; Kusuma, H. S. Potential of activated carbon from various sources as a low-cost adsorbent to remove heavy metals and synthetic dyes. *Results Chem.* 2023, 5, 100711.

32. Ouyang, H.; Safaeipour, N.; Othman, R. S.; Otadi, M.; Sheibani, R.; Kargaran, F.; Van Le, Q.; Khonakdar, H. A.; Li, C. Agricultural waste-derived (nano) materials for water and wastewater treatment: Current challenges and future perspectives. *J. Clean. Prod.* 2023, 421, 138524.

33. Rasha, A.; Mohamed, A.; Magdy, B.; Mohamed, I. Heavy metals removal from industrial wastewater using bio-adsorbent materials based on agricultural solid waste through batch and continuous flow mechanisms. *J. Water Process Eng.* 2024, 27, 104665.

34. Xu, K.; He, T.; Li, L.; Iqbal, J.; Tong, Y.; Hua, L.; Tian, Z.; Zhao, L.; Li, H. DOTA functionalized adsorbent DOTA@sludge@chitosan derived from recycled shrimp shells and sludge and its application for lead and chromium removal from water. *Int. J. Biol. Macromol.* 2024, 255, 128263.

35. Salleh, M. A. M.; Mahmoud, D. K.; Karim, W. A. W. A.; Idris, A. Cationic and anionic dye adsorption by agricultural solid waste: A comprehensive review. *Desalination* 2011, 280 (1), 1–13.

36. Simonescu, C. M.; Tătăruş, A.; Culiță, D. C.; Stănică, N.; Ionescu, I. A.; Butoi, B.; Banici, A. Comparative study of $CoFe_2O_4$ nanoparticles and $CoFe_2O_4$-chitosan composite for congo red and methyl orange removal by adsorption. *Nanomaterials* 2021, 11 (3), 711.

37. Hevira, L.; Ighalo, J. O.; Sondari, D. Chitosan-based polysaccharides for effective synthetic dye adsorption. *J. Mol. Liq.* 2024, 393, 123604.

38. Gkika, D. A.; Liakos, E. V.; Vordos, N.; Kontogoulidou, C.; Magafas, L.; Bikiaris, D. N.; Bandekas, D. V.; Mitropoulos, A. C.; Kyzas, G. Z. Cost estimation of polymeric adsorbents. *Polymers* 2019, 11 (5), 925.

39. Rehan, A. I.; Rasee, A. I.; Awual, M. E.; Waliullah, R. M.; Hossain, M. S.; Kubra, K. T.; Salman, M. S.; Hasan, M. M.; Hasan, M. N.; Sheikh, M. C.; Marwani, H. M.; Khaleque, M. A.; Islam, A.; Awual, M. R. Improving toxic dye removal and remediation using novel nanocomposite fibrous adsorbent. *Colloids Surf. A* 2023, 673, 131859.

40. Franco, D. S. P.; Georgin, J.; Netto, M. S.; Allasia, D.; Oliveira, M. L. S.; Foletto, E. L.; Dotto, G. L. Highly effective adsorption of synthetic phenol effluent by a novel activated carbon prepared from fruit waste of the *Ceiba speciosa* forest species. *J. Environ. Chem. Eng.* 2021, 9 (5), 105927.

41. Jain, M.; Khan, S. A.; Sahoo, A.; Dubey, P.; Pant, K. K.; Ziora, Z. M.; Blaskovich, M. A. T. Statistical evaluation of cow-dung derived activated biochar for phenol adsorption: Adsorption isotherms, kinetics, and thermodynamic studies. *Bioresour. Technol.* 2022, 352, 127030.

42. Abdel-Gawwad, H. A.; Ahmed, M. S.; Mohammed, A. H.; Badawi, M.; Bonilla-Petriciolet, A.; Lima, E. C.; Salama, Y. F.; Mobarak, M.; Seliem, M. K. Utilization of red clay brick waste in the green preparation of an efficient porous nanocomposite for phenol adsorption: Characterization, experiments and statistical physics treatment. *Sustain. Chem. Pharm.* 2023, 32, 101027.

43. Zbair, M.; Ainassaari, K.; El Assal, Z.; Ojala, S.; El Ouahedy, N.; Keiski, R. L.; Bensitel, M.; Brahmi, R. Steam activation of waste biomass: Highly microporous carbon, optimization of bisphenol A, and diuron adsorption by response surface methodology. *Environ. Sci. Pollut. Res.* 2018, 25 (35), 35657–35671.

44. Mpatani, F. M.; Han, R.; Aryee, A. A.; Kani, A. N.; Li, Z.; Qu, L. Adsorption performance of modified agricultural waste materials for removal of emerging micro-contaminant bisphenol A: A comprehensive review. *Sci. Total Environ.* 2021, 780, 146629.

45. Ali, H. Biodegradation of synthetic dyes – A review. *Water. Air. Soil. Pollut.* 2010, 213 (1), 251–273.

46. Sophia A, C.; Lima, E. C. Removal of emerging contaminants from the environment by adsorption. *Ecotoxicol. Environ. Saf.* 2018, 150, 1–17.

47. Ateia, M.; Alsbaiee, A.; Karanfil, T.; Dichtel, W. Efficient PFAS removal by amine-functionalized sorbents: Critical review of the current literature. *Environ. Sci. Technol. Lett.* 2019, 6 (12), 688–695.

48. Hassan, M.; Liu, Y.; Naidu, R.; Du, J.; Qi, F. Adsorption of perfluorooctane sulfonate (PFOS) onto metal oxides modified biochar. *Environ. Technol. Innovation* 2020, 19, 100816.

49. Yuan, C.; Huang, Y.; Cannon, F. S.; Zhao, Z. Adsorption mechanisms of PFOA onto activated carbon anchored with quaternary ammonium/epoxide-forming compounds: A combination of experiment and model studies. *J. Environ. Sci.* 2020, 98, 94–102.

50. Xiao, F.; Zhang, X.; Penn, L.; Gulliver, J. S.; Simcik, M. F. Effects of monovalent cations on the competitive adsorption of perfluoroalkyl acids by kaolinite: Experimental studies and modeling. *Environ. Sci. Technol.* 2011, 45 (23), 10028–10035.

51. Saeidi, N.; Kopinke, F.; Georgi, A. Understanding the effect of carbon surface chemistry on adsorption of perfluorinated alkyl substances. *Chem. Eng. J.* 2020, 381, 122689.

52. Karoyo, A. H.; Wilson, L. D. Tunable macromolecular-based materials for the adsorption of perfluorooctanoic and octanoic acid anions. *J. Colloid Interface Sci.* 2013, 402, 196–203.

53. Zhang, Q.; Deng, S.; Yu, G.; Huang, J. Removal of perfluorooctane sulfonate from aqueous solution by crosslinked chitosan beads: Sorption kinetics and uptake mechanism. *Bioresour. Technol.* 2011, 102 (3), 2265–2271.

54. Olusegun, S. J.; Mohallem, N. D. S. Comparative adsorption mechanism of doxycycline and congo red using synthesized kaolinite supported $CoFe_2O_4$ nanoparticles. *Environ. Pollut.* 2020, 260, 114019.

55. Wang, T.; Pan, X.; Ben, W.; Wang, J.; Hou, P.; Qiang, Z. Adsorptive removal of antibiotics from water using magnetic ion exchange resin. *J. Environ. Sci.* 2017, 52, 111–117.

56. Pašalić, H.; Aquino, A. J. A.; Tunega, D.; Haberhauer, G.; Gerzabek, M. H.; Lischka, H. Cation–π interactions in competition with cation microhydration: A theoretical study of alkali metal cation–pyrene complexes. *J. Mol. Model.* 2017, 23 (4), 131.

57. Zhong, J.; Cui, X.; Guan, W.; Lu, C. Direct observation of adsorption kinetics on clays by cation–π interaction-triggered aggregation luminescence. *J. Mater. Chem. C* 2018, 6 (48), 13218–13224.

58. Singh, S. K.; Das, A. The n → π* interaction: A rapidly emerging non-covalent interaction. *Phys. Chem. Chem. Phys.* 2015, 17 (15), 9596–9612.

59. Olusegun, S. J.; Larrea, G.; Osial, M.; Jackowska, K.; Krysinski, P. Photocatalytic degradation of antibiotics by superparamagnetic iron oxide nanoparticles. Tetracycline case. *Catalysts* 2021, 11 (10), 1243.

60. Smith, D. A. A brief history of the hydrogen bond. *Modeling the Hydrogen Bond* 1994, 569 (569), 1–5.

61. Qu, Z.; Sun, F.; Qie, Z.; Gao, J.; Zhao, G. The change of hydrogen bonding network during adsorption of multi-water molecules in lignite: Quantitative analysis based on AIM and DFT. *Mater. Chem. Phys.* 2020, 247, 122863.

62. Ahmed, I.; Jhung, S. H. Applications of metal-organic frameworks in adsorption/separation processes via hydrogen bonding interactions. *Chem. Eng. J.* 2017, 310, 197–215.

63. Weng, X.; Cai, W.; Owens, G.; Chen, Z. Magnetic iron nanoparticles calcined from biosynthesis for fluoroquinolone antibiotic removal from wastewater. *J. Clean. Prod.* 2021, 319, 128734.

64. Ji, J.; Yuan, X.; Zhao, Y.; Jiang, L.; Wang, H. Mechanistic insights of removing pollutant in adsorption and advanced oxidation processes by sludge biochar. *J. Hazard. Mater.* 2022, 430, 128375.

65. Pan, X.; Gu, Z.; Chen, W.; Li, Q. Preparation of biochar and biochar composites and their application in a Fenton-like process for wastewater decontamination: A review. *Sci. Total. Environ.* 2021, 754, 142104.

66. Thomas, N.; Dionysiou, D.; Pillai, S. Heterogeneous Fenton catalysts: A review of recent advances. *J. Hazard. Mater.* 2021, 404, 124082.

67. Wang, R.; Zhang, S.; Chen, H.; He, Z.; Cao, G.; Wang, K.; Li, F.; Ren, N.; Xing, D.; Ho, S. H. Enhancing biochar-based nonradical persulfate activation using data-driven techniques. *Environ. Sci. Technol.* 2023, 57, 4050–4059.

68. Yang, B.; Dai, J.; Zhao, Y.; Wu, J.; Ji, C.; Zhang, Y. Advances in preparation, application in contaminant removal, and environmental risks of biochar-based catalysts: A review. *Biochar* 2022, 4, 51.

69. Song, G.; Qin, F.; Yu, J.; Tang, L.; Pang, Y.; Zhang, C.; Wang, J.; Deng, L. Tailoring biochar for persulfate-based environmental catalysis: Impact of biomass feedstocks. *J. Hazard. Mater.* 2022, 424, 127663.

70. Chen, Y.; Wang, R.; Duan, X.; Wang, S.; Ren, N.; Ho, S. Production, properties, and catalytic applications of sludge derived biochar for environmental remediation. *Water Res.* 2020, 187, 116390.

71. Alkurdi, S.; Herath, I.; Bundschuh, J.; Juboori, R.; Vithanage, M.; Mohan, D. Biochar versus bone char for a sustainable inorganic arsenic mitigation in water: What needs to be done in future research? *Environ. Int.* 2019, 127, 52–69.

72. Tong, Q.; Li, C.; Wang, G.; Wang, Y.; Peng, S.; Wang, J.; Lai, B.; Guo, Y. Enhanced peroxymonosulfate activation by Co-bHAP catalyst for efficient degradation of sulfamethoxazole. *J. Environ. Chem. Eng.* 2023, 11, 109499.

73. Liu, H.; Liu, Y.; Tang, L.; Wang, J.; Yu, J.; Zhang, H.; Yu, M.; Zou, J.; Xie, Q. Egg shell biochar-based green catalysts for the removal of organic pollutants by activating persulfate. *Sci. Total. Environ.* 2020, 745, 141095.

74. Fanaei, F.; Moussavi, G.; Srivastava, V.; Sillanpää, M. The enhanced catalytic potential of sulfur-doped MgO (S-MgO) nanoparticles in activation of peroxysulfates for advanced oxidation of acetaminophen. *Chem. Eng. J.* 2019, 371, 404–413.

75. Chen, S.; Xiong, P.; Zhan, W.; Xiong, L. Degradation of ethylthionocarbamate by pyrite-activated persulfate. *Miner. Eng.* 2018, 122, 38–43.

76. Hanci, T.; Alaton, I. Comparison of sulfate and hydroxyl radical based advanced oxidation of phenol. *Chem. Eng. J.* 2013, 224, 10–16.

77. Solís, R.; Mena, I.; Nadagouda, M.; Dionysiou, D. Adsorptive interaction of peroxymonosulfate with graphene and catalytic assessment via non-radical pathway for the removal of aqueous pharmaceuticals. *J. Hazard. Mater.* 2020, 384, 121340.

78. Chueca, J.; Giannakis, S.; Marjanovic, M.; Kohantorabi, M.; Gholami, M.; Grandjean, D.; Alencastro, L.; Pulgarín, C. Solar-assisted bacterial disinfection and removal of contaminants of emerging concern by Fe^{2+}-activated HSO_5^- vs. $S_2O_8^{2-}$ in drinking water. *Appl. Catal. B: Environ.* 2019, 248, 62–72.

79. Li, N.; Li, R.; Duan, X.; Yan, B.; Liu, W.; Cheng, Z.; Chen, G.; Hou, L.; Wang, S. Correlation of active sites to generated reactive species and degradation routes of organics in peroxymonosulfate activation by Co-loaded carbon. *Environ. Sci. Technol.* 2021, 55, 16163–16174.

80. Xu, Y.; Lin, H.; Li, Y.; Zhang, H. The mechanism and efficiency of MnO_2 activated persulfate process coupled with electrolysis. *Sci. Total. Environ.* 2017, 609, 644–654.

81. Deng, J.; Cheng, Y.; Lu, Y.; Crittenden, J. C.; Zhou, S.; Gao, N.; Li, J. Mesoporous manganese cobaltite nanocages as effective and reusable heterogeneous peroxymonosulfate activators for carbamazepine degradation. *Chem. Eng. J.* 2017, 330, 505–517.

82. Duan, X.; Su, C.; Miao, J.; Zhong, Y.; Shao, Z.; Wang, S.; Sun, H. Insights into perovskite-catalysed peroxymonosulfate activation: Maneuverable cobalt sites for promoted evolution of sulfate radicals. *Appl. Catal. B: Environ.* 2018, 220, 626–634.

83. Zhu, K.; Wang, X.; Chen, D.; Ren, W.; Lin, H.; Zhang, H. Wood-based biochar as an excellent activator of peroxydisulfate for Acid Orange 7 decolorization. *Chemosphere* 2019, 231, 32–40.

84. Ouyang, D.; Chen, Y.; Yan, J.; Qian, L.; Han, L.; Chen, M. Activation mechanism of peroxymonosulfate by biochar for catalytic degradation of 1,4-dioxane: Important role of biochar defect structures. *Chem. Eng. J.* 2019, 370, 614–624.

85. Wu, Y.; Guo, J.; Han, Y.; Zhu, J.; Zhou, L.; Lan, Y. Insights into the mechanism of persulfate activated by rice straw biochar for the degradation of aniline. *Chemosphere* 2018, 200, 73–379.

86. Wang, B.; Li, Y.; Wang, L. Metal-free activation of persulfates by corn stalk biochar for the degradation of antibiotic norfloxacin: Activation factors and degradation mechanism. *Chemosphere* 2019, 237, 124454.

87. Ho, S.; Chen, Y.; Li, R.; Zhang, C.; Ge, Y.; Cao, G.; Ma, M.; Duan, X.; Wang, S.; Ren, N. N-doped graphitic biochars from C-phycocyanin extracted *Spirulina* residue for catalytic persulfate activation toward nonradical disinfection and organic oxidation. *Water Res.* 2019, 159, 77–86.

88. Yu, J.; Tang, L.; Pang, Y.; Zeng, G.; Feng, H.; Zou, J.; Wang, J.; Feng, C.; Zhu, X.; Ouyang, X.; Tan, J. Hierarchical porous biochar from shrimp shell for persulfate activation: A two-electron transfer path and key impact factors. *Appl. Catal. B: Environ.* 2020, 260, 118160.

89. Zhou, X.; Zeng, Z.; Zeng, G.; Lai, C.; Xiao, R.; Liu, S.; Huang, D.; Qin, L.; Liu, X.; Li, B.; Yi, H.; Fu, Y.; Li, L.; Wang, Z. Persulfate activation by swine bone char-derived hierarchical porous carbon: Multiple mechanism system for organic pollutant degradation in aqueous media. *Chem. Eng. J.* 2020, 383, 123091.

90. Shi, C.; Li, Y.; Feng, H.; Jia, S.; Xue, R.; Li, G.; Wang, G. Removal of p-nitrophenol using persulfate activated by biochars prepared from different biomass materials. *Chem. Res. Chinese Univ.* 2018, 34, 39–43.

91. Wang, J.; Shen, M.; Gong, Q.; Wang, X.; Cai, J.; Wang, S.; Chen, Z. One-step preparation of ZVI-sludge derived biochar without external source of iron and its application on persulfate activation. *Sci. Total. Environ.* 2020, 714, 136728.

92. Huang, B.; Jiang, J.; Huang, G.; Yu, H. Sludge biochar-based catalysts for improved pollutant degradation by activating peroxymonosulfate. *J. Mater. Chem. A* 2018, 6, 8978–8985.

93. Wang, S.; Wang, J. Activation of peroxymonosulfate by sludge-derived biochar for the degradation of triclosan in water and wastewater. *Chem. Eng. J.* 2019, 356, 350–358.

94. Yang, Z.; Yang, H.; Liu, Y.; Hu, C.; Jing, H.; Li, H. Heterogeneous catalytic ozonation for water treatment: Preparation and application of catalyst. *Ozone Sci. Eng.* 2023, 45 (2), 147–173.

95. Agudelo, E. A.; Cardona, G., S. Selection of catalysts for use in a heterogeneous catalytic ozonation system. *Ozone Sci. Eng.* 2020, 42 (2), 146–156.

96. Rocca, D.; Peralta, R.; Peralta, R.; Rodríguez-Castellón, E.; Moreira, R. Adding value to aluminosilicate solid waste to produce adsorbents, catalysts and filtration membranes for water and wastewater treatment. *J. Mater. Sci.* 2021, 56 (2), 1039–1063.

97. Zhang, Y.; Han, Z.; He, P.; Chen, H. Geopolymer-based catalysts for cost-effective environmental governance: A review based on source control and end-of-pipe treatment. *J. Clean. Prod.* 2020, 263, 121556.

98. Rocca, D.; Sousa, F.; Ardisson, J.; Peralta, R.; Rodríguez, E.; Moreira, R. Magnetic mining waste based-geopolymers applied to catalytic reactions with ozone. *Heliyon* 2023, 9 (6), e17097.

99. Chen, C.; Yoza, B.; Chen, H.; Li, Q.; Guo, S. Manganese sand ore is an economical and effective catalyst for ozonation of organic contaminants in petrochemical wastewater. *Water Air Soil Pollut.* 2015, 226 (6), 182.

100. Chakraborty, R.; Vilya, K.; Pradhan, M.; Nayak, A. Recent advancement of biomass-derived porous carbon based materials for energy and environmental remediation applications. *J. Mater. Chem. A* 2022, 10 (13), 6965–7005.

101. Gunarathne, V.; Ashiq, A.; Ramanayaka, S.; Wijekoon, P. Vithanage, M. Biochar from municipal solid waste for resource recovery and pollution remediation. *Environ. Chem. Lett.* 2019, 17 (3), 1225–1235.

102. Mortazavi, S.; Asgari, G.; Hashemian, S.; Moussavi, G. Degradation of humic acids through heterogeneous catalytic ozonation with bone charcoal. *React. Kinet. Mech. Catal.* 2010, 100 (2), 471–485.

103. Asgari, G.; Mohammadi, A.; Mortazavi, S.; Ramavandi, B. Investigation on the pyrolysis of cow bone as a catalyst for ozone aqueous decomposition: Kinetic approach. *J. Anal. Appl. Pyrol*sis 2013, 99, 149–154.

104. Moussavi, G.; Khosravi, R. Preparation and characterization of a biochar from pistachio hull biomass and its catalytic potential for ozonation of water recalcitrant contaminants. *Bioresour. Technol.* 2012, 119, 66–71.

105. Tian, S.; Qi, J.; Wang, Y.; Liu, Y.; Wang, L.; Ma, J. Heterogeneous catalytic ozonation of atrazine with Mn-loaded and Fe-loaded biochar. *Water Res.* 2021, 193, 116860.

106. Luo, Z.; Wang, D.; Zeng, W.; Yang, J. Removal of refractory organics from piggery bio-treatment effluent by the catalytic ozonation process with piggery biogas residue biochar as the catalyst. *Sci. Total Environ.* 2020, 734, 139448.

107. Li, X.; Fu, L.; Chen, F.; Zhao, S.; Zhu, J.; Yin, C. Application of heterogeneous catalytic ozonation in wastewater treatment: An overview. *Catalysts* 2023, 13 (2), 342.

108. Rocca, D.; De Noni Júnior, A.; Rodríguez-Aguado, E.; Peralta, R.; Rodríguez-Castellón, E.; Puma, G.; Moreira, R. Mechanistic insights on the catalytic ozonation of trimethoprim in aqueous phase using geopolymer catalysts produced from mining waste. *J. Environ. Chem. Eng.* 2023, 11 (6), 111163.

109. Nguyen, L.; Nguyen, X.; Van Thai, N.; Le, H.; Thi, T.; Thi, K.; Nguyen, H.; Le, M.; Van, H.; Nguyet, D. Promoted degradation of ofloxacin by ozone integrated with Fenton-like process using iron-containing waste mineral enriched by magnetic composite as heterogeneous catalyst. *J. Water Process Eng.* 2022, 49, 103000.

110. Yong, K.; Wu, J.; Andrews, S. Heterogeneous catalytic ozonation of aqueous reactive dye. *Ozone Sci. Eng.* 2005, 27 (4), 257–263.

111. Asgari, G.; Akbari, S.; Mohammadi, A.; Poormohammadi, A.; Ramavandi, B. Preparation and catalytic activity of bone-char ash decorated with MgO–FeNO$_3$ for ozonation of reactive black 5 dye from aqueous solution: Taguchi optimization data. *Data in Brief* 2017, 13, 132–136.

112. Wang, Y.; Cao, H.; Chen, L.; Chen, C.; Duan, X.; Xie, Y.; Song, W.; Sun, H.; Wang, S. Tailored synthesis of active reduced graphene oxides from waste graphite: Structural defects and pollutant-dependent reactive radicals in aqueous organics decontamination. *Appl. Catal. B: Environ.* 2018, 229, 71–80.

113. Ryu, S.; Lee, J.; Kannapu, H.; Jang, S.; Kim, Y.; Jang, H.; Ha, J.; Jung, S.; Park, Y. Acid-treated waste red mud as an efficient catalyst for catalytic fast copyrolysis of lignin and polyproylene and ozone-catalytic conversion of toluene. *Environ. Res.* 2020, 191, 110149.

114. Bing, J.; Hu, C.; Nie, Y.; Yang, M.; Qu, J. Mechanism of catalytic ozonation in Fe$_2$O$_3$/Al$_2$O$_3$@SBA-15 aqueous suspension for destruction of ibuprofen. *Environ. Sci. Technol.* 2015, 49 (3), 1690–1697.

115. Yu, G.; Wang, Y.; Cao, H.; Zhao, H.; Xie, Y. Reactive oxygen species and catalytic active sites in heterogeneous catalytic ozonation for water purification. *Environ. Sci. Technol.* 2020, 54 (10), 5931–5946.

116. Zhang, T.; Li, W.; Croue, J. Catalytic ozonation of oxalate with a cerium supported palladium oxide: An efficient degradation not relying on hydroxyl radical oxidation. *Environ. Sci. Technol.* 2011, 45 (21), 9339–9346.

117. Ma, J.; Sui, M.; Zhang, T.; Guan, C. Effect of pH on MnO$_x$/GAC catalysed ozonation for degradation of nitrobenzene. *Water Res.* 2005, 39 (5), 779–786.

118. Bing, J.; Hu, C.; Zhang, L. Enhanced mineralization of pharmaceuticals by surface oxidation over mesoporous γ-Ti-Al$_2$O$_3$ suspension with ozone. *Appl. Catal. B: Environ.* 2017, 202, 118–126.

119. Rodríguez-Padrón, D.; Luque, R.; Muñoz-Batista, M. J. Waste-derived materials: Opportunities in photocatalysis. *Top. Curr. Chem.* 2020, 378, 1–28.

120. Khan, A.; Bhoi, R. G.; Saharan, V. K.; George, S. Green calcium-based photocatalyst derived from waste marble powder for environmental sustainability: A review on synthesis and application in photocatalysis. *Environ. Sci. Pollut. Res.* 2022, 29, 86439–86467.

121. Zhang, Q.; Wu, Y.; Zuo, T. Green recovery of titanium and effective regeneration of TiO_2 photocatalysts from spent selective catalytic reduction catalysts. *ACS Sustain. Chem. Eng.* 2018, 6 (3), 3091–3101,

122. Mohamed, H. H.; Alsanea, A. A.; Alomair, N. A.; Akhtar, S.; Bahnemann, D. W. ZnO@ porous graphite nanocomposite from waste for superior photocatalytic activity. *Environ. Sci. Pollut. Res.* 2019, 26, 12288–12301.

123. Zuliani, A.; Muñoz-Batista, M. J.; Luque, R. Microwave-assisted valorization of pig bristles: Towards visible light photocatalytic chalcocite composites. *Green Chem.* 2018, 20, 3001–3007.

124. Sree, G. V.; Nagaraaj, P.; Kalanidhi, K.; Aswathy, C.; Rajasekaran, P. Calcium oxide a sustainable photocatalyst derived from eggshell for efficient photo-degradation of organic pollutants. *J. Clean. Prod.* 2020, 270, 122294.

125. Lancet, D.; Pecht, I. Spectroscopic and immunochemical studies with nitrobenzoxadiazolealanine, a fluorescent dinitrophenyl analog. *Biochemistry* 1977, 16, 5150–5157.

126. Zhong, Y.; Zhang, X.; Wang, Y.; Zhang, X.; Wang, X. Carbon quantum dots from tea enhance Z-type $BiOBr/C_3N_4$ heterojunctions for RhB degradation: Catalytic effect, mechanisms, and intermediates. *Appl. Surf. Sci.* 2023, 639, 158254.

127. Li, H.; Hu, J.; Zhou, X.; Li, X.; Wang, X. An investigation of the biochar-based visible-light photocatalyst via a self-assembly strategy. *J. Environ. Manage.* 2018, 217, 175–182.

128. Yang, B.; Dai, J.; Zhao, Y.; Wang, Z.; Wu, J.; Ji, C.; Zhang, Y.; Pu, X. Synergy effect between tetracycline and Cr (VI) on combined pollution systems driving biochar-templated $Fe_3O_4@SiO_2/TiO_2/g-C_3N_4$ composites for enhanced removal of pollutants. *Biochar* 2023, 5, 1.

129. Jamil, T S.; Sharaf El-Deen, S E A. Removal of persistent tartrazine dye by photodegradation on TiO2 nanoparticles enhanced by immobilized calcinated sewage sludge under visible light. Separation Science and Technology. 2016, 51 (10), 1744–1756.

130. Mian, M. M.; Liu, G.; Sewage sludge-derived $TiO_2/Fe/Fe_3C$-biochar composite as an efficient heterogeneous catalyst for degradation of methylene blue. *Chemosphere* 2019, 215, 101–114.

131. Dong, G.; Chi, W.; Chai, D.-f.; Zhang, Z.; Li, J.; Zhao, M.; Zhang, W.; Lv, J.; Chen, S. A novel Ag$_3$BiO$_3$/ZnO/BC composite with abundant defects and utilizing hemp BC as charge transfer mediator for photocatalytic degradation of levofloxacin. *Appl. Surf. Sci.* 2023, 619, 156732.

132. Tang, R.; Gong, D.; Deng, Y.; Xiong, S.; Zheng, J.; Li, L.; Zhou, Z.; Su, L.; Zhao, J. π-π stacking derived from graphene-like biochar/g-C$_3$N$_4$ with tunable band structure for photocatalytic antibiotics degradation via peroxymonosulfate activation. *J. Hazard. Mater.* 2022, 423, 126944.

133. Martínez-Huitle, C. A.; Ferro, S. Electrochemical oxidation of organic pollutants for the wastewater treatment: Direct and indirect processes. *Chem. Soc. Rev.* 2006, 35 (12), 1324–1340.

134. Martínez-Huitle, C. A.; Panizza, M. Electrochemical oxidation of organic pollutants for wastewater treatment. *Curr. Opin. Electrochem.* 2018, 11, 62–71.

135. Qiao, J.; Xiong, Y. Electrochemical oxidation technology: A review of its application in high-efficiency treatment of wastewater containing persistent organic pollutants. *J. Water Process Eng.* 2021, 44, 102308.

136. Deng, F.; Olvera-Vargas, H.; Garcia-Rodriguez, O.; Zhu, Y.; Jiang, J.; Qiu, S.; Yang, J. Waste-wood-derived biochar cathode and its application in electro-Fenton for sulfathiazole treatment at alkaline pH with pyrophosphate electrolyte. *J. Hazard. Mater.* 2019, 377, 249–258.

137. Li, X.; Yang, H.; Ma, H.; Zhi, H.; Li, D.; Zheng, X. Carbonaceous materials applied for cathode electro-Fenton technology on the emerging contaminants degradation. *Process Saf. Environ. Protect.* 2023, 169, 186–198.

138. Luo, Z.; Liu, M.; Tang, D.; Xu, Y.; Ran, H.; He, J.; Chen, K.; Sun, J. High H$_2$O$_2$ selectivity and enhanced Fe^{2+} regeneration toward an effective electro-Fenton process based on a self-doped porous biochar cathode. *Appl. Catal. B: Environ.* 2022, 315, 121523.

139. Gong, Y.; Wan, J.; Zhou, P.; Wang, X.; Chen, J.; Xu, K. Oxygen and nitrogen-enriched hierarchical MoS$_2$ nanospheres decorated cornstalk-derived activated carbon for electrocatalytic degradation and supercapacitors. *Mater. Sci. Semicond. Process.* 2021, 123, 105533.

140. Ji, J.; Li, X.; Xu, J.; Yang, X.; Meng, H.; Yan, Z. Zn-Fe-rich granular sludge carbon (GSC) for enhanced electrocatalytic removal of bisphenol A (BPA) and rhodamine B (RhB) in a continuous-flow three-dimensional electrode reactor (3DER). *Electrochim. Acta* 2018, 284, 587–596.

141. Xing, L.; Wei, J.; Zhang, Y.; Xu, M.; Pan, G.; Li, J.; Li, J.; Li, Y. Boosting active sites of protogenetic sludge-based biochar by boron doping for electro-Fenton degradation towards emerging organic contaminants. *Sep. Purif. Technol.* 2022, 294, 121160.

142. Zhou, W.; Li, F.; Su, Y.; Li, J.; Chen, S.; Xie, L.; Wei, S.; Meng, X.; Rajic, L.; Gao, J.; Alshawabkeh, A. N. O-doped graphitic granular biochar enables pollutants removal via simultaneous H_2O_2 generation and activation in neutral Fe-free electro-Fenton process. *Sep. Purif. Technol.* 2021, 262, 118327.

143. Wu, D.; Yi, X.; Tang, R.; Feng, C.; Wei, C. Single microbial fuel cell reactor for coking wastewater treatment: Simultaneous carbon and nitrogen removal with zero alkaline consumption. *Sci. Total Environ.* 2018, 621, 497–506.

144. Zhang, T.; Liu, Y.; Yang, L.; Li, W.; Wang, W.; Liu, P. Ti–Sn–Ce/bamboo biochar particle electrodes for enhanced electrocatalytic treatment of coking wastewater in a three-dimensional electrochemical reaction system. *J. Clean. Prod.* 2020, 258, 120273.

145. Wei, T.; Meng, Y.; Ai, D.; Zhu, C.; Wang, B. Ball milling Fe_3O_4@biochar cathode coupling persulfate for the removal of sulfadiazine from water: Effectiveness and mechanisms. *J. Environ. Chem. Eng.* 2022, 10 (6), 108879.

146. Wang, T.; Ta, M.; Guo, J.; Liang, L.; Bai, C.; Zhang, J.; Ding, H. Insight into the synergy between rice shell biochar particle electrodes and peroxymonosulfate in a three-dimensional electrochemical reactor for norfloxacin degradation. *Sep. Purif. Technol.* 2023, 304, 122354.

147. Wang, W.; Li, W.; Li, H.; Xu, C.; Zhao, G.; Ren, Y. Kapok fiber derived biochar as an efficient electro-catalyst for H_2O_2 in-situ generation in an electro-Fenton system for sulfamethoxazole degradation. *J. Water Process Eng.* 2022, 50, 103311.

148. Ren, Y.; Lu, P.; Qu, G.; Ning, P.; Ren, N.; Wang, J.; Wu, F.; Chen, X.; Wang, Z.; Zhang, T.; Cheng, M.; Chu, X. Study on the mechanism of rapid degradation of rhodamine B with Fe/Cu@antimony tailing nano catalytic particle electrode in a three dimensional electrochemical reactor. *Water Res.* 2023, 244, 120487.

149. Zhang, C.; Li, H.; Yang, X.; Tan, X.; Wan, C.; Liu, X. Characterization of electrodes modified with sludge-derived biochar and its performance of electrocatalytic oxidation of azo dyes. *J. Environ. Manage.* 2022, 324, 116445.

150. Temur Ergan, B.; Aydin, E. S.; Gengec, E. Improving electro-Fenton degradation performance using waste biomass-derived-modified biochar electrodes: A real environment textile water treatment. *J. Environ. Chem. Eng.* 2023, 11 (6), 111439.

151. Jiang, H.; Chen, H.; Duan, Z.; Huang, Z.; Wei, K. Research progress and trends of biochar in the field of wastewater treatment by electrochemical

advanced oxidation processes (EAOPs): A bibliometric analysis. *J. Hazard. Mater. Adv.* 2023, 10, 100305.

152. Wang, X.; Zhao, Z.; Wang, H.; Wang, F.; Dong, W. Decomplexation of Cu-1-hydroxyethylidene-1,1-diphosphonic acid by a three-dimensional electrolysis system with activated biochar as particle electrodes. *J. Environ. Sci. – China* 2023, 124, 630–643.

CHAPTER 5

Application in Contaminated Soil Remediation

Liming Sun, Fengbo Yu, Chao Jia,
Shaobin Wang, and Xiangdong Zhu

5.1 INTRODUCTION

Soil contamination has emerged as a critical global environmental issue, posing significant risks to human health and ecological systems.[1, 2] Rapid industrialization and urbanization have resulted in the discharge of diverse pollutants, including heavy metals, organic compounds, and pesticides, into the soil environment.[3, 4] These contaminants can accumulate in the soil, leading to soil degradation, reduced agricultural productivity, and threats to food safety. Furthermore, the migration of pollutants from soil to groundwater and the food chain poses far-reaching implications to human health.[5] Therefore, effective remediation strategies are urgently needed to address soil contamination and restore soil quality.

Solid waste-based materials, derived from industrial, municipal, and agricultural activities, offer a versatile and eco-friendly arsenal for the remediation of contaminated soils.[6] These materials, possessing unique structural and physicochemical properties, provide a versatile platform for removing and transforming a wide range of contaminants through adsorption/passivation, catalytic oxidation, and catalytic reduction.

The application of solid waste-based materials in soil remediation is a multifaceted endeavour. In the context of adsorption/passivation, these

DOI: 10.1201/9781003535409-5

materials exhibit a strong affinity for pollutants, facilitated by their extensive surface areas, complex pore structures, and reactive surface functional groups. Contaminants are effectively immobilized through physical adsorption, electrostatic interactions, and complexation, thereby reducing their bioavailability and environmental mobility.[7] The effectiveness of this process is contingent upon the interplay between the properties of the solid waste materials, the characteristics of the contaminants, and the prevailing soil conditions.

In contrast, catalytic oxidation utilises the oxidative capacity of solid waste-based materials to degrade organic pollutants into less hazardous substances. This advanced oxidation process involves the generation of Reactive Oxygen Species (ROS), which are produced by the catalytic action of the materials.[8] The effectiveness of solid waste-based materials in this context depends on their catalytic activity, which can be enhanced through structural modifications, heteroatom doping, or the incorporation of metal nanoparticles to promote the generation and stability of reactive species.

Catalytic reduction employs solid waste-based materials as catalysts,[9, 10] thereby facilitating the transformation of hazardous pollutants into by products under mild reaction conditions involving the materials. This approach is particularly advantageous for the treatment of persistent organic pollutants and heavy metals, offering high efficiency and cost-effectiveness. The catalytic efficiency of these materials depends on their composition, structure, and surface properties, which determine the dynamics of electron transfer and the activation of pollutants for reduction.

The pursuit of optimizing the performance of solid waste-based materials for soil remediation is an ongoing endeavour. Through rigorous evaluation, characterisation, and modification strategies, the catalytic properties of these materials can enhance soil remediation. Innovations in material synthesis, the elucidation of structure-activity relationships to enhance catalytic performance, and the incorporation of additional substances or energy inputs include approaches being explored to improve the efficacy of these materials in real-world applications.

This chapter provides a comprehensive review of the application of solid waste-based materials in the remediation of contaminated soil. The adsorption/passivation, catalytic oxidation, and catalytic reduction mechanisms are discussed, highlighting the potential of these materials as sustainable and cost-effective solutions in soil remediation. The chapter also explores the challenges and prospects in this field, highlighting the need for further optimization and integration of solid waste-based materials in soil remediation strategies.

5.2 ADSORPTION/PASSIVATION

Soil pollution is one of the serious environmental problems facing contemporary society, posing a considerable threat to ecosystems and to human health. In the soil remediation process, solid waste-based materials have attracted much attention because of their unique structure and excellent physical and chemical properties. Various types of contaminants present in the soil, including organic and inorganic contaminants, can be removed by physical adsorption, pore filling, π-π bond interactions, electrostatic interaction, precipitation, complexation, ion exchange, and so on.[11, 12]

5.2.1 Pollutant Adsorption Mechanism on Solid Waste-Based Materials in Soil

The pore structure of solid waste-based materials can provide a large number of adsorption sites, which can allow contaminants to be adsorbed on the surface of waste-based materials or diffuse into their internal pore structure through physical interactions such as van der Waals forces or hydrogen bonds.[13] This adsorption is a non-chemical process that is usually affected by various factors such as the physical and chemical properties of solid waste-based materials, pollutant properties and soil environmental conditions. Generally, the larger the average pore size of solid waste-based materials, the more contaminants that can penetrate the internal pores. Physical adsorption is usually divided into single-layer and multi-layer adsorption, which can be fitted by Langmuir and Freundlich models.[14]

As one of the essential mechanisms for waste-based materials to adsorb soil contaminants, electrostatic adsorption mainly refers to the uneven distribution of charges due to the difference in electrical properties between waste-based materials and soil contaminants, resulting in electrostatic attraction.[15] Specifically, waste materials are negatively charged, while many soil contaminants may be positively charged or polar molecules, and this electrical difference causes them to attract each other on contact. Therefore, this effect can effectively capture and immobilize contaminants in the soil on the surface of waste-based materials, thereby reducing their mobility and biological toxicity in the soil.

The precipitation reaction usually refers to the removal of part of the contaminants in the form of precipitation after contact with waste-based materials. This process may be due to the chemical properties or physical structure of waste-based materials surface interacting with the contaminants, causing the contaminants to precipitate on the waste-based materials surface, thereby reducing the contaminant mobility in the soil. In general, the surface of waste-based materials has some active functional groups, such as -OH, -COOH or -NH$_2$, which promote chemical adsorption reactions with soil contaminants.[16] This chemical adsorption causes contaminant molecules to precipitate on the waste-based materials surface, thereby removing them from the soil. In addition, the pore structure of waste-based materials also promotes the precipitation of soil contaminants.

The complexation of waste-based materials with soil contaminants is an important immobilization mechanism that plays an important role in soil remediation and contaminant control. This process typically involves interactions between pollutant molecules and waste-based material surface functional groups (such as -OH and -COOH). This effect often results in contaminant molecules being firmly anchored on the waste-based material surfaces in the form of complexes, thereby reducing their migration and release in the soil.[17] Notably, this process is affected by many factors, including the soil environmental conditions and the properties of solid waste-based materials and pollutant. Different types of waste-based materials have different content and types of functional groups, so complexation with various contaminants may also differ.

The ion exchange between solid waste-based materials and soil contaminants usually involves the exchange between acidic oxygen-containing functional groups on the solid waste-based materials surface and heavy metal ions or cationic organic contaminants. In this reaction, the charged functional groups on the solid waste-based material surfaces can adsorb the ions formed by ionization and release ions of the same charge.[15] For example, the -OH on the surface of solid waste-based materials can form complexes with cations formed by soil heavy metals and release negative ions on the surface of solid waste-based materials.

Notably, the removal mechanism of solid waste-based materials for contaminants in soil is relatively complex. It is often the joint action of multiple mechanisms that can promote the effective removal of contaminants. When removing the same contaminant from soil, differences in the physical and chemical properties of solid waste-based materials may also lead to different adsorption mechanisms.

5.2.2 Effect of Solid Waste-Based Materials on Adsorption of Soil Contaminants

The adsorption effect of solid waste-based materials on soil contaminants is affected by many factors, including the properties of solid waste-based materials and contaminants, and soil environmental.

The properties of solid waste-based materials, such as pore structure, specific surface area, pH value, and microstructure, are important for their ability to adsorb contaminants.[18, 19] Generally, solid waste-based materials have a porous structure, including micropores, mesopores and macropores. These pore structures provide a large number of adsorption sites, which is beneficial to improving the adsorption capacity and rate of the material.[20] The microporous structure can effectively limit the movement of pollutant molecules and provide more specific surface area, thereby improving the material's adsorption capacity for soil contaminants. Then, the microstructure of solid waste-based materials, such as crystal structure and graphitisation, may also affect its surface activity and stability, ultimately changing its adsorption performance for soil contaminants. In addition, the surface functional groups (such as -OH and -COOH) of solid waste-based materials may have specific interactions with contaminants, which lead to chemical adsorption, complexation or ion exchange reactions, thus affecting the adsorption efficiency and selectivity of the materials.

The chemical properties of contaminants, including molecular structure, functional groups, and polarity, determine the type and intensity of interactions between them and solid waste-based materials.[21] For example, contaminants containing functional groups may hydrogen-bond or ionic-bond with the surface functional groups of solid waste-based materials, making them more likely to be adsorbed. The molecular size of the contaminant also affects its interaction with solid waste-based materials. Smaller molecules can more easily enter the micropores of solid waste-based materials for adsorption, while larger molecules may be more prone to physical adsorption or complexation on the surface of the material. In addition, the high concentration of contaminants in the soil will increase the chance of contact with solid waste-based materials, which is more conducive to their removal from the soil. However, when multiple pollutants are present in the soil, it is not beneficial to the adsorption process of the material because competition between pollutants may occur. In addition, the charge state of the contaminant will also affect its interaction with the charge of the solid waste-based materials. Contaminants with opposite charges are more likely to be adsorbed to material surfaces through electrostatic interactions.

The influence of soil properties on the adsorption of contaminants in soil by solid waste-based materials are complex and diverse. Different types of soil vary greatly in structure, particle size distribution, and organic matter content, which can affect the interaction between solid waste-based materials and contaminants in the soil.[22] For example, soil rich in organic matter may interact with organic pollutants, while soil with smaller particle sizes increases the contact surface area with solid waste-based materials, promoting the adsorption of contaminants. In addition, pH is an important factor affecting the chemical form of contaminants in the soil and the surface charge of solid waste-based materials, which in turn affects the interaction between them.[23] The soil moisture also affects the rate of dispersion and dispersion between solid waste-based materials and contaminants in the soil. Generally speaking, higher soil moisture will increase the contact opportunities between materials and contaminants in the soil, which is conducive to the occurrence of adsorption. The Cation Exchange Capacity (CEC) of soil reflects the ability of the surface of soil particles to adsorb ions.[24] Generally speaking, soil particles with higher CEC may compete with solid waste-based materials for adsorption sites, thereby reducing the adsorption efficiency of the material. In summary, the effects of soil properties on the adsorption of contaminants in soil by solid waste-based materials are multifaceted and may interact with each other.

5.2.3 Current Status and Development Trends of Solid Waste-Based Materials for Adsorption of Contaminants in Soil

At present, the application of solid waste-based materials in soil contaminant remediation has made some progress and has been widely used in various contaminated sites. Guo et al.[25] confirmed the advantages of solid waste-based materials in adsorbing organic matter in soil. Benefit from the high specific surface area, well-developed pore structure, and controllable surface charge and functional groups, solid waste-based materials show strong resistance to some common organic pollutants, including petroleum hydrocarbons, pesticides, and organic solvents, which are mainly achieved by physical or chemical adsorption.[26, 27] Wu et al.[28] effectively removed bisphenol A from the soil by adding biochar and reduced its migration and diffusion in the soil. Some organic contaminants usually have high polarity and can form hydrogen bonds or other interactions with the functional groups on the surface of solid waste-based materials, which can greatly improve the materials' adsorption capacity.[29]

Trace metals or heavy metals in soil are different from organic matter and cannot be degraded by microorganisms or mineralized into other forms, and can only persist in the environment.[30] For these inorganic pollutants, their removal by solid waste-based materials are mainly achieved by physical adsorption and ion exchange.[31] For example, solid waste-based materials can effectively reduce Cd content in soil and reduce its toxic effects on plants. Zhang et al.[32] modified biochar with Fe-based materials to increase the pH value of acidic soil and improve the Cation Exchange Capacity (CEC) of soil and organic matter environment, thereby completing the rapid and effective immobilization of metal ions (Cu, Pb, Zn) in the soil. In addition, some heavy metal ions usually have multiple coordination states and can form coordination bonds or complexes with functional groups of solid waste-based materials, being thereby adsorbed on the material surface. In addition, to facilitate the recycling of solid waste-based materials, Liu et al.[33] used $FeCl_3$ as a magnetic material and combined this with gelation to form a magnetic macro-porous biochar sphere with a layered structure. This easily magnetically recyclable composite material can effectively remove heavy metals (including Cd(II) and As(V)) from contaminated agricultural soils.

In summary, as important soil remediation materials, solid waste-based materials have good adsorption properties and have shown great potential in the removal of soil contaminants. Further, the preparation process and properties of solid waste-based materials can be optimized to improve their adsorption efficiency and selectivity to adapt to different types of contaminants and soil environments. In addition, new technologies such as machine learning can be combined with other remediation technologies (such as phytoremediation and bioremediation) to establish a multi-dimensional system based on solid waste-based materials to accelerate the soil remediation process. With the continuous advancement of relevant research, the application of solid waste-based materials in soil contamination remediation will be more extensive, making greater contributions to protecting the global environment and human health.

5.3 CATALYTIC OXIDATION

5.3.1 Catalytic Oxidation Mechanism

With the rapid progress in industrialization and urbanization, a substantial quantity of organic contaminants have been discharged into the environment.[34, 35] The primary reservoir for these organic pollutants is environmental soil. Simultaneously, a considerable volume of solid waste, encompassing municipal, industrial, electronic, biomedical, and biomass waste, is generated.[36] This solid waste contains numerous valuable components, such as recyclable metals and organic compounds, which can be utilised for synthesising catalytic materials. Consequently, to address soil pollution and facilitate resource recycling, various methodologies have been under development for an extended period. These methods include absorption, solidification, catalytic oxidation, and passivation, utilising solid waste materials.[37] Among these approaches, catalytic oxidation by solid waste materials assumes a significant role in organic pollutant degradation due to their potent oxidation capabilities.

Catalytic oxidation, also known as the Advanced Oxidation Process (AOP), emerged around 1980 as an innovative approach for treating high-concentration toxic and hazardous pollutants in wastewater.[38] After a lot of research and development, AOPs have been also an effective means of soil remediation. AOPs typically consist of organic pollutants, oxidizing agents, catalytic materials, and external conditions (such as ultrasound and illumination). Currently, the commonly used oxidants include four types: permanganate, hydrogen peroxide, persulfate, and ozone. These oxidants undergo reactions via two different mechanisms. Permanganate oxidation degradation primarily occurs through electron transfer of the oxidant, wherein it directly reacts with pollutants in the soil through a conventional redox reaction. The oxidation degradation process of hydrogen peroxide and persulfate involves the formation of free radicals or non-free radicals. Free radicals are typically unstable and highly reactive species, capable of strong interactions with pollutants. Both mechanisms are also involved in ozone oxidation degradation.

Based on their compositions, AOPs can also be categorized into various types, including Fenton oxidation, Fenton-like oxidation, peroxydisulfate-based oxidation, and photochemical oxidation, and so forth.[38] The main catalytic processes of AOPs involve three steps: (1) Adsorption of the oxidizing agent onto the catalytic materials; (2) Activation of the oxidizing

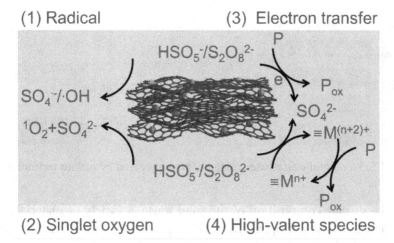

FIGURE 5.1 Persulfate activation mechanism by heterogeneous materials[40]. Note: P represents organic pollutants, P_{ox} represents the degraded products, and M represents metal.

agent post-adsorption, leading to the generation of ROS; (3) Oxidation of organic pollutants into inorganic substances by ROS, owing to their high oxidation-reduction potential.[39] Additionally, there are four reaction pathways based on the ROS species involved, including radical (OH and SO_4^{-}), non-radical ($^1O_2$), electron transfer, and high-valent metal species. For instance, the activation mechanism of persulfate using heterogeneous materials is illustrated in Figure 5.1.[40] Therefore, the selection of catalytic materials significantly influences the generation of ROS for organic pollutant removal. In this chapter, we focus solely on introducing solid waste-based materials for the catalytic oxidation of soil organic pollutants.

5.3.2 Solid Waste-Based Materials for Catalytic Oxidation

Solid waste-based materials are derived from the solid waste generated through diverse production processes, daily life, and other activities. The transformation of solid waste into solid waste-based materials is pivotal for sustainable development, as it mitigates waste generation, preserves resources, and advances circular economy. As a significant application, solid waste-based materials can be used in catalytic oxidation for the removal of soil organic pollutants.

Some kinds of typical solid waste are municipal solid waste, industrial solid waste, electronic waste, biomedical waste, and biomass waste. Based

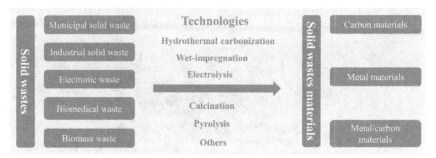

FIGURE 5.2   Solid waste-based materials were prepared by various technologies.

on the diverse composition of the solid waste, a variety of materials can be prepared for soil applications. Zhang and colleagues[41] reported the synthesis of cordierite using incineration fly ash from municipal solid waste as an additive to enhance the catalytic oxidation of volatile organic compounds. Due to the high content of metal or carbon elements in a wide range of raw solid waste, solid waste-based materials can be prepared using various technologies, such as hydrothermal carbonization, wet impregnation, electrolysis, calcination, and pyrolysis. These materials mainly encompass carbon materials, metal materials, and metal/carbon composites (Figure 5.2).

Carbon materials include hydrochar, biochar, activated carbon, graphene, oxidized graphene, and carbon nanotubes, which are typically synthesised through hydrothermal carbonization or pyrolysis from solid waste with high carbon, oxygen, and hydrogen content. Subsequent treatment at high temperatures and in anaerobic atmospheres leads to carbonization and graphitisation of solid waste, along with the generation of surface oxygen functional groups. The degree of graphitisation and the presence of surface oxygen functional groups are key factors influencing catalytic oxidation reactions. Numerous studies have indicated that a higher degree of graphitisation can facilitate electron transfer processes, while surface oxygen functional groups, such as -C=O, can serve as the primary active sites for generating reactive oxygen species. Balancing the degree of graphitisation and surface oxygen functional groups is crucial and remains an area of active research.

Metal materials typically consist of transition metal elements, such as Fe-based, Cu-based, Mn-based, and Co-based catalysts, which exhibit low oxidation-reduction potentials. Metal active sites can activate oxygen agents into reactive oxygen species by providing electrons. Following electron transfer, the valence state of the metal element increases, and reactivity diminishes. Therefore, maintaining the stability of catalytic materials remains a challenge for achieving high-efficiency applications. Moreover, the low utilisation efficiency of metal materials due to bulk particles necessitates the further development of metal/carbon materials. Carbon substrates can offer numerous anchoring sites for loading nanoparticles, clusters, or single atoms. These anchoring sites on carbon substrates typically include nitrogen, sulfur, phosphorus, oxygen, and defects. Through modification methods, catalytic efficiency and stability can be rapidly enhanced. Consequently, the properties of solid waste-based materials play a crucial role in determining their catalytic ability.

5.3.3 Solid Waste-Based Materials in Catalytic Oxidation

We delineated three approaches for soil remediation using catalytic oxidation methods: (1) Direct synthesis of solid waste-based materials for soil application; (2) Synthesis of solid waste-based materials by incorporating additional substances to enhance catalytic properties. For instance, some nitrogen-doping agents were added during synthesis to produce single-atom materials; (3) Addition of extra energy in catalytic oxidation processes to stimulate solid waste-based materials to generate ROS. To elucidate the structure-properties relationships, detailed reports from researchers were referenced.

Polycyclic Aromatic Hydrocarbons (PAHs) are prevalent in soil ecosystems and pose significant threats to human health. Mahmoud Mazarji and colleagues[42] utilised biochar derived from sunflower husks as a catalytic material to degrade PAHs using the Fenton reaction activated by H_2O_2. Additionally, biochar can synergize with Fe iron to enhance PAH removal efficiency by immobilizing heavy metals due to their pore structure and surface oxygen functional groups. Given the high Fe content in soil,[43] biochar serves as an effective modifier for practical soil applications. To enhance catalytic oxidation properties, various modified materials were prepared from solid waste with additional substances. For example, Xu and co-workers[44] developed ABD-Fe/N co-doped biochar to oxidize petroleum hydrocarbons in soil. The ABD-Fe/N co-doped biochar exhibited a strong capacity to activate H_2O_2, generating ·OH with excellent oxidation potential, resulting in maximum petroleum hydrocarbon oxidation amounts of 18.5–19.6 g/kg. The "adsorption-oxidation" mechanism played a pivotal role in mitigating the inhibitory effects of soil organic matter. Moreover, oxidation reactions can be enhanced by additional energy sources such as light or electron-induced reactions.[45] Some researchers utilised sunlight to induce ROS production, offering an environmentally friendly and sustainable approach. For instance, Kong and colleagues[46] synthesised zinc-iron layered double hydroxide for As(III) removal, demonstrating a high removal capacity of 134.5 mg/g towards As(III) with the assistance of light.

5.3.4 Improvement of Catalytic Oxidation

Currently, the soil remediation sector in our nation is in its early stages, characterised by various imperfections. The advancement and application of diverse remediation technologies are still undergoing engineering trials, and relevant standards and regulations require continuous refinement. Various methods, including physical, chemical, and biological approaches, are employed. However, in terms of efficacy and cost-effectiveness, these methods pose challenges for treating polluted soil. To mitigate treatment costs, catalytic materials are often derived from solid waste. Nevertheless, during oxidation processes, solid waste-based materials may introduce secondary environmental pollution due to the release of significant amounts of oxidizing agents and hazardous substances. Low catalytic efficiency necessitates the addition of substantial oxidizing agents to augment oxidation reactions, potentially leading to soil contamination by inactive oxidizing agents and their byproducts. Moreover, solid waste-based materials may release metals into the soil after reactions or under acidic conditions, posing further environmental risks. Consequently, researchers are exploring novel materials to enhance treatment efficiency and reduce oxidizing agents and metal leaching.

Single-atom materials present a promising solution to these challenges, offering high utilisation efficiency and excellent stability due to strong metal-substrate interactions.[47] Wang and colleagues[48] synthesised a novel cobalt–carbon-based single-atom catalyst derived from mild pyrolysis of spent coffee grounds soaked in a cobalt solution. Spent coffee grounds contain abundant nitrogen and sulfur atoms that serve as anchoring sites for cobalt metal, resulting in a low-cost synthesis process and efficient treatment efficacy and stability of the cobalt single-atom material. Furthermore, some integrated techniques have been employed to address inherent limitations. Zhou and co-workers utilised single-atom iron on three-dimensional nitrogen-doped carbon supports for microwave soil remediation, achieving 99.9% removal of chloramphenicol within 5 minutes through the synergistic effects of hot spots and ROS.[49] As the soil remediation industry in our country is still emerging, strategies for efficiently remediating polluted soil are continuously evolving. Apart from synthesising solid waste materials, sustainable and cost-effective technologies, particularly in-situ repair methods, need to be developed.

5.4 CATALYTIC REDUCTION

5.4.1 Mechanism of Catalytic Reduction

Catalytic reduction is a widely applied remediation technology for contaminated soil,[50] utilising catalysts to facilitate the reduction reaction under mild conditions and transform toxic and hazardous pollutants into innocuous substances. In comparison with conventional physicochemical remediation techniques, catalytic reduction offers several advantages, including milder reaction conditions, high efficiency, and low costs, thus exhibiting extensive application potential in the realm of soil remediation.[51]

The catalytic reduction process involves three stages: adsorption, activation, and reduction.[52] Initially, pollutant molecules are adsorbed and concentrated on the catalyst surface. This step typically entails reversible physical adsorption, with the adsorption force predominantly arising from weak interactions such as van der Waals forces.[53] Following the adsorption of pollutant molecules on the catalyst surface, their molecular configuration and electron cloud distribution undergo certain alterations, establishing favourable conditions for the subsequent activation stage. Then, the adsorbed pollutant molecules undergo activation, a crucial step in catalytic reduction. The activation process fundamentally involves electron transfer between the catalyst and pollutant molecules, weakening the chemical bonds within the pollutant molecules and lowering the activation energy of the reduction reaction. Active sites on the catalyst surface, such as unoccupied or partially occupied d orbitals of transition metals, can form coordination bonds with pollutant molecules.[54] This promotes the polarization of the pollutants, reduces their electron cloud density, and facilitates electron transfer and chemical bond cleavage.[55] Finally, in the presence of reducing agents like hydrogen (H_2),[56] the activated pollutant molecules are reduced and converted into harmless smaller molecules. The catalyst serves as an electronic intermediary in the reduction reaction, accepting electrons from the reducing agent and subsequently transferring them to the pollutant molecules, enabling their reduction and transformation.

The catalyst plays a crucial role in catalytic reduction reactions. The composition and structure of the catalyst have a decisive effect on the catalytic reduction efficiency. Transition metals, oxides, and sulfides are excellent catalytic reduction catalysts due to their variable valence states and strong coordination ability.[57] In a catalytic reduction reaction, the active sites on the catalyst surface provide electrons to the contaminant molecules, weakening their internal chemical bonds and making them easier to reduce. Furthermore, the lattice structure and pore structure of the catalyst also play significant roles in the catalytic reduction reaction. The lattice structure provides a favourable reaction site, while the pore structure offers a larger surface area, increasing the contact area with pollutants and thereby enhancing the efficiency of the catalytic reduction reaction.[58]

The mechanism of catalytic reduction reaction can be summarized as follows: (1) adsorption of pollutant molecules on the catalyst surface; (2) activation of the catalyst and adsorption of pollutant molecules; (3) reduction and transformation of pollutants by reducing agents. The composition and structure of the catalyst have a decisive effect on the catalytic reduction efficiency, with transition metals, oxides, and sulfides being excellent catalytic reduction catalysts due to their variable valence states and strong coordination ability.

5.4.2 Solid Waste-Based Materials for Catalytic Reduction

Traditional catalytic reduction technology usually uses precious metals (such as Pt and Pd) or metal oxides as catalysts.[59] Although these catalysts have good catalytic effects, high costs and large resource consumption limit their large-scale applications. Meanwhile, China generates a large amount of industrial and agricultural solid waste annually.[60] Therefore, properly disposing of and recycling this solid waste has become an urgent issue that needs to be addressed. Solid waste has a wide range of sources, low prices, diverse components. The development of solid waste-based catalytic reduction materials offers multiple benefits. It not only realises the concept of "waste into treasure" and reduces the environmental load, but also lowers the catalyst cost and improves the economic feasibility of catalytic reduction technology.

As the primary source of electricity in China, coal-fired power plants generate a substantial amount of solid waste annually, including coal waste slag, fly ash, and furnace slag.[61] These waste residues are abundant in iron, calcium, silicon, aluminium, and other elements. After proper modification, they can be endowed with the redox properties necessary for catalytic reduction. Water-washing fly ash significantly enhances the catalytic reduction efficiency of nitrophenol pollutants, achieving a removal rate close to 90%.[62] This finding offers a new perspective on the high-value utilisation of fly ash. Another major source of solid waste is the iron and steel industry. Blast-furnace slag, steel slag, and other waste slags contain a large quantity of transition metal oxides and exhibit excellent catalytic reduction properties. It has been discovered that iron-modified steel slag can effectively catalyse the reduction of trichloroethylene and other organic pollutants,[63] showing promising application prospects in soil remediation.

Developing a circular economy through source substitution and process enhancement is an important measure to realise resource recycling and pollution control simultaneously. Using solid waste to prepare catalytic materials has two main advantages. First, it can reduce the consumption of primary resources and alleviate the environmental load. Second, solid waste-based catalytic materials can significantly improve the degradation rate and removal efficiency of pollutants, reduce emissions, and enhance environmental quality.[64] Therefore, vigorously developing and promoting solid waste-based catalytic materials is crucial for advancing the coupled development of environmental governance and circular economy.

5.4.3 Material Properties and Evaluation of Solid Waste-Based Materials

Performance evaluation is a crucial step in the development and application of solid waste-based catalytic reduction raw materials. It is necessary to comprehensively characterise the physical and chemical properties, and test catalytic activities of these materials at both micro and macro levels. At the microscopic level, advanced characterisation methods such as X-Ray Diffraction (XRD), X-ray Photoelectron Spectroscopy (XPS), Scanning Electron Microscopy (SEM), and Transmission Electron Microscopy (TEM) can be employed to thoroughly analyse the crystalline structure, elemental composition, surface morphology, pore structure, and other microscopic properties of the materials. These analyses provide an essential basis for understanding the catalytic mechanism of the materials. Specifically, XRD can determine the crystal type and grain size, while XPS can analyse the elemental valence and chemical environment on the surface of the materials. Moreover, SEM and TEM can directly observe the surface morphology and microstructure of the materials, offering direct evidence for material modification and optimization.

At the macroscopic level, parameters such as specific surface area, pore volume, pH, and redox potential of the materials have significant effects on their catalytic performance and need to be systematically measured and evaluated. The low-temperature N_2 adsorption-desorption technique can determine the specific surface area and pore size distribution of the materials, providing a basis for evaluating their adsorption properties.[65] The NH_3/CO_2 temperature-programmed desorption technique can characterise the acid-base properties on the surface of materials and elucidate the interaction mechanism between materials and pollutants.[66] The redox properties of materials can be investigated by the H_2 temperature-programmed reduction technique, which can provide guidance for optimizing the electron transport and catalytic reduction processes of the materials.[67] Through comprehensive analysis of the physical, chemical, and surface properties of the materials, the catalytic mechanism can be thoroughly understood, and a scientific basis can be provided for the improvement and optimization of material properties.

Kinetic parameters such as conversion, rate constant, and apparent activation energy of the catalytic reduction reaction system are direct indicators for evaluating the catalytic activity of solid waste-based materials. A continuous flow reaction device is employed to systematically study the catalytic reduction performance of materials under different reaction conditions (for example, temperature, pH, pollutant concentration), which can comprehensively investigate the feasibility and effectiveness of materials in practical applications. By comparing the catalytic activity of different materials, the best-performing material system can be selected to provide technical support for engineering applications. Additionally, catalyst stability is another crucial index for evaluating its performance. Through repeated use experiments, the anti-poisoning ability and regeneration performance of the material can be investigated, and the service life and economic cost of the material can be assessed, providing an essential reference for practical applications.

The structure-activity relationship between material structure and properties is a crucial aspect of research on solid waste-based catalytic materials. Through systematic analysis of the intrinsic relationships amongst material composition, structure, and properties, the catalytic mechanism can be elucidated, and the determining factors linking material structure to properties can be identified. This analysis provides theoretical guidance for the design and synthesis of novel, efficient catalytic raw materials. With an in-depth understanding of the structure-activity relationship, targeted material modifications and optimizations can be performed, such as regulating the composition, crystal shape, morphology, and defects. These optimizations aim to enhance the performance of solid waste-based catalytic reduction materials and promote their widespread application in the field of environmental governance.

5.4.4 Improvement of Catalytic Reduction

To address their inherent shortcomings, the catalytic properties of solid waste-based materials can be significantly enhanced through various modification techniques, including composition optimization, structural regulation, and interface modification. Metal/metal oxide nanoparticles, such as nZVI, Fe_3O_4, TiO_2, and ZnO, in combination with element-doped biochar (for example, N, P, B, S), are widely pursued nanohybrids that exhibit promising properties and offer numerous benefits for soil remediation.[68] Acid treatment can optimize surface functional groups, modulate pore structure, increase specific surface area, and promote pollutant adsorption and enrichment.[69] Depositing nano-metal particles (for example, Pd, Pt, Ni) on the surface of the material can significantly enhance metal dispersion and increase the number of reactive sites.[70] Combining different kinds of solid waste can achieve complementary advantages and produce synergistic effects. Alternatively, compositing with other functional materials (for example, carbon, zeolite, clay) can endow the material with new functions (for example, adsorption, photocatalysis), expanding its application range.[71] Pretreatments such as heat treatment, water washing, and pickling can remove impurities from solid waste, modulate particle size, expose more active sites, thereby enhancing mass transfer performance.[72]

5.5 CONCLUSION

In conclusion, solid waste-based materials have shown great potential in the remediation of contaminated soil through various methods, including adsorption/passivation, catalytic oxidation, and catalytic reduction. In the adsorption/passivation process, solid waste materials can effectively remove both organic and inorganic contaminants from soil through physical adsorption, electrostatic interaction, precipitation, and complex sorption. The removal efficiency for contaminants of soil is influenced by the properties of the solid waste-based materials (for example, pore structure, surface area, surface functional groups), the properties of the contaminants (for example, molecular structure, polarity, charge), and the soil environmental conditions (for example, pH, organic matter content, moisture). Optimization of the preparation process and properties of solid waste-based materials can enhance their adsorption efficiency and selectivity for different contaminants and soil environments.

Catalytic oxidation, involving the activation of oxidizing agents by solid waste-based materials to generate ROS, is an effective approach for degrading organic pollutants in soil. Various solid waste-based materials, such as carbon materials, metal materials, and metal/carbon composites, have been developed for this application. The catalytic performance is determined by factors such as the degree of graphitisation, surface oxygen functional groups, and metal loading. Modification of solid waste-based materials with additional substances (for example, nitrogen-doping) or the application of extra energy (for example, light) can further enhance their catalytic oxidation efficiency.

Catalytic reduction, utilising solid waste-based materials as catalysts to facilitate the reduction of contaminants under mild conditions, is another promising strategy for soil remediation. The catalytic reduction process also involves adsorption, activation, and reduction stages. Solid waste-based materials derived from coal waste slag, fly ash, and steel slag have shown excellent catalytic reduction properties for the removal of organic pollutants and heavy metals. Comprehensive characterisation and evaluation of the physical, chemical, and surface properties of these materials are essential for understanding their catalytic mechanism and optimizing their performance.

REFERENCES

1. You, Q.; Yan, K.; Yuan, Z.; Feng, D.; Wang, H.; Wu, L.; Xu, J. Polycyclic aromatic hydrocarbons (PAHs) pollution and risk assessment of soils at contaminated sites in China over the past two decades. *J. Clean. Prod.* 2024, 450, 141876.

2. Yaashikaa, P. R.; Kumar, P. S. Bioremediation of hazardous pollutants from agricultural soils: A sustainable approach for waste management towards urban sustainability. *Environ. Pollut.* 2022, 312, 120031.

3. Tajik, S.; Beitollahi, H.; Nejad, F. G.; Sheikhshoaie, I.; Nugraha, A. S.; Jang, H. W.; Yamauchi, Y.; Shokouhimehr, M. Performance of metal-organic

frameworks in the electrochemical sensing of environmental pollutants. *J. Mater. Chem. A* 2021, 9 (13), 8195–8220.

4. Maqsood, Q.; Sumrin, A.; Waseem, R.; Hussain, M.; Imtiaz, M.; Hussain, N. Bioengineered microbial strains for detoxification of toxic environmental pollutants. *Environ. Res.* 2023, 227, 115665.

5. Zheng, S.; Wang, Q.; Yuan, Y.; Sun, W. Human health risk assessment of heavy metals in soil and food crops in the Pearl River Delta urban agglomeration of China. *Food Chem.* 2020, 316, 126213.

6. Ramírez Calderón, O. A.; Abdeldayem, O. M.; Pugazhendhi, A.; Rene, E. R. Current updates and perspectives of biosorption technology: An alternative for the removal of heavy metals from wastewater. *Curr. Pollut. Rep.* 2020, 6 (1), 8–27.

7. Huang, F.; Dong, F.; Chen, L.; Zeng, Y.; Zhou, L.; Sun, S.; Wang, Z.; Lai, J.; Fang, L. Biochar-mediated remediation of uranium-contaminated soils: Evidence, mechanisms, and perspectives. *Biochar* 2024, 6 (1), 16.

8. Wang, K.; Li, H.; Yu, W.; Ma, T. Insights into structural and functional regulation of chalcopyrite and enhanced mechanism of reactive oxygen species (ROS) generation in advanced oxidation process (AOP): A review. *Sci. Total Environ.* 2024, 919, 170530.

9. Ji, Z.; Zhang, G.; Liu, R.; Qu, J.; Liu, H. Potential applications of solid waste-based geopolymer materials: In wastewater treatment and greenhouse gas emission reduction. *J. Clean. Prod.* 2024, 443, 141144.

10. Xiao, S.; Ma, K.; Tang, X.; Shaw, H.; Pfeffer, R.; Stevens, J. G. The lean catalytic reduction of nitric oxide by solid carbonaceous materials. *Appl. Catal. B Environ.* 2001, 32 (1), 107–122.

11. Vithanage, M.; Herath, I.; Almaroai, Y. A.; Rajapaksha, A. U.; Huang, L.; Sung, J.-K.; Lee, S. S.; Ok, Y. S. Effects of carbon nanotube and biochar on bioavailability of Pb, Cu and Sb in multi-metal contaminated soil. *Environ. Geochem. Health* 2017, 39, 1409–1420.

12. He, L.; Zhong, H.; Liu, G.; Dai, Z.; Brookes, P. C.; Xu, J. Remediation of heavy metal contaminated soils by biochar: Mechanisms, potential risks and applications in China. *Environ. Pollut.* 2019, 252, 846–855.

13. Li, N.; He, M.; Lu, X.; Yan, B.; Duan, X.; Chen, G.; Wang, S.; Hou, L. a. Municipal solid waste derived biochars for wastewater treatment: Production, properties and applications. *Resour. Conserv. Recycl.* 2022, 177, 106003.

14. Qiu, Y.; Xiao, X.; Cheng, H.; Zhou, Z.; Sheng, G. D. Influence of environmental factors on pesticide adsorption by black carbon: pH and model dissolved organic matter. *Environ. Sci. Technol.* 2009, 43 (13), 4973–4978.

15. Reynier, N.; Coudert, L.; Blais, J.-F.; Mercier, G.; Besner, S. Treatment of contaminated soil leachate by precipitation, adsorption and ion exchange. *J. Environ. Chem. Eng.* 2015, 3 (2), 977–985.

16. Case, S. D. C.; McNamara, N. P.; Reay, D. S.; Whitaker, J. The effect of biochar addition on N₂O and CO₂ emissions from a sandy loam soil – The role of soil aeration. *Soil Biol. Biochem.* 2012, 51, 125–134.

17. Tong, X.-j.; Li, J.-y.; Yuan, J.-h.; Xu, R.-k. Adsorption of Cu(II) by biochars generated from three crop straws. *Chem. Eng. J.* 2011, 172 (2), 828–834.

18. Uchimiya, M.; Wartelle, L. H.; Klasson, K. T.; Fortier, C. A.; Lima, I. M. Influence of pyrolysis temperature on biochar property and function as a heavy metal sorbent in soil. *J. Agric. Food Chem.* 2011, 59 (6), 2501–2510.

19. Sen, T. K. Agricultural solid wastes based adsorbent materials in the remediation of heavy metal ions from water and wastewater by adsorption: A review. *Molecules* 2023, 28 (14), 5575.

20. Blenis, N.; Hue, N.; Maaz, T. M.; Kantar, M. Biochar production, modification, and its uses in soil remediation: A review. *Sustainability* 2023, 15 (4), 3442.

21. Zhang, Q.; Wang, C. Natural and human factors affect the distribution of soil heavy metal pollution: A review. *Water Air Soil Pollut.* 2020, 231 (7), 350.

22. Liang, M.; Lu, L.; He, H.; Li, J.; Zhu, Z.; Zhu, Y. Applications of biochar and modified biochar in heavy metal contaminated soil: A descriptive review. *Sustainability* 2021, 13 (24), 14041.

23. Hu, Y.; Zhang, P.; Yang, M.; Liu, Y.; Zhang, X.; Feng, S.; Guo, D.; Dang, X. Biochar is an effective amendment to remediate Cd-contaminated soils – A meta-analysis. *J. Soil. Sediment.* 2020, 20 (11), 3884–3895.

24. Bouajila, K.; Hechmi, S.; Mechri, M.; Jeddi, F. B.; Jedidi, N. Short-term effects of Sulla residues and farmyard manure amendments on soil properties: Cation exchange capacity (CEC), base cations (BC), and percentage base saturation (PBS). *Arabian J. Geosci.* 2023, 16 (7), 410.

25. Guo, X.-x.; Liu, H.-t.; Zhang, J. The role of biochar in organic waste composting and soil improvement: A review. *Waste Manage.* 2020, 102, 884–899.

26. Hu, B.; Ai, Y.; Jin, J.; Hayat, T.; Alsaedi, A.; Zhuang, L.; Wang, X. Efficient elimination of organic and inorganic pollutants by biochar and biochar-based materials. *Biochar* 2020, 2, 47–64.

27. Du, J.; Ma, A.; Wang, X.; Zheng, X. Review of the preparation and application of porous materials for typical coal-based solid waste. *Materials* 2023, 16 (15), 5434.

28. Wu, F.; Gong, X.; Meng, D.; Li, H.; Ren, D.; Zhang, J. Effective immobilization of bisphenol A utilising activated biochar incorporated into soil: Combined with batch adsorption and fixed-bed column studies. *Environ. Sci. Pollut. Res.* 2023, 30 (46), 103259–103273.

29. Qiu, M.; Liu, L.; Ling, Q.; Cai, Y.; Yu, S.; Wang, S.; Fu, D.; Hu, B.; Wang, X. Biochar for the removal of contaminants from soil and water: A review. *Biochar* 2022, 4 (1), 19.

30. Wang, M.; Zhu, Y.; Cheng, L.; Andserson, B.; Zhao, X.; Wang, D.; Ding, A. Review on utilisation of biochar for metal-contaminated soil and sediment remediation. *J. Environ. Sci.* 2018, 63, 156–173.

31. Liu, S.; Xie, Z.; Zhu, Y.; Zhu, Y.; Jiang, Y.; Wang, Y.; Gao, H. Adsorption characteristics of modified rice straw biochar for Zn and in-situ remediation of Zn contaminated soil. *Environ. Technol. Innovation* 2021, 22, 101388.

32. Zhang, X.; Xue, J.; Han, H.; Wang, Y. Study on improvement of copper sulfide acid soil properties and mechanism of metal ion fixation based on Fe-biochar composite. *Sci. Rep.* 2024, 14 (1), 247.

33. Liu, Y.; Huang, J.; Xu, H.; Zhang, Y.; Hu, T.; Chen, W.; Hu, H.; Wu, J.; Li, Y.; Jiang, G. A magnetic macro-porous biochar sphere as vehicle for the activation and removal of heavy metals from contaminated agricultural soil. *Chem. Eng. J.* 2020, 390, 124638.

34. Alharbi, O. M. L.; Basheer, A. A.; Khattab, R. A.; Ali, I. Health and environmental effects of persistent organic pollutants. *J. Mol. Liq.* 2018, 263, 442–453.

35. Xiang, L.; Harindintwali, J. D.; Wang, F.; Redmile-Gordon, M.; Chang, S. X.; Fu, Y.; He, C.; Muhoza, B.; Brahushi, F.; Bolan, N.; et al. Integrating biochar, bacteria, and plants for sustainable remediation of soils contaminated with organic pollutants. *Environ. Sci. Technol.* 2022, 56 (23), 16546–16566.

36. Bui, T.-D.; Tseng, J.-W.; Tseng, M.-L.; Lim, M. K. Opportunities and challenges for solid waste reuse and recycling in emerging economies: A hybrid analysis. *Resour. Conserv. Recycl.* 2022, 177, 105968.

37. Nascimento, Í. V. d.; Fregolente, L. G.; Pereira, A. P. d. A.; Nascimento, C. D. V. d.; Mota, J. C. A.; Ferreira, O. P.; Sousa, H. H. d. F.; Silva, D. G. G. d.; Simões, L. R.; Souza Filho, A. G.; et al. Biochar as a carbonaceous material to enhance soil quality in drylands ecosystems: A review. *Environ. Res.* 2023, 233, 116489.

38. Deng, Y.; Zhao, R. Advanced oxidation processes (AOPs) in wastewater treatment. *Curr. Pollut. Rep.* 2015, 1 (3), 167–176.

39. Huang, M.; Li, Y.-S.; Zhang, C.-Q.; Cui, C.; Huang, Q.-Q.; Li, M.; Qiang, Z.; Zhou, T.; Wu, X.; Yu, H.-Q. Facilely tuning the intrinsic catalytic sites of the spinel oxide for peroxymonosulfate activation: From fundamental investigation to pilot-scale demonstration. *Proc. Natl. Acad. Sci. U.S.A.* 2022, 119 (30), e2202682119.

40. Wu, X.; Kim, J.-H. Outlook on single atom catalysts for persulfate-based advanced oxidation. *ACS ES&T Eng.* 2022, 2 (10), 1776–1796.

41. Tian, D.; Yang, Y.; Zhang, J.; Yue, Y.; Qian, G. Synthesis of cordierite using municipal solid waste incineration fly ash as one additive for enhanced catalytic oxidation of volatile organic compounds. *Sci. Total Environ.* 2024, 906, 167420.

42. Mazarji, M.; Minkina, T.; Sushkova, S.; Mandzhieva, S.; Barakhov, A.; Barbashev, A.; Dudnikova, T.; Lobzenko, I.; Giannakis, S. Decrypting the synergistic action of the Fenton process and biochar addition for sustainable remediation of real technogenic soil from PAHs and heavy metals. *Environ. Pollut.* 2022, 303, 119096.

43. Yu, F.; Jia, C.; Wu, X.; Sun, L.; Shi, Z.; Teng, T.; Lin, L.; He, Z.; Gao, J.; Zhang, S.; et al. Rapid self-heating synthesis of Fe-based nanomaterial catalyst for advanced oxidation. *Nat. Commun.* 2023, 14 (1), 4975.

44. Li, X.; Xu, J.; Yang, Z. Efficient oriented interfacial oxidation of petroleum hydrocarbons by functionalized Fe/N co-doped biochar-mediated heterogeneous Fenton for heavily contaminated soil remediation. *Chem. Eng. J.* 2022, 450, 138466.

45. Brillas, E. Fenton, photo-Fenton, electro-Fenton, and their combined treatments for the removal of insecticides from waters and soils. A review. *Sep. Purif. Technol.* 2022, 284, 120290.

46. Liu, T. T.; Zheng, M. Q.; Hao, P. P.; Ji, K. Y.; Shao, M. F.; Duan, H. H.; Kong, X. G. Efficient photo-oxidation remediation strategy toward arsenite-contaminated water and soil with zinc-iron layered double hydroxide as amendment. *J. Environ. Chem. Eng.* 2023, 11 (1), 109233.

47. Shang, Y. N.; Xu, X.; Gao, B. Y.; Wang, S. B.; Duan, X. G. Single-atom catalysis in advanced oxidation processes for environmental remediation. *Chem. Soc. Rev.* 2021, 50 (8), 5281–5322.

48. Cui, P. X.; Yang, Q.; Liu, C.; Wang, Y.; Fang, G. D.; Dionysiou, D. D.; Wu, T. L.; Zhou, Y. Y.; Ren, J. X.; Hou, H. B.; et al. An N,S-anchored single-atom catalyst derived from domestic waste for environmental remediation. *ACS ES&T Eng.* 2021, 1 (10), 1460–1469.

49. Zhou, C.; Sun, Q. M.; Cao, Q.; He, J. H.; Lu, J. M. Synergistic effect of Fe single-atom catalyst for highly efficient microwave-stimulated remediation of chloramphenicol-contaminated soil. *Small* 2023, 19 (2), 2205341.

50. Zheng, W.; Cui, T.; Li, H. Combined technologies for the remediation of soils contaminated by organic pollutants. A review. *Environ. Chem. Lett.* 2022, 20 (3), 2043–2062.

51. Aparicio, J. D.; Raimondo, E. E.; Saez, J. M.; Costa-Gutierrez, S. B.; Álvarez, A.; Benimeli, C. S.; Polti, M. A. The current approach to soil remediation: A review of physicochemical and biological technologies, and the potential of their strategic combination. *J. Environ. Chem. Eng.* 2022, 10 (2), 107141.

52. Schoutteten, K.; Hennebel, T.; Dheere, E.; Bertelkamp, C.; De Ridder, D. J.; Maes, S.; Chys, M.; Van Hulle, S. W. H.; Vanden Bussche, J.; Vanhaecke, L.; et al. Effect of oxidation and catalytic reduction of trace organic contaminants on their activated carbon adsorption. *Chemosphere* 2016, 165, 191–201.

53. Li, W.; Pang, X.; Snape, C.; Zhang, B.; Zheng, D.; Zhang, X. Molecular simulation study on methane adsorption capacity and mechanism in clay minerals: Effect of clay type, pressure, and water saturation in shales. *Energ. Fuel.* 2019, 33 (2), 765–778.

54. Wu, D.; Wu, J.; Li, H.; Lv, W.; Song, Y.; Ma, D.; Jia, Y. Unlocking the potential of alkaline-earth metal active centers for nitrogen activation and ammonia synthesis: The role of s–d orbital synergy. *J. Mater. Chem. A* 2024, 12 (7), 4278–4289.

55. Wang, Z.; Chen, X.; Xie, X.; Yang, S.; Sun, L.; Li, T.; Chen, L.; Hua, D. Synthesis of aromatic monomers via hydrogenolysis of lignin over nickel catalyst supported on nitrogen-doped carbon nanotubes. *Fuel Process. Technol.* 2023, 248, 107810.

56. Li, S.; Dong, M.; Yang, J.; Cheng, X.; Shen, X.; Liu, S.; Wang, Z. Q.; Gong, X. Q.; Liu, H.; Han, B. Selective hydrogenation of 5-(hydroxymethyl)fur-fural to 5-methylfurfural over single atomic metals anchored on Nb_2O_5. *Nat. Commun.* 2021, 12 (1), 584.

57. Lu, F.; Astruc, D. Nanocatalysts and other nanomaterials for water remediation from organic pollutants. *Coord. Chem. Rev.* 2020, 408, 213180.

58. Pidko, E. A. Toward the balance between the reductionist and systems approaches in computational catalysis: Model versus method accuracy for the description of catalytic systems. *ACS Catal.* 2017, 7 (7), 4230–4234.

59. Goula, M. A.; Charisiou, N. D.; Papageridis, K. N.; Delimitis, A.; Papista, E.; Pachatouridou, E.; Iliopoulou, E. F.; Marnellos, G.; Konsolakis, M.; Yentekakis, I. V. A comparative study of the H_2-assisted selective catalytic reduction of nitric oxide by propene over noble metal (Pt, Pd, Ir)/γ-Al_2O_3 catalysts. *J. Environ. Chem. Eng.* 2016, 4 (2), 1629–1641.

60. Aravani, V. P.; Sun, H.; Yang, Z.; Liu, G.; Wang, W.; Anagnostopoulos, G.; Syriopoulos, G.; Charisiou, N. D.; Goula, M. A.; Kornaros, M.; et al. Agricultural and livestock sector's residues in Greece & China: Comparative qualitative and quantitative characterisation for assessing their potential for biogas production. *Renew. Sustain. Energ. Rev.* 2022, 154, 111821.

61. Ma, Z.; Tian, X.; Liao, H.; Guo, Y.; Cheng, F. Improvement of fly ash fusion characteristics by adding metallurgical slag at high temperature for production of continuous fiber. *J. Clean. Prod.* 2018, 171, 464–481.

62. Park, J.; Bae, S. Formation of Fe nanoparticles on water-washed coal fly ash for enhanced reduction of p-nitrophenol. *Chemosphere* 2018, 202, 733–741.

63. Gao, W.; Zhou, W.; Lyu, X.; Liu, X.; Su, H.; Li, C.; Wang, H. Comprehensive utilisation of steel slag: A review. *Powder Technol.* 2023, 422, 118449.

64. Wang, G. R.; Zhang, J.; Liu, L.; Zhou, J. Z.; Liu, Q.; Qian, G. R.; Xu, Z. P.; Richards, R. M. Novel multi-metal containing MnCr catalyst made from manganese slag and chromium wastewater for effective selective catalytic reduction of nitric oxide at low temperature. *J. Clean. Prod.* 2018, 183, 917–924.

65. Alhwaige, A. A.; Ishida, H.; Qutubuddin, S. Carbon aerogels with excellent CO_2 adsorption capacity synthesised from clay-reinforced biobased chitosan-polybenzoxazine nanocomposites. *ACS Sustain. Chem. Eng.* 2016, 4 (3), 1286–1295.

66. Zhang, D.; Luo, J.; Wang, J.; Xiao, X.; Liu, Y.; Qi, W.; Su, D. S.; Chu, W. Ru/FeO_x catalyst performance design: Highly dispersed Ru species for selective carbon dioxide hydrogenation. *Chin. J. Catal.* 2018, 39 (1), 157–166.

67. Bellakki, M. B.; Baidya, T.; Shivakumara, C.; Vasanthacharya, N. Y.; Hegde, M. S.; Madras, G. Synthesis, characterisation, redox and photocatalytic properties of $Ce_{1-x}Pd_xVO_4$ ($0 \leq x \leq 0.1$). *Appl. Catal. B Environ.* 2008, 84 (3), 474–481.

68. Zheng, T.; Ouyang, S.; Zhou, Q. Synthesis, characterisation, safety design, and application of NPs@BC for contaminated soil remediation and sustainable agriculture. *Biochar* 2023, 5 (1), 5.

69. Ternero-Hidalgo, J. J.; Rosas, J. M.; Palomo, J.; Valero-Romero, M. J.; Rodríguez-Mirasol, J.; Cordero, T. Functionalization of activated carbons by HNO_3 treatment: Influence of phosphorus surface groups. *Carbon* 2016, 101, 409–419.

70. Prats, H.; Stamatakis, M. Atomistic and electronic structure of metal clusters supported on transition metal carbides: Implications for catalysis. *J. Mater. Chem. A* 2022, 10 (3), 1522–1534.

71. Juela, D. M. Promising adsorptive materials derived from agricultural and industrial wastes for antibiotic removal: A comprehensive review. *Sep. Purif. Technol.* 2022, 284, 120286.

72. Di Noto, V.; Negro, E.; Patil, B.; Lorandi, F.; Boudjelida, S.; Bang, Y. H.; Vezzù, K.; Pagot, G.; Crociani, L.; Nale, A. Hierarchical metal–[carbon nitride shell/carbon core] electrocatalysts: A promising new general approach to tackle the ORR bottleneck in low-temperature fuel cells. *ACS Catal.* 2022, 12 (19), 12291–12301.

CHAPTER 6

Application in Air Pollution Control

Gaoqi Han, Han Yan, Xiaotong Zhao, and Rui Han

6.1 INTRODUCTION

Since human society entered industrialization, with the growing population and rapid economic development, energy consumption and environmental protection are usually considered to be in an antagonistic relationship, for instance, high energy consumption exacerbates environmental pollution. The combustion of fossil energy sources such as coal, oil and natural gas releases many harmful pollutants into the atmosphere, such as SO_2, VOCs, Hg^0, H_2S and other harmful gases, which causes serious air pollution and poses a great threat to human health and sustainable development. Therefore, environmental protection and clean use of energy is a great challenge we will face, in which the development of environmental functional materials with excellent performance, recyclable and rich sources have a pivotal position in the control of harmful gases, and is one of the focuses of researchers' attention at present.[1]

Since the 20th century, significant progress has been made in the areas of electrocatalysis, photocatalysis, non-homogeneous catalysis, and other related fields. Superior performance materials, such as activated carbon, carbon nanotubes, and two-dimensional graphene, have been developed and utilised to achieve effective control of harmful gases. However, most of

DOI: 10.1201/9781003535409-6

the materials are based on the precursors of fossil fuel derivatives, and even synthesised under more severe or high-energy-consuming conditions, which not only have high costs but also cause hazards to the environment.

Agricultural and industrial wastes such as crop residues and slag are not only widely available, but can also be processed to become environmentally functional materials with excellent performance. Therefore, the use of cheap solid waste resources as precursors to prepare low-cost, excellent surface properties and structurally controllable environmentally functional materials is receiving more and more attention.[2] In this chapter, the application of solid waste-based materials, such as biomass carbon, fly ash, red mud, steel slag, and the like., in the adsorption and catalytic reaction of harmful gases, such as CO_2, SO_2, VOCs, NO_x, and so forth., will be introduced in detail by taking biomass carbon, fly ash, red mud, steel slag, and so on, as examples.

6.2 ADSORPTION

6.2.1 Overview

6.2.1.1 Basic Concepts and Principles

Adsorption is the process of separating a specific component by contacting a mixed component system with a porous solid material so that a specific component in the mixed system selectively attaches to the surface of the porous solid, thus achieving the separation of the specific component. In this case, the component adsorbed onto the solid surface is the adsorbate, and the porous solid on which the adsorbent is adsorbed is the adsorbent.

The adsorption process can be divided into two categories based on the forces performing on the adsorbate and adsorbent: physical adsorption, which is brought on by van der Waals forces between molecules, and chemical adsorption, which is caused on by a chemical reaction between both substances.[3] For example, the adsorption of CO_2 gas by calcium-based materials is chemisorption, while the adsorption of waste gas by activated carbon is physisorption. The adsorption process can also be divided into Temperature-Variable Adsorption (TSA) and Pressure-Variable Adsorption (PSA) depending on how the adsorbent is regenerated.

6.2.1.2 The Kinetic Models

The study of adsorption kinetic modelling of gases by solid porous materials is of great theoretical significance for understanding their mass transfer mechanisms and laws. The adsorption process of gaseous pollutants by porous solid materials can be divided into three processes: external diffusion, internal diffusion and active site adsorption. External diffusion is the process in which gas molecules diffuse to the surface of the porous solid and are immobilized on the surface, internal diffusion is the process in which gas molecules diffuse from the surface of the adsorbent to the inner surface in the micropores, and active site adsorption is the process in which gas molecules are adsorbed by the adsorption sites and undergo a chemical reaction, and the three processes can be fitted with pseudo first-order kinetic, internal diffusion, and pseudo-second-order kinetic models,[3] respectively.

The mathematical expression for the pseudo-primary kinetic model is equation (1):

$$\frac{dq_t}{dt} = k_1 \left(q_e - q_t \right) \tag{1}$$

where q_t (mg/g) and q_e (mg/g) refer to the unit adsorbed amount of gas analysed by the adsorbent at t and the adsorbed amount at adsorption equilibrium, respectively, and k_1 is the pseudo-primary rate constant.

The internal diffusion model can be represented by the Weber-Moris equation, which is expressed as equation (2):

$$q_t = k_{id} t^{\frac{1}{2}} + C_i \tag{2}$$

where q_t (mg/g) represents the unit adsorbed amount of gas analysed by the adsorbent at time t, k_{id} is the rate constant for the internal diffusion model, and C_i (mg/g) is the boundary layer effect constant.

The mathematical expression for the pseudo-secondary kinetic model is equation (3):

$$\frac{dq_t}{dt} = k_2 \left(q_e - q_t \right)^2 \tag{3}$$

where q_t(mg/g) and q_e(mg/g) represent the unit adsorption amount of the gas analysed by the adsorbent at time t and the adsorption amount at adsorption equilibrium, respectively, and k_2 is the pseudo-secondary rate constant.

In addition, there are some mechanism study models that are often used, such as the Yoon-Nelson model for predicting the rate and saturation of gas adsorption, the Bangham model for inferring the rate of adsorption, and the Thomas model for predicting the saturation amount of gas adsorption.

6.2.1.3 The Adsorption Thermodynamic Models

The adsorption activation energy of the adsorbent can be calculated from the Arrhenius equation, the expression of which is shown in equation (4):

$$\ln k_2 = -\frac{E_a}{RT} + \ln k_0 \tag{4}$$

where k_2 is the reaction rate constant of the pseudo-secondary kinetic model, g/µg-min; k_0 is the finger-prior factor of the Arrhenius equation; E_a is the activation energy of adsorption, kJ/mol; R is the ideal gas constant, 8.314 J/(mol·K); and T is the temperature of the reaction, K. Generally, E_a in the range of -5 to -40 kJ/mol represents physical adsorption, while E_a in the range of -40 to 800 kJ/mol represents chemical adsorption, while in the range of -40 to -800 kJ/mol, it represents chemisorption.

The thermodynamic behaviour of the adsorbent during adsorption can be analysed on the basis of the changes in thermodynamic parameters such as standard enthalpy ΔH, standard free energy ΔG and standard entropy ΔS. In general, ΔG can represent the spontaneity of the reaction, and ΔH and ΔS can be used to judge the heat absorption or exotherm of the reaction, and the above parameters can be calculated by Equation (5) and Equation (6):

$$\Delta G = RT \ln K \tag{5}$$

$$\Delta G = \Delta H - T\Delta S \tag{6}$$

where K is the partition coefficient of the adsorbent, $K = q_e/C_e$; q_e is the cumulative adsorbed amount of the adsorbent at equilibrium, µg/g; C_e is the equilibrium concentration at the equilibrium state of adsorption, µg/m³; and ΔG, ΔH, and ΔS represent the standard enthalpy (kJ/mol), the standard free energy (kJ/mol), and the standard entropy (J·mol⁻¹·K⁻¹), respectively.

6.2.1.4 The Adsorption Isotherm Models

The adsorption characteristics of the adsorbent can be represented by the Langmuir model, the Freundlich adsorption isotherm model, and the Slips adsorption isotherm model,[3, 4] whose expressions are shown in Eq. (7), Eq. (8), and Eq. (10).

$$q_e = \frac{q_m K_L C_e}{1 + K_L C_e} \tag{7}$$

$$q_e = K_f C_e^{1/n} \tag{8}$$

Where, q_m and q_e are the theoretical maximum adsorption amount and equilibrium adsorption amount, mg/g, respectively; C_e is the concentration of the adsorbent at equilibrium, mg/L; K_L and K_f are Langmuir's constant and Freundlich's constant, respectively.

The adsorbent's adsorption capacity is denoted by K_f in the Freundlich equation, whereas the adsorbent's inhomogeneity or the adsorption reaction's intensity is indicated by the n value of 1. The adsorption performance improves with a decreasing n value.

In the Langmuir equation, the factor less equilibrium parameter (R_L) characterizes the adsorption constant of the adsorption capacity. When $0 < R_L < 1$, it indicates that adsorption can occur easily; when $R_L > 1$, it indicates that adsorption is unlikely; when $R_L = 1$, it is a linear distribution. When R_L tends to 0, it indicates that the adsorption is irreversible.

The factor less equilibrium parameter R_L is calculated as shown in Equation (9):

$$R_L = \frac{1}{1 + C_0 k_L} \tag{9}$$

where K_L is the Langmuir isotherm equation constant; C_0 is the initial concentration of adsorbate, mg/L.

The combined effect of Freundlich and Langmuir isotherms is presented by the Slips model, which describes the typical monolayer adsorption characteristics of Freundlich isotherms at relatively low pressures while predicting the typical monolayer adsorption characteristics of Langmuir isotherms at higher pressures, which is given by:

$$q_e = q_m \frac{K_s \left(C_e\right)^{1/n}}{1+K_s \left(C_e\right)^{1/n}} \tag{10}$$

where K_s is the affinity constant, L/M; C_e is the equilibrium concentration, mg/L; n is a parameter that characterises the system heterogeneity $(0 < \frac{1}{n} < 1)$.

6.2.2 CO_2

Carbon dioxide has emerged as a primary constituent of greenhouse gases that are linked to global warming since the industrial revolution. 65% of the radiative forcing resulting from greenhouse gases comes from CO_2. The atmospheric concentration of CO_2 had been maintained at 280 ppm for thousands of years prior to the industrial revolution, and the atmospheric concentration of CO_2 reached 367 ppm in 1999, whereas in 2019, this value grew to 409.8 ppm, showing an even faster growth trend. Accelerating the development of decarbonisation technologies across all sectors, including energy, industry, and transportation, is necessary to meet the goals set forth in the Paris Agreement, which includes striving to limit the temperature increase to 1.5°C over pre-industrial levels and limiting the average increase in global temperature to a value lower than 2°C beyond pre-industrial levels. Despite the increasing share of low-carbon emitting clean energy and biomass power generation in recent years, 65% of electricity production in China, the United States and the Asia-Pacific region still comes directly from the combustion of fossil fuels.

At present, the capture technology of CO_2 in combustion flue gas mainly includes the liquid amine absorption method, the zeolite adsorption method and the limestone circulating calcination method. The liquid amine absorption method has the advantages of mature technology, convenient operation and fast reaction rate, which is the most widely used CO_2 capture technology, and solid adsorption technology has the advantages of low cost of raw materials and strong adsorption capacity, which is a very promising CO_2 capture technology. However, for solid adsorption materials, both physical and chemical adsorption materials, there are certain problems to be solved, such as high energy consumption for regeneration and serious sintering, which makes it difficult for this technology to meet the requirements of large-scale and low energy consumption in practical applications in the short term.[5] In this section, the use of two solid waste-based materials, biochar material and steel slag, in CO_2 capture will be introduced.

6.2.2.1 Biochar

At present, the main low-temperature solid adsorbents for CO_2 are Metal-Organic Frameworks (MOFs), zeolites, mesoporous silica materials and carbon-based adsorbents, and so forth, among which carbon-based porous materials are widely used because of their high CO_2 adsorption capacity, good regeneration performance, lower cost and higher utilisation rate, and the comparison of different low-temperature CO_2 solid adsorbents is shown in Table 6.1.

With the gradual deepening of people's understanding of biomass materials, the technology of using cheap and readily available materials such as rice husk, sugarcane bagasse, palm shell, coconut shell, coffee grounds, weeds, and the like, as precursors, modified and converted into activated carbon adsorbent for CO_2 adsorption by various ways is receiving more and more attention,[3] and Table 6.2 lists some examples of the research on CO_2 adsorption using biochar as adsorbent.

In order to maximize the CO_2 capture efficiency, it is necessary to comprehensively consider several influencing factors such as the treatment

TABLE 6.1   Comparison of low temperature CO_2 adsorbents

Adsorbents	Advantages	Disadvantages
MOFs Materials	Good adsorption performance under high pressure	Poor adsorption performance at low pressure
	Regular pore distribution with adjustable structure	Poor cyclic stability and complex synthesis
Zeolite Molecular Sieve	Uniform pore size and large specific surface area	Poor regeneration performance
	Mild conditions and high selectivity	High CO_2 adsorption heat
Silicon-based Materials	Mesoporous structure	Poor selectivity
	Easy to modify	Very sensitive to water
	Mild adsorption conditions	
Carbon-based Porous Materials	Wide range of sources and low cost	Orifices are prone to collapse
	Large specific surface area and high stability	Poor mechanical stability

TABLE 6.2 The reported CO_2 adsorption and capture performance by biochar

Biomass Precursors	Activating Conditions	Adsorption Conditions	Typical Adsorption Capacity	Refs
Olive Stone	CO_2: 800°C. NH_3: 400, 600, 800, 900°C.	TGA. CO_2; 25°C, 100°C. 15% CO_2/85% N_2; 40°C, 1 bar.	biochar CO_2/NH_3; 10.7 wt%, 25°C; 3.2 wt%, 100°C. biochar CO_2/NH_3; ~3.0 wt%, 40°C, 15% CO_2.	6
Almond Shell	CO_2: 700°C. NH_3: 400, 600, 800, 900°C.	TGA. CO_2; 25°C, 100°C. 15% CO_2/85% N_2; 25°C, 1 bar.	biochar CO_2/NH_3; 11.7 wt%, 25°C; 4.7 wt%, 100°C. 15% CO_2; 5.2 wt%, 25°C, 1 bar.	7
Olive Stones, Almond Shell	CO_2: 700, 800°C. NH_3: 800°C.	FBR. 17% CO_2/83% N_2; 30°C, 130 kPa.	biochar CO_2/NH_3; 0.82 mmol/g, 30°C.	8
Olive Stones, Almond Shell	CO_2: 700, 800°C.	TGA. CO_2; 0, 25, 50°C, 0–120 kPa. FBR. 14% CO_2/86% N_2; 30, 50°C; 0–120 kPa.	biochar CO_2: 3.1 mmol/g, 25°C; 1.0 mmol/g, 100°C. biochar CO_2: 1.08 mmol/g, 25°C, 15 kPa, 14% CO_2; 0.68 mmol/g, 50°C, 15 kPa, 14% CO_2.	9
Olive Stone	CO_2: 800°C	SVA*. H_2O: 25, 50, 70°C. FBR: CO_2 (12.8–15.6%), H_2O (0.1–1.9%), O_2 (3.6–3.9%), N_2.	biochar CO_2: ~1.4 mmol/g, 25°C, 150 kPa.	10
Olive Stones, Almond Shell	air: 400, 450, 500°C. 3, 5% O_2: 500, 650°C.	SVA. 0, 30, 50, 70°C, 0–120 kPa. FBR: CO_2 (8, 14, 30%), 25, 50, 70°C; 140 kPa.	biochar O_2: ~2.1 mmol/g, 30°C, 120 kPa. biochar O_2: 1.48 mmol/g, 25°C, 140 kPa, 30% CO_2.	11
Olive Stones, Almond Shell	O_2 (3, 5, 21%): 400, 500, 650°C.	TGA. CO_2; 25, 100°C, 101 kPa; SVA. 0, 25, 50°C, 0–120 kPa.	biochar O_2: 2.11 mmol/g, 25°C, 101 kPa; 0.74 mmol/g, 100°C, 101 kPa.	12
Soybean	CO_2: 600°C	FBR. CO_2 15.4%, O_2 4%, N_2; 30, 75, 120°C.	biochar $ZnCl_2$: 41 mg/g, 30°C; 22.4 mg/g, 75°C; 23 mg/g, 120°C.	13

Feedstock	Activation Conditions	Method/Conditions	Performance	Ref.
Eucalyptus Sawdust	KOH/C = 2, 4; 600, 650, 700, 800°C.	SVA: CO_2; 0, 25, 50°C.	Biochar KOH: 4.8 mmol/g, 25°C; 3.6 mmol/g, 50°C.	14
Algae Adding With Glucose	KOH/C = 1, 2, 4; 650, 700, 750°C.	Ditto	Biochar KOH: 4.5 mmol/g, 25°C; 2.8 mmol/g, 50°C.	15
Palm Shell-Based Activated Carbon	NH_3; 400, 800°C; Air: 400°C.	TGA. CO_2; 30–105°C.	biochar NH_3: 73.5 mg/g, 30°C; 30.1 mg/g, 105°C.	16
Bamboo	KOH/C = 1, 2, 3, 4, 5; Temperature: 500, 600, 700, 800, 850°C	SVA. CO_2; 0, 25°C; 0.01–1 bar.	Biochar KOH: 4.5 mmol/g, 25°C, 1 bar.	17
Pine Nut Shell	KOH: 500–900°C	SVA. CO_2; 0, 25, 50, 75°C; 0.01–1 bar.	biochar$_{KOH}$: 2.0 mmol/g, 25°C, 0.15 bar; 5.0 mmol/g, 25°C, 1 bar.	18
Grass Cuttings, Bio-Sludge, Beer Waste, Horse Manure	CO_2; 800°C. H_3PO_4; 600°C.	SVA. CO_2; 0, 25, 50, 100°C; 0–101kPa.	biochar $4H_3PO$: 1.45 mmol/g, 0°C, 10kPa.	19

and activation conditions of biochar and adsorption working conditions. The thermal treatment conditions of biochar, such as processing temperature and gas flow rate, may have a large impact on the structural evolution and pore structure of biochar, thus affecting the CO_2 adsorption performance of the material. Generally speaking, when the pyrolysis temperature reaches 600-900°C, the pores of biochar are developed and the CO_2 adsorption capacity gradually increases, but as the temperature further increases, the pores further increase, accelerating the diffusion and escape of CO_2 gas molecules, which will adversely affect the CO_2 adsorption. In addition, compared with traditional pyrolysis, microwave pyrolysis can save cost, time and energy consumption,[3] and is also a promising treatment method.

For physical adsorption, the number of active adsorption sites gradually decreases with the gradual increase of temperature, and the CO_2 adsorption capacity gradually decreases, while for chemical adsorption, the increase of temperature promotes CO_2 adsorption to some extent. For adsorbents where physical and chemical adsorption coexist, such as ammonia-treated biocarbon materials, the number of micropores is the main limiting condition at lower temperatures, whereas the number of chemical adsorption sites will replace the number of micropores as the main limiting condition as the temperature increases. Therefore, the development of biocarbon materials with more ultra-micropores and stronger chemisorption capacity will be an effective way to further enhance the adsorption performance.

6.2.2.2 Steel Slag

As a major by-product of the steelmaking industry, the accumulation of large quantities of steel slag can have a bad impact on the atmosphere, soil and groundwater environment. The main composition of steel slag is shown in Table 6.3, and it can be seen that CaO, SiO_2, Fe_2O_3 and Al_2O_3 are the most important constituent elements of steel slag, and the sum of their mass percent content exceeds 80% of the total mass of steel slag. Among them, CaO has the highest content of 42.328%, providing a large number of alkaline sites for CO_2 adsorption.[5] In addition, Al_2O_3 and MgO can be used as stabilizers to prevent the sintering of the adsorbent at high temperatures, providing favourable conditions for the stable adsorption of CO_2.[20]

TABLE 6.3 The composition of steel slag

Component	MgO	Al_2O_3	SiO_2	CaO	MnO	Fe_2O_3	CuO	Na_2O
Ratios (%)	7.391	7.405	17.64	42.328	3.526	14.043	0.006	0.106

The reaction of calcium-based CO_2 adsorbent with CO_2 occurs as Eq. (11) and the thermodynamic equation of the reaction can be described by Eq. (12):

$$CaO(s) + CO_2(g) = CaCO_3(s), \Delta H < 0 \qquad (11)$$

$$LogP_{CO_2}[atm] = 7.079 - (8308 / T[K]) \qquad (12)$$

The basic principle of steel slag CO_2 capture is as follows: in the mineralization furnace at a temperature of 650-700°C, CaO and CO_2 in the flue gas occurs in the mineralization reaction in Equation (11), thus realising the fixation of CO_2; the generated $CaCO_3$ enters the calciner at a temperature of > 900°C to thermally decompose, and the required heat is supplied by the combustion of fuel, and the exhaust gas can be condensed to obtain a purity higher than 95% of the CO_2 gas stream. 95% purity of CO_2 can be obtained after condensation of the tail gas, the calcined regenerated CaO continues to enter the mineralization furnace for CO_2 capture, thus realising the cycle of CO_2 adsorption. When the CaO is deactivated, it is discharged, and new adsorbent is added at the same time.

Many researchers have conducted in-depth studies on the reaction conditions for CO_2 fixation by steel slag, which greatly improved the adsorption efficiency. For example, Polettini et al investigated the effects of pressure, CO_2 concentration, and temperature on the carbon fixation rate in the process of CO_2 fixation by converter slag (BOF) mineralization. The results showed that the operational factors had a large effect on CO_2 uptake, and the carbon fixation rates ranged from 6.7 to 53.6 g CO_2/100g slag. Under mild experimental conditions, namely, P = 5 bar, C = 40% vol. CO_2, T = 50°C, and t = 4 h, carbon sequestration was superior to previous studies.[21] In addition, adding high temperature or pressure to the system through a specific reactor or improving the mass transfer efficiency during the reaction is another way to increase the carbon sequestration rate of steel slag mineralization. For example, Chang et al. investigated CO_2 mineralization with steel slag in a supergravity rotary filled bed and obtained the optimum process parameters of 750 rpm, 65°C, 30 min, and 2.5 L/min of pure CO_2, with a maximum conversion of 93.5% for converter steel slag.[22]

6.2.3 SO$_2$

In order to meet the requirements of rapid industrial development, the combustion of fossil energy such as coal will remain as one of the main forms of energy supply for some time to come, which will bring about a large amount of SO$_2$ emissions, which will bring about serious harm to the ecological environment and human health. SO$_2$ gas is easily soluble in water, and is highly toxic, which will cause different degrees of damage to respiratory mucous membranes, lungs, and other organs of humans and animals after inhalation. Furthermore, acid rain is mostly caused by SO$_2$, which also kills a lot of aquatic life, depletes soil, and seriously corrodes structures, all of which have a negative impact on social and economic advancement.

At present, low concentration SO$_2$ flue gas treatment technology generally uses the adsorption method, the microbial method and the membrane separation method. The adsorption method is a method that utilises a porous solid adsorbent to adsorb pollutants in the exhaust gas in order to remove or recycle the pollutants so that the exhaust gas can be purified. Microbial wet desulfurisation is the oxidation-reduction of sulfur oxides and sulfur-containing salts by microorganisms in the presence of water. The membrane separation method is through the different gas molecules in the membrane material mass transfer rate difference, so as to realise the selective separation of gas. At present, the harsh operating conditions of microbial desulfurisation and the high cost of membrane materials for the membrane separation method limit its application in industry. The preparation of cheap adsorbent materials from solid waste has the advantages of simple process, good adsorption effect, low energy consumption, renewable reuse, and so forth, and has a better application prospect.[23, 24]

6.2.3.1 Fly Ash

Fly ash is a solid waste produced after coal combustion, and its main chemical composition includes SiO$_2$ and Al$_2$O$_3$, which account for more than 70%~80% of the fly ash, and the remaining 20%~30% are FeO, CaO, Mn$_3$O$_4$, TiO$_2$ and other substances, in addition to unburned charcoal particles and rare elements such as germanium and gallium.[5, 25] China is a large coal-burning country, a large amount of fly ash is generally piled up in the ash storage pond, through various factors such as rainwater scouring, fly ash will jeopardize the surface and groundwater, thus polluting the environment, and through the bio-enrichment, so that the heavy metal substances in the fly ash jeopardize the survival of plants and animals, and the fine fly

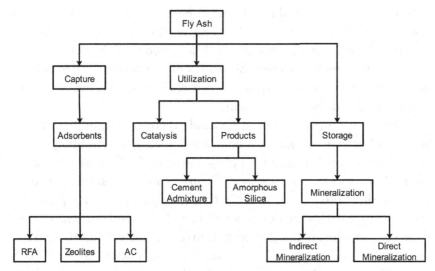

FIGURE 6.1   Application path of fly ash in flue gas treatment.[26]

ash substances diffuse into the atmosphere, and enter into the human body through the respiratory tract, triggering a variety of diseases. It will jeopardize human health and affect the living environment.

Since fly ash contains more alkaline oxides and is characterized by multiple micropores, multiple adsorption sites, and large specific surface area, it can be used as an adsorbent for the treatment of flue gas pollutants through corresponding modifications. Figure 6.1 shows the application path of fly ash in flue gas treatment.

Fly ash adsorption of SO_2 utilises the rich microporous structure contained in fly ash to adsorb SO_2 gas molecules, which are then converted into non-toxic and harmless components, with the adsorption principle (* represents the adsorption state):[27]

$$SO_2 \rightarrow SO_2^* \tag{13}$$

$$O_2 \rightarrow O_2^* \tag{14}$$

$$H_2O \rightarrow H_2O^* \tag{15}$$

$$SO_2^* + 1/2O_2^* \rightarrow SO_3^* \tag{16}$$

$$SO_3^* + H_2O^* \rightarrow H_2SO_4^* \tag{17}$$

$$H_2SO_4^* + 2NH_3 \rightarrow \left(NH_4\right)_2 SO_4^* \tag{18}$$

Since fly ash is a product of coal at high temperature, its activity is low at room temperature, and the effect of directly using it for flue gas desulfurisation is not obvious. In the fly ash used for flue gas purification treatment to improve the chemical activity of fly ash is needed, to enhance its flue gas purification ability, the modification method is mainly divided into physical and chemical methods.

The primary physical strategies include thermal modification, microwave modification, ultrasonic modification, mechanical grinding modification, and so on. The idea is to physically alter the fly ash's surface and internal structure to increase its specific surface area, porosity, and exposure of active sites. This will increase the fly ash's capacity for adsorption. The two primary chemical processes are alkali and acid modification, amongst others. The purpose of this procedure is to further increase the fly ash's specific surface area and porosity while simultaneously destroying the Si-O-Si and Si-O-Al network structure in the fly ash with an acid or alkali to boost the fly ash's reactivity and improve its adsorption effect. The modification of fly ash is typically combined with two or more modification methods, which can make up for the shortcomings of a single modification method and better improve the adsorption capacity of modified fly ash. The modification of fly ash by physical method alone can only get a limited modification effect, and the modification of fly ash by chemical method alone can get an improved modification effect, but it takes a longer time. The adsorption capacity of modified fly ash can be improved. For example, some scholars calcined the fly ash with Na_2CO_3, and then prepared fly ash adsorbent by hydration reaction with $Ca(OH)_2$ as the calcium-based absorbent, and the study showed that the SO_2 adsorption capacity of the modified fly ash was increased by 10.5 times compared with that before modification. There are also scholars who modified the fly ash by mechanical grinding, and then chemically modified the fly ash with $Ca(OH)_2$ aqueous solution, the specific surface area and activity of the fly ash had a more obvious improvement than the mechanical grinding method.

6.2.3.2 Calcium Carbide Slag

The process of creating calcium carbide slag is seen below. Calcium carbide slag is an alkaline byproduct that is created when calcium carbide is hydrolised during the preparation of acetylene gas in the PVC industry.

$$CaCO_3 \rightarrow CaO + CO_2 \tag{19}$$

$$CaO + 3C \rightarrow CaC_2 \left(Calcium\,Carbide\right) \tag{20}$$

$$CaC_2 + 2H_2O \rightarrow C_2H_2 + Ca(OH)_2 \, (Calcium \, Carbide \, Slag) \quad (21)$$

Calcium carbide slag can be used in the dry flue gas desulfurisation process, the principle is similar to limestone dry flue gas desulfurisation, the desulfurising agent will be ground into powder and then sprayed into the desulfurisation equipment, and then high temperature calcium carbide slag is decomposed into CaO, absorbing SO_2 in the flue gas to generate calcium sulfite, and in the oxygen-rich conditions converted to calcium sulfate. Compared with limestone, calcium carbide slag calcination temperature is lower, with better pore structure and desulfurisation activity, some studies show that calcium carbide slag at 700°C has a higher SO_2 adsorption capacity. However, compared with the dry flue gas desulfurisation process, semi-dry and wet desulfurisation are more widely used, and higher desulfurisation efficiency can be achieved by dissolving the calcium carbide slag to form a slurry to react with SO_2.[28]

6.2.4 VOCs

According to the U.S. Environmental Protection Agency (EPA), a class of organic compounds known as Volatile Organic Compounds (VOCs) includes ammonium carbonate, metal carbonates, carbonic acid, and CO in addition to other organic compounds that are involved in the atmospheric photochemical reaction of carbon compounds; The World Health Organization (WHO) will define volatile organic compounds as those with room temperature saturated vapor pressure higher than 133.322 Pa, the boiling point between 50°C and 260°C, and the European Union (EU) on the definition of volatile organic compounds at standard atmospheric pressure (101.3 kPa) under the boiling point of 250°C or less than the organic compounds; The boiling point between 50°C and 260°C, and room temperature saturated vapor pressure greater than 133.322 Pa are the definitions used by the World Health Organization (WHO). VOCs are described by the World Health Organization (WHO) as organic compounds having a boiling point between 50 and 260°C and a saturated vapor pressure of more than 133.322 Pa at room temperature. VOCs are complex in composition and highly toxic, and when they enter the atmosphere, they not only cause the greenhouse effect and environmental problems such as photochemical smog, but also produce strong irritation to the human eye and throat, causing dizziness, nausea and other symptoms and even inducing respiratory problems, nausea and other symptoms, even inducing respiratory diseases.[29]

According to the final form of the substance after treatment, the treatment technology of VOCs can currently be categorized into recycling technology and elimination technology. Recycling technology mainly utilises selective absorbents or selective permeable membranes and other means to separate VOCs, and typical methods include adsorption, absorption, membrane separation,[30] and the like. Elimination technology utilises light, heat, catalysts and microorganisms to completely transform organic matter into H_2O and CO_2. There are combustion methods, biological methods and photocatalytic emerging treatment technologies.[31] Currently, adsorption technology is the predominant means of treating Volatile Organic Compounds (VOCs). It performs by adsorbing harmful components and achieving the goal of removing harmful pollution using specific materials with adsorption capacity, including activated carbon, activated alumina, the zeolite molecular sieve, and other adsorbent compounds.

6.2.4.1 Fly Ash Based Zeolite

Zeolite is a cage-like structure of aluminosilicate minerals, light grey or flesh red. Its structural skeleton is formed by the TO_4 tetrahedron (T for Si, Al) sharing the top of an oxygen atom. The zeolite cage-like porous structure has uses in a wide range of petrochemical, environmental protection and other industries. Zeolite in the adsorption of waste gas is one of the main directions of the application of volatile organic compounds.

Reagents are now the primary raw material used in the production of synthetic zeolite, and they are very expensive. In addition to addressing the environmental issues raised by fly ash, the process of synthesising zeolite from fly ash can lower the cost of the raw materials used to create synthetic zeolite. As a result, researchers both domestically and overseas have focused a great deal of attention on the synthesis of fly ash zeolite. The chemical composition of fly ash and related applications have been introduced in Section 3. Based on previous studies, it can be concluded that the following approaches are commonly used to synthesise zeolite from fly ash: the one-step, the two-step, the alkali melting-hydrothermal, the microwave-assisted, and the ultrasonic treatment hydrothermal methods. For example, K. S. Hui et al. synthesised high-quality 4A-type zeolites by controlling the temperature of the hydrothermal reaction.[32] G. G. Hollman et al. synthesised X-type zeolites by the two-step hydrothermal method using fly ash as a raw material, 4A-type zeolite and P-type zeolite, and studied the adsorption properties of zeolite on heavy metal ions and ammonium root ions in wastewater.[33] However, there are many difficulties in the process of synthesising zeolite from fly ash. For example, fly ash is a small bead-shaped particle, its main material composition is quartz and mullite, the physical structure and chemical properties are very stable, it is reluctant to participate in a hydrothermal reaction, the zeolite synthesis process of fly ash silica-aluminium utilisation rate is low; the silica-aluminium ratio of the fly ash is limited, it is difficult to prepare high silica-aluminium ratio zeolite without the addition of reagents as a source of silica, and so on.

6.2.4.2 Biochar

As mentioned in Section 2, biochar is a cheap adsorbent that is frequently employed for the adsorption of organic pollutants in air and water because it is rich in surface functional groups and has a porous structure. It can be made from a variety of sources at a low preparation temperature. The adsorption of VOCs on biochar is a physicochemical interaction, which is usually affected by van der Waals' force, the oxygen-containing functional groups, and the hydrophilicity energy of the VOCs themselves, and so forth. Some researchers have proposed an adsorption mechanism diagram as shown in Figure 6.2. To determine the absorption capacity of typical VOCs, it is therefore required to take into account the balance between the pore characteristics, chemical functional groups, and VOC properties of biochar.

The adsorption temperature has an impact on the adsorption capacity of Volatile Organic Compounds (VOCs), much like it does on the adsorption of carbon dioxide. Research has indicated that increasing the adsorption temperature from 20 to 40°C results in a 73.1-98.2% drop in the adsorption capacity of acetone by biochar. Furthermore, in contrast to CO_2 adsorption, the size, molecular weight, and boiling points of Volatile Organic Compounds (VOCs) are among the intrinsic properties that affect how well they adsorb substances. For example, benzene showed a faster rate of diffusion than other Volatile Organic Compounds (VOCs) derived from xylene. This could be attributed to its tiny molecular size of 0.58 nm, which enables it to diffuse into micropores and reach the sites of adsorption. The adsorption process was aided by the high molecular weight Volatile Organic Compounds' (VOCs) substantially larger diffusion coefficients than those of the lower molecular weight VOCs and with stronger intermolecular interactions between the biochar and the high boiling point VOCs. Table 6.4 demonstrates the application examples of adsorption of VOCs using biochar materials as adsorbents.

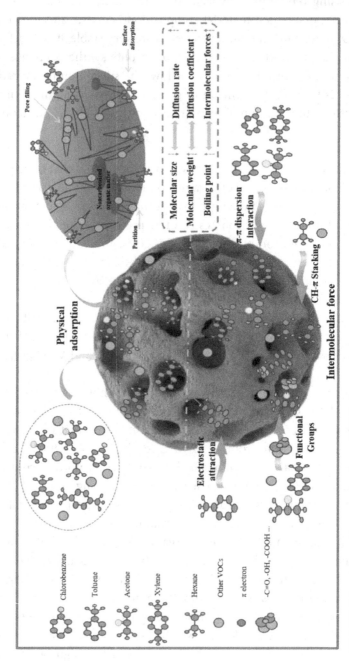

FIGURE 6.2 Adsorption mechanism of VOCs.[3]

TABLE 6.4 The reported gaseous VOC adsorption and removal performance by biochar

Biomass Precursors	Activating Conditions	Adsorption Conditions	Typical Adsorption Capacity	Refs
Sludge	-	Kinetic adsorption. Acetone (80–6900 ppm), chloroform (90–7800 ppm), acetonitrile (43–2700 ppm); 10, 30, 50, 80°C.	Biochar $ZnCl_2$: Acetone: ~150 mg/g, 30°C. Chloroform: 244 mg/g, 30°C. Acetonitrile: 41 mg/g, 30°C.	34
Sewage Sludge	KOH, NaOH: acid washing.	FBR. Air, VOCs <100 ppm, H_2O 20%; 20°C.	biochar KOH: Toluene: 350 mg/g, 20°C. MEK: 220 mg/g, 20°C. Limonene: 650 mg/g, 20°C.	35
Leather Waste	KOH, NaOH, K_2CO_3:0.33, 1; 750, 900°C.	FBR.	biochar KOH: Toluene: 700 mg/g.	30
Bamboo, Pepper Wood, Sugarcane Bagasse, Hickory Wood, Sugar Beet Tailings	-	TGA. VOC (50 ml/min), N_2; 20, 30, 40°C.	Toluene: 62.91 mg/g, 20°C. Acetone: 91.16 mg/g, 20°C. Cyclohexane: 69.33 mg/g, 20°C.	36
Hickory Wood, Peanut Hull	H_3PO_4 or KOH:C = 1:1, 600°C.	TGA. VOC (50 ml/min), N_2; 20°C.	Biochar $4H_3PO$: Acetone: 147.77 mg/g, 20°C. Cyclohexane: 159.66 mg/g, 20°C.	37
Milled Hickory, Wood Chips, Peanut Shell	CO_2; 600, 700, 800, 900°C.	TGA. VOC (50 ml/min), N_2; 20, 40, 60°C.	Biochar CO_2: Acetone 121.74 mg/g, 20°C. Cyclohexane: 155.41 mg/g, 40°C.	38
Hickory Wood	ball milling, 1:100.	TGA. VOC (50 ml/min), N_2; 20°C.	Toluene: ~70 mg/g, 20°C. acetone: 103.4 mg/g, 20°C. Cyclohexane: 72.5 mg/g, 20°C. Ethanol: ~65 mg/g, 20°C.	39
Corn Stalk	ball milling, 3:200; H_2O_2/NH_4OH.	TGA. VOC (50 ml/min), N_2; 25°C.	Chloroform: 87.0 mg/g, 20°C. biochar $H_2O_2NH_4)H/NH_4OH$: Benzene: ~118 mg/g, 25°C. M-xylene: ~125 mg/g, 25°C. O-xylene: ~120 mg/g, 25°C. P-xylene: 130.21 mg/g, 25°C.	40

6.2.5 Other Gaseous Pollutants

Mercury is a highly volatile, dense, low melting point, liquid at room temperature, toxic heavy metal, and one of several metallic pollutants that are difficult to control. Mercury is characterised by persistence, bioaccumulation and biotoxicity, and it mainly causes very serious harm to the human nervous system, embryonic development, kidneys and liver. As one of the trace elements, although the concentration of monomercury emitted from coal-fired power plants is not high, monomercury has also become one of the main pollutants controlled in coal-fired flue gases because of its own strong accumulation and high toxicity. A third of the emissions of anthropogenic mercury come from the burning of fossil fuels, particularly in coal-fired power plants. Red mud is a byproduct of aluminium manufacturers' alumina manufacturing. Approximately 1-2 tonnes of red mud waste are created for every tonne of alumina produced. Red mud can be reused in addition to lowering the cost of removing mercury from flue gas because it has a high concentration of metal oxides like CuO, Fe_2O_3, Al_2O_3, and so on in addition to having a highly developed pore structure. In addition, some scholars have also studied the mercury removal from biochar, for example, Zhu et al. prepared biochar using ammonium halide-modified rice husk-based, proving the role of halogens, and experimentally investigated the effect of the combined removal of NO, Hg^0 and SO_2, and the removal rate of mercury by the adsorption catalyst was more than 80%.[41] Klasson K T et al. proposed the feasibility of mercury removal from biochar of poultry manure fertilizers, and found that hydrochloric acid rinsing of bio charcoal increased the removal of mercury.[42]

H_2S is a colourless, highly toxic gas with a strong irritating rotten egg odour, which can cause serious harm to human health. H_2S exists in many scenarios such as oil mine wells, landfills, sewage treatment plants, and so on, so it is of great practical significance to effectively remove H_2S from polluted sites. At present, the main technologies for the removal of hydrogen sulfide are wet desulfurisation, dry desulfurisation and other technological desulfurisation. Among them, dry desulfurisation is to utilise the reductive characteristics of H_2S and use solid particles or powders for selective desulfurisation. Commonly used materials for dry desulfurisation include molecular sieves, metal oxides, activated carbon and biochar, and the metal-organic skeleton. It has the advantages of high precision, simple and low-cost equipment and low energy consumption. Many researchers have tried to use agricultural waste, such as peanut shells and corn stalks, as raw materials for H_2S adsorption and obtained excellent H_2S adsorption results. For example, Guo et al tested the adsorption performance of KOH-activated palm shell biochar, and found that the adsorption of H_2S can reach up to 169 mg/g.[43] Wang et al studied the H_2S removal performance of Cu-loaded cocoa-based biochar, and the results showed that the copper loading possesses good physicochemical properties, and the adsorption amount can reach 220.37 mg/g.[44] The above ideas can broaden the use of adsorbents and reduce the cost, which can help to realise the industrialized application.

6.3 CATALYTIC REDUCTION

6.3.1 NO_x

6.3.1.1 Blast Furnace Slag

One common solid waste produced during the manufacturing of iron and steel is Blast Furnace Slag (BFS). Even though BFS is abundant in SiO_2, Al_2O_3 and Fe_2O_3, it is not well used. Therefore, from the standpoints of waste utilisation and environmental preservation, the synthesis of catalysts utilising BFS is quite important.

For the production of zeolites, using solid waste with high SiO_2 and Al_2O_3 percentages is a viable approach. Chen[45] used a combination of ion-exchange and BFS impregnation to create CuX zeolite catalysts doped with Mn or Ce oxides (Figure 6.3). At relatively low temperatures, the catalysts Mn/CuX, Ce/CuX, and Mn-Ce/CuX all showed around 100% NO conversion.

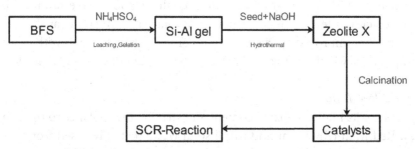

FIGURE 6.3 Preparation of CuX Zeolite catalysts.

By loading V_2O_5 and WO_3 onto pretreatment Ti containing BFS using the impregnation technique, Yang[46] created V_2O_5-WO_3/Ti containing BFS catalysts. This catalyst's nitrogen removal effectiveness was significantly higher than that of commercial TiO_2-loaded catalysts, with a NOx conversion of around 77% at 250-400°C. Tran[47] used commercial Ti/Si precursors and Ti containing BFS to create four V_2O_5-WO_3/-TiO_2-SiO_2 samples. Among these, two catalysts, designated BFS-1 and BFS-2, were made from titanium-containing BFS with varying Fe_2O_3, Al_2O_3, and SO_4^{2-} contents. The BFS-1 catalyst demonstrated the highest NO conversion when a little quantity of SO_4^{2-} was applied. The findings demonstrated that a moderate level of Al_2O_3/Fe_2O_3/SO_4^{2-}/SiO_2 doping promoted the production of Ti-O-Si, generated more acidic site dots, and enhanced the material's high dispersibility and specific surface area. Impurities may not decrease the activity of V_2O_5-WO_3/TiO_2 catalysts when Ti containing BFS is employed as a carrier; rather, they may improve the catalysts' performance by producing mesoporous structures with greater BET specific surface area and pore volume.

High SCR catalytic activity CeZrTiAl catalysts can be processed from BFS containing Ti. H_2SO_4 was used to leach the pretreated BFS, yielding an Al and Ti-containing leach solution. Afterwards, CeZrTiAl catalysts were prepared by doping Ce and Zr species, and γ-Al_2O_3 nanosheet structures were constructed by hydrothermally adjusting the pH value of the leach solution. It is noted that at 260-440°C, the NOx conversion can exceed 90%, and the γ-Al_2O_3 nanosheets of the CeZrTiAl catalyst show strong SCR catalytic activity. The specific surface area of CeZrTiAl catalysts was increased to approximately three times that of CeZrTiAl nanoparticles by using the nanosheet structure of γ-Al_2O_3 as a skeleton carrier. Additionally, the γ-Al_2O_3 nanosheets offered a large number of surface-active oxygen vacancies, which led to good reduction qualities and a large number of surface acidic site dots. Furthermore, the catalyst's reduction capabilities were further improved by the interaction of γ-Al_2O_3 nanosheets with other components, which led to an increase in the amount of Ce^{3+} and defects.

6.3.1.2 Red Mud

Because of its high iron oxide content and striking resemblance to reddish clay, Red Mud (RM) is an industrial solid waste that is released from the aluminium industry's alumina extraction process. Because RM has a high concentration of SiO_2, Al_2O_3, Fe_2O_3, and TiO_2, it can be used as a raw material or loaded directly into zeolite carriers, which can then be doped with reactive metal oxides to create NH_3-SCR catalysts. Many alkaline oxides, primarily Na, K, and Ca oxides, are present in RM.

These alkaline oxides have a negative impact on the catalyst structure and decrease the amount of acidic sites available for NH_3 adsorption, which lowers RM's NH_3-SCR activity. It is possible to activate RM's NH_3-SCR activity by eliminating these basic oxides. Consequently, the principal techniques for mitigating alkali metals are a range of pretreatment procedures such as calcination, acid washing, and water washing. RM-based catalysts dominated by active Fe_2O_3 and SiO_2-Al_2O_3-TiO_2 composite oxides were prepared using these pretreatment techniques.

Wang[48] studied how binders affected RM-based catalysts made by calcining and leaching hydrochloric acid. Sesbania powder's binder effect increased the number of oxygen vacancies and improved NH_3 adsorption. Furthermore, the catalysts' surface acidity, reducibility, and contact area with reactants all saw notable improvements, which supported the ionization performance.

6.3.1.3 Silica Fume

Because of its high silicon content and ability to track other metals, silica powder can be utilised as a source of silicon for the synthesis of zeolites. For instance, Chen[49] created ZSM-5 molecular sieves without the use of organic templates by combining SF with metakaolin. Consequently, 80% NO conversion and nearly 100% selectivity were shown by sf-derived ZSM-5 loaded with Ni/Cu/Zn in the 300-450°C range. This suggests that the Si/Al ratio of the sf-derived catalysts can be adjusted through the addition of additional Al-containing materials to alter the surface acidic sites.

6.3.1.4 Fly Ash

Fly Ash (FA) is an irregular substance that has a high elemental composition of Si, Al and Fe. It can be utilised as an SCR catalyst at low to medium temperatures. Enhancing the redox characteristics of the FA surface can be achieved through metal loading. The SCR activity of Fe, Cu, Ni and V supplied FA-based catalysts was compared with that of the wet impermeability approach by Xuan.[50] In the range of 270-310°C, 10% Cu/FA shows the maximum NO conversion among all. The NO conversion rate was significantly impacted by the roasting temperature throughout the preparation process, whereas the impact of the nitric acid pretreatment was minimal. This implies that acid is not able to significantly alter the FA surface characteristics on its own. Furthermore, Wang[51] studied the SCR mechanism at low temperatures of FA-derived porous cordierite ceramics (Ce/FPSS) loaded with CeOx. Ce/FPSS outperformed commercial Ce/FPSS in terms of SCR performances and had excellent oxygen storage and release characteristics while adhering to the E-R and L-H mechanisms. Additionally, it was discovered that the stimulation of electron transfer through numerous redox pairs in binary metal loading results in a greater enhancement of the SCR activity at the low temperature compared to only one metal. Atomic filter carriers for SCR processes are produced from FA in a large number of examples because of its high concentration of crystalline Si. Li[52] used SBA-15 produced from FA to make mesoporous particle sieves. Wet impregnation (IWI) was used to load Fe and Mn into the FA-derived SBA-15. In comparison to the single laden metal SBA-15, the loaded SBA-15 showed improved metallic dispersion, an increased Mn^{4+} molar proportion, and more bound oxygen molecules. By creating N-intermediates with Lewis acidic site and Brønsted acidic site point formation, the addition of Fe enhanced NH_3 adsorption even more. Excellent water thermal stability was also demonstrated by the Fe-Mn/FA synthesised SBA-15.

Furthermore, Zhang[53] used a hydrothermal process to manufacture ZSM-5 carriers from fly ash and red mud. The range of 100-250°C produced the maximum NO conversion (>85%) for mass fractions of 5% and 10% for Co and Mn, respectively. A stronger Mn^{4+} ratio in the Co-Mn/ZSM-5 catalyst resulted in better NOx conversion at low temperatures, whereas stronger N_2 selectivity was demonstrated by Co^{3+} compared to no Co loading. Additional slags have also been created as inexpensive catalysts.

Han[54] used the impregnation approach to manufacture SCR catalysts utilising Coal Gasification Slag (CGS) as a carrier. The findings demonstrated that, in the 180-290°C temperature range, the maximum NOx conversion efficiency was achieved with 1% vanadium loading in CGS because of the significant dispersion of vanadium species. This suggests that CGS can function as a carrier for the SCR catalyst at low temperatures. Additionally, 1% V/CGS displayed good water resistance at 260°C. The catalytic activity was boosted in the presence of SO_2 at 240-290°C because V_2O_5 was oxidized to produce sulfate, which facilitated NH_3 adsorption.

To create NH_3-SCR catalysts, a variety of solid waste, including BFS, RM, FA, and the like, were utilised. The majority of them are superior carriers with high specific surface area, which enhances the redox and NH_3 adsorption characteristics and makes it easier for the active components to be uniformly loaded onto the carriers. The denitrification performance of the catalysts made from solid waste was improved by appropriately doping them with metal oxides. Certain impurities found in the carriers made from solid waste, however, performed even better than those found in other carriers made from more costly materials. Because of the two distinct loading metal fractions, the titanium slag displayed two distinct temperature windows.

In the low and medium temperature windows, the V-W/TiO$_2$ catalyst with titanium slag as the titanium source demonstrated comparatively poor NO conversion. On the other hand, Mn-based Ti-BFS catalysts showed increased SCR activity at low temperatures (< 250°C), indicating that attention should be given when choosing the loading metal component. Furthermore, because of the number of oxygen vacancies and the dispersion of MnO$_x$ species, manganese slag also demonstrated good low-temperature SCR activity. High Fe$_2$O$_3$ red mud-based SCR catalysts showed a medium to high temperature window, while fly ash zeolites and fly ash carriers obtained a wide temperature window in the 150-350°C range. These examples show how, depending on its phase structure, certain solid waste can be used in various contexts. Notwithstanding their favourable preparation cost, environmental protection, and denitrification effectiveness, NH$_3$-SCR catalysts made from solid waste materials continue to encounter numerous obstacles.

6.3.2 CO$_2$

Reaching neutrality in carbon emissions depends on CO$_2$ outputs; Carbon Capture and Storage (CCS) alone is insufficient. By transforming CO$_2$ into advantageous chemical compounds, the greenhouse effect can be significantly reduced. It is critical to view CO$_2$ as a significant carbon source that is capable of being effectively converted to create syngas or C1 and C2 compounds. Nickel-based catalysts have garnered significant interest due to their strong hydrogenation process performance as well as low price in comparison to platinum group metals. As a result, numerous researchers have employed Nickel-loaded catalysts made from solid waste to catalyse the removal of CO$_2$.

6.3.2.1 Pulverized Coal

Dong[55] used wet impregnation to create 15%Ni-1%Re/coal coke catalysts for the methanization of carbon dioxide. The H$_2$-TPR as well as H$_2$-TPD outcomes showed that the addition of 1%Re led to a considerable reduction ability, enhanced nickel metal dispersion, and a shift in the reduction peak location to low temperatures. Furthermore, with the addition of Re, the particle size decreased from 24.0 nm to 19.8 nm and the nickel dispersion rose from 9.6% for Nickel/coal meal to 14.7% for Ni-Re/coal meal. In the airspeed range of 350-450°C and 1000-3000 h-1, the Ni-Re/coal meal catalysts continuously outperformed the Ni-Re/coal meal in terms of the conversion of CO_2 and CH_4 selectivity. According to the in-situ DRIFTS data, formate is produced when hydrogen atoms are separated from one another by metal Ni clusters and Re. Formate is then further hydrogenated by hydrogen atoms to produce methane.

6.3.2.2 BFS

Chen[56] introduced CeO_2 to Ni-CeO_2/hBFS, which are BFS carriers treated with HNO_3. The extra CeO_2 improved the particular area of surface and the Nickel dispersion in comparison to Ni/hBFS without it, resulting in smaller Nickel nanoparticles (from a range 8.95 nm to 3.75 nm). Moreover, adding CeO_2 resulted in a different reaction pathway, validating the in-situ DRIFTS. The catalysts demonstrated a greater rate of intermediary formation than the Ni/hBFS catalysts due to the process of hydrogenation mixed bicarbonate and carbonate. Notably, the in-situ DRFITS does not exhibit a gaseous CO product, suggesting that formate did not dissociate to CO prior to hydrogenating to CH_4. The Ni-CeO_2/hBFS catalyst demonstrated 81.6% CO_2 conversion and 99.8% CH_4 selectivity at 350°C, according to the data. After 100 hours, there was no discernible deactivation or carbon accumulation. Compared to the Ni/hBFS catalyst, the Ni-CeO_2/hBFS catalyst achieved a CO_2 conversion that was almost 40% greater.

Because of the improved catalytic impact that BFS carriers brought, Wan[57] investigated the effects of KOH-activated BFS carriers (S-SGS) and simulated KOH-activated BFS carriers for the CO_2 methanation reaction. The Ni-P-SGS catalysts had greater particular area of surface, smaller nickel size of particles, and stronger reducibility. As a result, Ni-P-SGS had a 7.4% greater CO_2 conversion than Ni-S-SGS, whilst Ni-S-SGS had a 72.8% CO_2 conversion. On both catalysts, the chemical pathway for CO_2 methanation was direct formate hydrogenation to CH_4. By providing sufficient sites for metallic dispersion, the imitated BFS carrier (P-SGS) may have exhibited greater activity due to its higher initial specific surface area.

6.3.2.3 UGSO

Cu-UGSO reaches the reaction equilibrium[58] in the reverse water gas shift process with excellent CO_2 conversion and 96% CO selectivity. This shows that because the modified slag oxides are rich in spinel oxides like Fe_3O_4, they may possibly be supportive. Furthermore, UGSO has been employed as a transporter for the hydrogenation of CO_2 into methanol. Using the improved co-precipitation process followed by the addition of ethanol, the 10% Cu-7.5Zn/UGSO-EtOH catalyst showed superior CO_2 conversion and CH_4 selectivity compared to the 10Cu7.5Zn/UGSO-EtOH catalyst without the addition of ethanol. The existence of oxygen defects, the enhancement of strong basic centres that assured increased reducibility, the reduction of ZnO and CuO particle sizes, and the increase in CuO ratio are all responsible for the improved catalytic performance. At 260°C and 20 bars, the produced solid waste catalysts demonstrated a greater recovery of methanol. This indicates that at high temperatures (> 600°C), metallurgical solid waste with rich spinel oxides has good reducibility and oxidation resistance, making a suitable carrier for the hydrogenation of CO_2 to C2+ products.

6.3.2.4 Fly Ash

The process of Dry Reforming Methane (DRM) is a useful technique for turning CO_2 into carbon fuels like syngas. Nonetheless, it is typically necessary to pretreat or load solid waste transporters with metal. Huang[59] used the sol-gel approach to create FA catalysts doped with Ni and MgO. By applying NaOH treatment and doping with MgO, the specific surface area was significantly enhanced from 12.3 m^2/g to 20.9 m^2/g. The dispersion of NiO was enhanced by the interaction between NiO and MgO. Additionally, they discovered that adding 20% MgO to fly ash enhanced DRM's catalytic activity while simultaneously preventing carbon deposition.

Gao[60] treated the FA carriers in two steps using NaOH and HNO_3, after which they were wet impregnated with Ni to load them. The XRD results indicated that Al-Ni alloys were present in Ni/A-A-FA, but no diffraction peaks of Al-Ni alloys were found in Ni/FA that had not been treated or in Ni/AFA that had just been treated with alkali. Furthermore, in comparison to Ni/A-FA and Ni/FA, the decrease peaks in the H_2-TPR data were pushed to higher temperatures, indicating a significant metal-carrier interaction.

Consequently, the Ni/A-A-FA catalyst shown more activity than the conventional SiO_2 catalyst loaded with Ni, with 95% conversion of CH_4 and CO_2 at 850°C.

6.3.2.5 LHA

In order to recover aluminium from aluminium salt slag, Torrez-Herrera[61] produced Hibonite-type lanthanum hexaaluminate (Ni/LHA) by HCl leaching. With an H_2/CO selectivity of 0.99 and CO_2 and CH_4 conversions of 80% and 81%, respectively, the catalytic performance outperformed that of the pure catalysts, Ni/SiO_2 and Ni/Al_2O_3. Moreover, adding Pt to Ni / LHA can lower the particle size of Ni, enhance the metal Pt's ability to prevent hydrogen spillover, and boost its resistance to carbon accumulation. Consequently, the 10% Ni-Pt (0-1%) catalysts' deactivation rate decreased in comparison to those without the Pt addition.[62]

6.4 CATALYTIC OXIDATION

6.4.1 NO/SO_2

6.4.1.1 Fly Ash

The elimination of NO and SO_2 can be accomplished using the photocatalytic oxidation of catalysts generated from FA. Xu[63] used gel composites combining FA and TiO_2 to create photocatalysts. They discovered that changing the FA content could increase the NOx removal efficiency. Additionally, Yu[64] created an FA photocatalyst coated with anatase and TiO_2. They discovered that the crystal size and the Fe-containing phase in FA had an impact on the efficacy of nitrogen removal. While the photocatalytic activity of roasted FA hematite was higher than that of magnetite, the larger the crystal size, the lower the NO removal.

6.4.1.2 Red Mud

RM is capable of acting as a plasma system catalyst. Bhattacharyya[65] treated NOx in biodiesel exhaust using a high frequency AC plasma and RM. He found that at a specific energy of 250 J·L^{-1}, NO was oxidized to NO_2 as a result of the synergistic impact of the plasma-RM, and that NO_2 was then adsorbed by the RM with an efficiency of 72%. Similar to this, Mohapatro removed NO using a two-stage plasma system. In the pre-oxidation zone, the Dielectric Barrier Discharge (DBD) plasma oxidized NO to NO_2, and in the post-oxidation area, it converted NOx to N_2.

6.4.1.3 Steel Slag

Industrial solid waste can be used as a cost-effective replacement for conventional wet adsorbents/catalysts in the moist desulfurisation of flue gases (WFGD) process. Furthermore, leached transition metal ions may be essential for the reaction's liquid-phase catalytic oxidation of SO_2 and NOx,[66] which makes the approach appropriate for small-sized boilers to medium-sized boilers or facilities in the non-ferrous, thermal, or chemical sectors.[67] It is often believed that pairing ozone peroxidation with slag from steel slurry adsorption is an effective technique to remove SO_2 and NOx simultaneously.[68] In addition to using the waste's high alkalinity to remove SO_2 and NOx, solid waste with a high metal oxide concentration can be used to remove air pollutants by liquid-phase catalytic combustion of the metal ions. A high iron to metal ratio copper slag is the best way to enhance liquid-phase catalytic oxidation efficiency. Tao[69] discovered that using copper tailing can recover iron sulfate in addition to removing significant quantities of SO_2. The primary constituents of copper tailing are Fe_2SiO_4 and Fe_3O_4, and the addition of Mn^{2+} enhanced the copper slag slurry's ability to remove SO_2 because of the relationship between Fe^{2+}/Fe^{3+} and Mn^{2+} on S(IV).[70] In the WFGD process, sulfur(IV) is catalytically oxidized by Fe^{2+} through a sequence of free radical chain reactions that have high SO_2 absorption. We discovered in our earlier research that SO_2 and NOx could be eliminated using $KMnO_4$ and copper slag. On the other hand, high $KMnO_4$ consumption as a result of the leaching of Fe^{2+} ions from Fe_3O_4 and Fe_2SiO_4 decreased the effectiveness of NOx removal. This issue was resolved by thermally pre-treating the raw copper slag and using the CaO calcination method to transform the iron-containing phases (Fe_3O_4, Fe_2SiO_4, and FeS) into a-Fe_2O_3, which improved the elimination of NOx. Furthermore, Fenton-like processes involving this thermally altered copper slag can produce free radicals, increasing the efficiency of NO/SO_2 combustion.[71]

Moreover, NO and SO_2 were eliminated from water-quenched manganese slag by thermal pretreatment employing $KMnO_4$ as an oxidant.[72] MnS produced MnO_2 by the CaO calcination pretreatment, and NO was removed at a rate that was 3.6 times greater than that of pure $KMnO_4$ solution. It was discovered that the primary variable influencing the effectiveness of NOx removal was the calcination temperature. This suggests that the oxidizing capacity of NOx at various calcination temperatures is influenced by distinct phase structures.

During the denitrification process, high alkalinity steel slag, red mud, and phosphorus deposits (including limestone) are employed; these materials remove NOx at a high rate when they enter the reaction process with O_3 and P_4 produced O_3. The slurry's high alkalinity, which encourages the uptake of NO_2 or compounds with high valent nitrogen, is responsible for the improved removal of NOx. Additionally, when utilising copper as well as manganese slag, the effectiveness of nitrogen removal can be improved by adding ozone. This increase results from the interaction of O_3 with released Fe^{2+} as well as Mn^{2+}, which develops a sophisticated oxidizing process that produces particularly reactive forms of oxygen.

6.4.2 H_2S

Numerous studies have been conducted using either high calcium or high iron solid waste to remove H_2S from humidified air. Obis[73] showed that H_2S adsorption is significantly impacted by the iron concentration of bottom ash from municipal household waste. Furthermore, because of their comparable chemical compositions, converter steel and calcium carbide slag can be employed for H_2S adsorption in place of bottom ash.[74] Fe_2O_3 in the converter steel slag, which reacts with H_2S to generate iron sulfide or monosulfurisation, may be the essential element for H_2S removal. In conclusion, the reactivity of H_2S[75] is regulated by comparable iron oxide minerals found in converter steel slag and bottom ash.

6.4.2.1 Blast Furnace Slag

The removal method of H_2S by steel slag and abrasive soil treated with BFS was studied by Montrs and Xie,[76] respectively. They discovered that in a very alkaline setting with high humidity, Fe^{3+} hydrated oxides were essential for the oxidation of H_2S through redox reactions. Iron oxides and H_2S combine to generate singlet sulfur, or iron sulfide, which is often responsible for the majority of H_2S removal.

6.4.2.2 Fly Ash

Many teams of researchers have looked into how coal FA might be used to remove organic sulfur and H_2S. According to a prior study, coal FA is capable of catalysing the oxidation of ethanethiol (CH_3CH_2SH), methyl mercaptan (CH_3SH), and H_2S.[77] Bound wood ash, which contains mullite

$(3Al_2O_3 2SiO_2)$, magnetite (Fe_3O_4), and hematite (Fe_2O_3), has been suggested as an effective way to oxidize H_2S, methyl mercaptan (MT), DMS, and DMDS at room temperature by low-temperature catalytic ozonation. The findings have shown that while the elemental S generation rate is decreased in the presence of O_2, H_2S can be oxidized to SO_2 in the presence of O_3. O_3 causes DMS to solely convert to DMSO and $DMSO_2$, suggesting that ozone on wood ash surfaces undergoes a different catalytic oxidation pathway than oxygen. The L-H process, also known as free radical generation, is followed in the catalytic oxidation of reduced sulfur.[78]

6.4.3 VOCs

6.4.3.1 Fly Ash

FA can reduce VOC emissions from coal-fired power plants by oxidation and adsorption.[79] For example, its fuel additive cost (FA) is just 10% of activated carbon's, making it an ideal substance for use in thermodynamic power plants. By increasing the oxygen motion on the catalyst surface, adding the active ingredients to the FA surface can improve its capacity for catalytic oxidation. For example, at 180-280°C, multilayer VOx species primarily oxidized toluene to benzene and benzoic acid, respectively. On the other hand, MnOx seems to be more efficient in this sense; at 280-320°C, ball-milled MnO_2/FA converts toluene to CO and CO_2.

Moreover, DFT calculations show that Mn_2O_3 has a greater adsorption energy than MnO_2, which favours the oxidation of benzene following MnO_2 reduction. Additionally, iron appears to be a site that undergoes oxidation for volatile organic molecules. For instance, Fe-loaded bagasse that had been steam-activated demonstrated greater CO_2 selectivity and butanol oxidation than inactivated bagasse between 150 and 300°C. Higher CO_2 selectivity[80] and improved Fe_2O_3 particle dispersion can explain this.

6.4.3.2 Red Mud

Furthermore, a more effective method for iron enrichment on RM surfaces is multi-acid treatment. Higher toluene conversions were obtained from calcination on the feed RM after oxalic and l-ascorbic acid treatments than from single HCl treatments. In addition, calcined RM treatments have been used to break down light hydrocarbons through hydrodechlorination, showing greater conversion rates than Fe_2O_3/Al_2O_3 or pure Fe_2O_3 catalysts for combustion.

By improving lattice oxygen mobility and reducibility, MnO_x can efficiently oxidize volatile organic compounds (VOCs). Using a hydrothermal technique, a mixture of rice husk loaded with Mn_2O_3 and RM was able to oxidize paraxylene by 93% at 400°C. This result demonstrated good thermal stability because of the redox characteristics of MnO_x. Furthermore, a toluene conversion of 100% was observed in Pt-loaded iron sludge (iron oxide) that was calcined at 200°C, suggesting a greater oxidizing capacity supported by metallic parts.[81]

6.4.3.3 Steel Slag

Domínguez[82] used dust and reduced steel slag for preparing porcelain foam scaffolds. After colloidal cerium and alumina washing, Au was deposited on these scaffolds via ion exchange. The Au/CeO_2-coated monoliths showed a higher level of 2-propanol oxidation sensitivity than the Au/Al_2O_3-coated monoliths because CeO_2 has the capacity to store oxygen. It was found that the main factors influencing toluene's ability to oxidize after washing were the dispersion and surface area of particles of Pt in the Pt/steel slag. This suggests that widely dispersed active sites are preferable to VOC combustion.[83]

6.5 CATALYSED HYDROLYSIS

6.5.1 VOCs

Red mud was extracted from industrial waste produced by the Bayer process and used by In-Heon Kwak[84] to create a catalyst for the breakdown of the greenhouse gas hydrofluorocarbon (HFC-134a). By running the catalyst continuously at 650°C for 66 hours, the catalyst's durability was assessed. During the stability test, HFC-134a's conversion grew synchronously and reached 99% after 50 hours. Strong catalytic activity and durability are demonstrated by the red mud catalysts during the hydrolysis process, as per the findings of the HFC-134a disintegration test and catalyst analytical studies.

6.5.2 COS

Zhu[85] prepared KOH-modified calcium carbide slag by over-volume impregnation method using calcium carbide slag as raw material, investigated the effect of different preparation conditions and process conditions on the removal of carbonyl sulfide (COS), and found that with the increase in the concentration of COS imported KOH-modified calcium carbide slag penetration and adsorption of COS increased firstly and then decreased, and the effect of the airspeed increase in the removal of COS became worse, and the increase in temperature can promote the hydrolysis of COS to H_2S.

6.6 SUMMARY AND OUTLOOKS

In response to the increasing number of solid waste and the difficulty of handling them, the resource utilisation of solid waste has become an inevitable trend for future development. Solid waste as materials for removing polluted gases have the advantages of low cost, easy accessibility and renewable utilisation, which are in line with the principle of sustainable development. Therefore, the use of solid waste as materials for removing polluted gases has a broad application prospect and important significance. A series of solid waste such as Biochar, BFS, RM, FA, and the like, were utilised to prepare adsorbent materials or catalysts. Most of them act as excellent carriers with large specific surface area, which is favourable for uniform loading of active components and enhances the adsorption or catalytic properties of gases.

As was previously indicated, a lot of research has been performed ono using industrial solid waste as materials with environmental benefits. These materials still have a lot of obstacles to overcome, and additional study is required to completely comprehend their active ingredients and eliminate various impurities using a variety of modification or synthesis techniques. The chemistry of industrial solid waste-based material combinations needs to be optimized in order to achieve the dual goals of exhaust gas treatment and solid waste utilisation because these materials have different metal compositions and active sites, which may limit the direct or indirect utilisation of industrial solid waste to a particular reaction.

In summary, the use of solid waste as materials for removing pollutant gases is an effective, environmentally friendly and sustainable method that will provide important technical support for improving the quality of the atmospheric environment. In future research, the adsorption and catalytic properties of solid waste can be further explored to optimize the material structure and improve the removal efficiency to make a greater contribution to the realisation of clean air.

REFERENCES

1. Ahmad, W.; Sethupathi, S.; Kanadasan, G.; Lau, L. C.; Kanthasamy, R. A review on the removal of hydrogen sulfide from biogas by adsorption using sorbents derived from waste. *Rev. Chem. Eng.* 2021, 37 (3), 407-431.

2. Cho, S.-H.; Lee, S.; Kim, Y.; Song, H.; Lee, J.; Tsang, Y. F.; Chen, W.-H.; Park, Y.-K.; Lee, D.-J.; Jung, S.; Kwon, E. E. Applications of agricultural residue biochars to removal of toxic gases emitted from chemical plants: A review. *Sci. Total Environ.* 2023, 868, 161655.

3. Wen, C.; Liu, T.; Wang, D.; Wang, Y.; Chen, H.; Luo, G.; Zhou, Z.; Li, C.; Xu, M. Biochar as the effective adsorbent to combustion gaseous pollutants: Preparation, activation, functionalization and the adsorption mechanisms. *Prog. Energ. Combust. Sci.* 2023, 99, 101098.

4. Somayajula, A.; Aziz, A. A.; Saravanan, P.; Matheswaran, M. Adsorption of mercury (II) ion from aqueous solution using low-cost activated carbon prepared from mango kernel. *Asia-Pac. J. Chem. Eng.* 2013, 8 (1), 1-10.

5. Bao, J.; Sun, X.; Ning, P.; Li, K.; Yang, J.; Wang, F.; Shi, L.; Fan, M. Industrial solid waste to environmental protection materials for removal of gaseous pollutants: A review. *Green Energ. Environ.* 2024.

6. Plaza, M. G.; Pevida, C.; Arias, B.; Fermoso, J.; Casal, M. D.; Martín, C. F.; Rubiera, F.; Pis, J. J. Development of low-cost biomass-based adsorbents for postcombustion CO_2 capture. *Fuel* 2009. 88 (12), 2442-2447.

7. Plaza, M. G.; Pevida, C.; Martín, C. F.; Fermoso, J.; Pis, J. J.; Rubiera, F. Developing almond shell-derived activated carbons as CO_2 adsorbents. *Sep. Purif. Technol.* 2010, 71 (1), 102-106.

8. Plaza, M. G.; García, S.; Rubiera, F.; Pis, J. J.; Pevida, C. Evaluation of ammonia modified and conventionally activated biomass based carbons as CO_2 adsorbents in postcombustion conditions. *Sep. Purif. Technol.* 2011, 80 (1), 96-104.

9. González, A. S.; Plaza, M. G.; Rubiera, F.; Pevida, C. Sustainable biomass-based carbon adsorbents for post-combustion CO_2 capture. *Chem. Eng. J.* 2013, 230, 456-465.

10. Plaza, M. G.; González, A. S.; Pevida, C.; Rubiera, F. Influence of water vapor on CO_2 adsorption using a biomass-based carbon. *Ind. Eng. Chem. Res.* 2014, 53 (40), 15488-15499.

11. Plaza, M. G.; Durán, I.; Querejeta, N.; Rubiera, F.; Pevida, C. Experimental and simulation study of adsorption in postcombustion conditions using a microporous biochar. 1. CO_2 and N_2 adsorption. *Ind. Eng. Chem. Res.* 2016, 55 (11), 3097-3112.

12. Plaza, M. G.; González, A. S.; Pis, J. J.; Rubiera, F.; Pevida, C. Production of microporous biochars by single-step oxidation: Effect of activation conditions on CO_2 capture. *Appl. Energ.* 2014, 114, 551-562.

13. Thote, J. A.; Iyer, K. S.; Chatti, R.; Labhsetwar, N. K.; Biniwale, R. B.; Rayalu, S. S. In situ nitrogen enriched carbon for carbon dioxide capture. *Carbon* 2010, 48 (2), 396-402.

14. Sevilla, M.; Fuertes, A. B. Sustainable porous carbons with a superior performance for CO_2 capture. *Energ. Environ. Sci.* 2011, 4 (5), 1765-1771.

15. Sevilla, M.; Falco, C.; Titirici, M.-M.; Fuertes, A. B. High-performance CO_2 sorbents from algae. *RSC Adv.* 2012, 2 (33), 12792-12797.

16. Shafeeyan, M. S.; Daud, W. M. A. W.; Houshmand, A.; Arami-Niya, A. Ammonia modification of activated carbon to enhance carbon dioxide adsorption: Effect of pre-oxidation. *Appl. Surf. Sci.* 2011, 257 (9), 3936-3942.

17. Wei, H.; Deng, S.; Hu, B.; Chen, Z.; Wang, B.; Huang, J.; Yu, G. Granular bamboo-derived activated carbon for high CO_2 adsorption: The dominant role of narrow micropores. *ChemSusChem* 2012, 5 (12), 2354-2360.

18. Deng, S.; Wei, H.; Chen, T.; Wang, B.; Huang, J.; Yu, G. Superior CO_2 adsorption on pine nut shell-derived activated carbons and the effective micropores at different temperatures. *Chem. Eng. J.* 2014, 253, 46-54.

19. Hao, W.; Björkman, E.; Lilliestråle, M.; Hedin, N. Activated carbons prepared from hydrothermally carbonised waste biomass used as adsorbents for CO_2. *Appl. Energ.* 2013, 112, 526-532.

20. Tian, S.; Jiang, J.; Yan, F.; Li, K.; Chen, X. Synthesis of highly efficient CaO-based, self-stabilizing CO_2 sorbents via structure-reforming of steel slag. *Environ. Sci. Technol.* 2015, 49 (12), 7464-7472.

21. Polettini, A.; Pomi, R.; Stramazzo, A. CO_2 sequestration through aqueous accelerated carbonation of BOF slag: A factorial study of parameters effects. *J. Environ. Manage.* 2016, 167, 185-195.

22. Chang, E.-E.; Pan, S.-Y.; Chen, Y.-H.; Tan, C.-S.; Chiang, P.-C. Accelerated carbonation of steelmaking slags in a high-gravity rotating packed bed. *J. Hazard. Mater.* 2012, 227-228, 97-106.

23. Zhang, X.; Gao, B.; Creamer, A. E.; Cao, C.; Li, Y. Adsorption of VOCs onto engineered carbon materials: A review. *J. Hazard. Mater.* 2017, 338, 102-123.

24. Sun, C.; Yuan, J.; Xu, H.; Huang, S.; Wen, X.; Tong, N.; Zhang, Y. Simultaneous removal of nitric oxide and sulfur dioxide in a biofilter under micro-oxygen thermophilic conditions: Removal performance, competitive relationship and bacterial community structure. *Bioresour. Technol.* 2019, 290, 121768.

25. Wang, S. Application of solid ash based catalysts in heterogeneous catalysis. *Environ. Sci. Technol.* 2008, 42 (19), 7055-7063.

26. Dindi, A.; Quang, D. V.; Vega, L. F.; Nashef, E.; Abu-Zahra, M. R. M. Applications of fly ash for CO_2 capture, utilisation, and storage. *J. CO2 Util.* 2019, 29, 82-102.

27. Mochida, I.; Korai, Y.; Shirahama, M.; Kawano, S.; Hada, T.; Seo, Y.; Yoshikawa, M.; Yasutake, A. Removal of SO$_x$ and NO$_x$ over activated carbon fibers. *Carbon* 2000, 38 (2), 227-239.

28. Huang, Z.; Long, J.; Deng, L.; Che, D. Feasibility study of using carbide slag as in-bed desulfuriser in circulating fluidized bed boiler. *Appl. Sci.* 2019, 9 (21), 4517.

29. Li, Y.; Fan, Z.; Shi, J.; Liu, Z.; Shangguan, W. Post plasma-catalysis for VOCs degradation over different phase structure MnO$_2$ catalysts. *Chem. Eng. J.* 2014, 241, 251-258.

30. Gil, R. R.; Ruiz, B.; Lozano, M. S.; Martín, M. J.; Fuente, E. VOCs removal by adsorption onto activated carbons from biocollagenic waste of vegetable tanning. *Chem. Eng. J.* 2014, 245, 80-88.

31. Guieysse, B.; Hort, C.; Platel, V.; Munoz, R.; Ondarts, M.; Revah, S. Biological treatment of indoor air for VOC removal: Potential and challenges. *Biotechnol. Adv.* 2008, 26 (5), 398-410.

32. Hui, K. S.; Chao, C. Y. H. Effects of step-change of synthesis temperature on synthesis of zeolite 4A from coal fly ash. *Micropor. Mesopor. Mater.* 2006, 88 (1), 145-151.

33. Hollman, G. G.; Steenbruggen, G.; Janssen-Jurkovičová, M. A two-step process for the synthesis of zeolites from coal fly ash. *Fuel.* 1999, 78 (10), 1225-1230.

34. Tsai, J.-H.; Chiang, H.-M.; Huang, G.-Y.; Chiang, H.-L. Adsorption characteristics of acetone, chloroform and acetonitrile on sludge-derived adsorbent, commercial granular activated carbon and activated carbon fibers. *J. Hazard. Mater.* 2008, 154 (1), 1183-1191.

35. Anfruns, A.; Martin, M. J.; Montes-Morán, M. A. Removal of odourous VOCs using sludge-based adsorbents. *Chem. Eng. J.* 2011, 166 (3), 1022-1031.

36. Zhang, X.; Gao, B.; Zheng, Y.; Hu, X.; Creamer, A. E.; Annable, M. D.; Li, Y. Biochar for volatile organic compound (VOC) removal: Sorption performance and governing mechanisms. *Bioresour. Technol.* 2017, 245, 606-614.

37. Zhang, X.; Gao, B.; Fang, J.; Zou, W.; Dong, L.; Cao, C.; Zhang, J.; Li, Y.; Wang, H. Chemically activated hydrochar as an effective adsorbent for volatile organic compounds (VOCs). *Chemosphere* 2019, 218, 680-686.

38. Zhang, X.; Xiang, W.; Wang, B.; Fang, J.; Zou, W.; He, F.; Li, Y.; Tsang, D. C. W.; Ok, Y. S.; Gao, B. Adsorption of acetone and cyclohexane onto CO$_2$ activated hydrochars. *Chemosphere* 2020, 245, 125664.

39. Xiang, W.; Zhang, X.; Chen, K.; Fang, J.; He, F.; Hu, X.; Tsang, D. C. W.; Ok, Y. S.; Gao, B. Enhanced adsorption performance and governing mechanisms of ball-milled biochar for the removal of volatile organic compounds (VOCs). *Chem. Eng. J.* 2020, 385, 123842.

40. Zhang, X.; Miao, X.; Xiang, W.; Zhang, J.; Cao, C.; Wang, H.; Hu, X.; Gao, B. Ball milling biochar with ammonia hydroxide or hydrogen peroxide enhances its adsorption of phenyl volatile organic compounds (VOCs). *J. Hazard. Mater.* 2021, 403, 123540.

41. Zhu, C.; Duan, Y.; Wu, C.-Y.; Zhou, Q.; She, M.; Yao, T.; Zhang, J. Mercury removal and synergistic capture of SO_2/NO by ammonium halides modified rice husk char. *Fuel* 2016, 172, 160-169.

42. Klasson, K. T.; Lima, I. M.; Boihem, L. L.; Wartelle, L. H. Feasibility of mercury removal from simulated flue gas by activated chars made from poultry manures. *J. Environ. Manage.* 2010, 91 (12), 2466-2470.

43. Guo, J.; Luo, Y.; Lua, A. C.; Chi, R.; Chen, Y.; Bao, X.; Xiang, S. Adsorption of hydrogen sulfide (H_2S) by activated carbons derived from oil-palm shell. *Carbon* 2007, 45 (2), 330-336.

44. Wang, S.; Nam, H.; Nam, H. Utilisation of cocoa activated carbon for trimethylamine and hydrogen sulfide gas removals in a confined space and its techno-economic analysis and life-cycle analysis. *Environ. Prog. Sustain. Energ.* 2019, 38 (6), e13241.

45. Chen, L.; Ren, S.; Jiang, Y.; Liu, L.; Wang, M.; Yang, J.; Chen, Z.; Liu, W.; Liu, Q. Effect of Mn and Ce oxides on low-temperature NH_3-SCR performance over blast furnace slag-derived zeolite X supported catalysts. *Fuel* 2022, 320, 123969.

46. Yang, J.; Lei, S.; Yu, J.; Xu, G. Low-Cost V-W-Ti SCR catalyst from titanium-bearing blast furnace slag. *J. Environ. Chem. Eng.* 2014, 2 (2), 1007-1010.

47. Tran, T.-S.; Yu, J.; Li, C.; Guo, F.; Zhang, Y.; Xu, G. Structure and performance of a V_2O5-WO_3/TiO_2-SiO_2 catalyst derived from blast furnace slag (BFS) for $DeNO_x$. *RSC Adv.* 2017, 7 (29), 18108-18119.

48. Wang, B.; Ma, J.; Wang, D.; Gao, Z.; Shi, Q.; Gao, C.; Lu, J. Acid-pretreated red mud for selective catalytic reduction of NO_x with NH_3: Insights into inhibition mechanism of binders. *Catal. Today.* 2021, 376, 247-254.

49. Chen, Y.; Liu, C.; Guo, S.; Mu, T.; Wei, L.; Lu, Y. CO_2 capture and conversion to value-added products promoted by MXene-based materials. *Green Energ. Environ.* 2022, 7 (3), 394-410.

50. Xuan, X.; Yue, C.; Li, S.; Yao, Q. Selective catalytic reduction of NO by ammonia with fly ash catalyst. *Fuel* 2003, 82 (5), 575-579.

51. Wang, T.; Zheng, J.; Liu, H.; Peng, Q.; Zhou, H.; Zhang, X. Adsorption characteristics and mechanisms of Pb^{2+} and Cd^{2+} by a new agricultural waste - *Caragana korshinskii* biomass derived biochar. *Environ. Sci. Pollut. Res.* 2021, 28 (11), 13800-13818.

52. Li, G.; Wang, B.; Wang, H.; Ma, J.; Xu, W. Q.; Li, Y.; Han, Y.; Sun, Q. Fe and/or Mn oxides supported on fly ash-derived SBA-15 for low-temperature NH_3-SCR. *Catal. Commun.* 2018, 108, 82-87.

53. Chen, H.; Zhang, Y. J.; He, P. Y.; Li, C. J. Cost-effective and facile one step synthesis of ZSM-5 from silica fume waste with the aid of metakaolin and its NO_x removal performance. *Powder Technol.* 2020, 367, 558-567.

54. Han, F.; Gao, Y.; Huo, Q.; Han, L.; Wang, J.; Bao, W.; Chang, L. Characteristics of vanadium-based coal gasification slag and the NH_3-selective catalytic reduction of NO. *Catalysts* 2018, 8 (8).

55. Dong, X.; Jin, B.; Cao, S.; Meng, F.; Chen, T.; Ding, Q.; Tong, C. Facile use of coal combustion fly ash (CCFA) as Ni-Re bimetallic catalyst support for high-performance CO_2 methanation. *Waste Manage.* 2020, 107, 244-251.

56. Chen, X.; He, Y.; Cui, X.; Liu, L. High value utilisation of waste blast furnace slag: New Ni-CeO$_2$/hBFS catalyst for low temperature CO_2 methanation. *Fuel* 2023, 338, 127309.

57. Wan, H.; He, Y.; Su, Q.; Liu, L.; Cui, X. Slag-based geopolymer microspheres as a support for CO_2 methanation. *Fuel* 2022, 319, 123627.

58. Desgagnés, A.; Iliuta, M. C. Intensification of CO_2 hydrogenation by in-situ water removal using hybrid catalyst-adsorbent materials: Effect of preparation method and operating conditions on the RWGS reaction as a case study. *Chem. Eng. J.* 2023, 454, 140214.

59. Huang, Y.; Li, Q.; Zhao, T.; Zhu, X.; Wang, Z. The dry reforming of methane over fly ash modified with different content levels of MgO. *RSC Adv.* 2021, 11 (23), 14154-14160.

60. Gao, Y.; Jiang, J.; Meng, Y.; Aihemaiti, A.; Ju, T.; Chen, X.; Yan, F. A novel nickel catalyst supported on activated coal fly ash for syngas production via biogas dry reforming. *Renew. Energ.* 2020, 149, 786-793.

61. Torrez-Herrera, J. J.; Korili, S. A.; Gil, A. A comparative study of the catalytic performance of nickel supported on a hibonite-type La-hexaaluminate synthesised from aluminum saline slags in the dry reforming of methane. *Int. J. Hydrogen Energ.* 2022, 47 (93), 39678-39686.

62. Torrez-Herrera, J. J.; Korili, S. A.; Gil, A. Bimetallic (Pt-Ni) La-hexaaluminate catalysts obtained from aluminum saline slags for the dry reforming of methane. *Chem. Eng. J.* 2022, 433, 133191.

63. Xu, M.; Clack, H.; Xia, T.; Bao, Y.; Wu, K.; Shi, H.; Li, V. Effect of TiO_2 and fly ash on photocatalytic NO_x abatement of engineered cementitious composites. *Constr. Build. Mater.* 2020, 236, 117559.

64. Yu, Y. Preparation of nanocrystalline TiO_2-coated coal fly ash and effect of iron oxides in coal fly ash on photocatalytic activity. *Powder Technol.* 2004, 146 (1), 154-159.

65. Bhattacharyya, A.; Rajanikanth, B. S. Biodiesel exhaust treatment with HFAC plasma supported by red mud: Study on $DeNO_x$ and power consumption. *Energ. Procedia* 2015, 75, 2371-2378.

66. Bao, J.; Yang, J.; Song, X.; Han, R.; Ning, P.; Lu, X.; Fan, M.; Wang, C.; Sun, X.; Li, K. The mineral phase reconstruction of copper slag as Fenton-like catalysts for catalytic oxidation of NO_x and SO_2: Variation in active site and radical formation pathway. *Chem. Eng. J.* 2022, 450, 138101.

67. Meng, Z.; Wang, C.; Wang, X.; Chen, Y.; Li, H. Simultaneous removal of SO_2 and NO_x from coal-fired flue gas using steel slag slurry. *Energ. Fuels* 2018, 32 (2), 2028-2036

68. Pei-Shi, S.; Ping, N.; Wen-Biao, S. Liquid-phase catalytic oxidation of smelting-gases containing SO_2 in low concentration. *J. Clean. Prod.* 1998, 6 (3), 323-327.

69. Tao, L.; Wang, X.; Ning, P.; Wang, L.; Fan, W. Removing sulfur dioxide from smelting flue and increasing resource utilisation of copper tailing through the liquid catalytic oxidation. *Fuel Process. Technol.* 2019, 192, 36-44.

70. Grgić, I.; Berčič, G. A simple kinetic model for autoxidation of S(IV) oxides catalyzed by iron and/or manganese ions. *J. Atmospheric Chem.* 2001, 39 (2), 155-170.

71. Bao, J.; Li, K.; Ning, P.; Wang, C.; Song, X.; Luo, Y.; Sun, X. Study on the role of copper converter slag in simultaneously removing SO_2 and NO_x using $KMnO_4$/copper converter slag slurry. *J. Environ. Sci.* 2021, 108, 33-43.

72. Luo, Y.; Ning, P.; Wang, C.; Wang, F.; Ma, Y.; Bao, J.; Sun, X.; Li, K. Pretreated water-quenched-manganese-slag slurry for high-efficiency one-step desulfurisation and denitrification. *Sep. Purif. Technol.* 2020, 250, 117164.

73. Fontseré Obis, M.; Germain, P.; Bouzahzah, H.; Richioud, A.; Benbelkacem, H. The effect of the origin of MSWI bottom ash on the H_2S elimination from landfill biogas. *Waste Manage.* 2017, 70, 158-169.

74. Sarperi, L.; Surbrenat, A.; Kerihuel, A.; Chazarenc, F. The use of an industrial by-product as a sorbent to remove CO_2 and H_2S from biogas. *J. Environ. Chem. Eng.* 2014, 2 (2), 1207-1213.

75. Hu, Y.; Wu, S.; Li, Y.; Zhao, J.; Lu, S. H_2S removal performance of $Ca_3Al_2O_6$-stabilized carbide slag from CO_2 capture cycles using calcium looping. *Fuel Process. Technol.* 2021, 218, 106845.

76. Xie, M.; Leung, A. K.; Ng, C. W. W. Mechanisms of hydrogen sulfide removal by ground granulated blast furnace slag amended soil. *Chemosphere* 2017, 175, 425-430.

77. Kastner, J. R.; Das, K. C.; Melear, N. D. Catalytic oxidation of gaseous reduced sulfur compounds using coal fly ash. *J. Hazard. Mater.* 2002, 95 (1), 81-90.

78. Kastner, J. R.; Das, K. C.; Buquoi, Q.; Melear, N. D. Low temperature catalytic oxidation of hydrogen sulfide and methanethiol using wood and coal fly ash. *Environ. Sci. Technol.* 2003, 37 (11), 2568-2574.

79. Cheng, J.; Liu, J.; Wang, T.; Sui, Z.; Zhang, Y.; Pan, W.-P. Reductions in volatile organic compound emissions from coal-fired power plants by combining air pollution control devices and modified fly ash. *Energ. Fuels* 2019, 33 (4), 2926-2933.

80. Halász, J.; Hodos, M.; Hannus, I.; Tasi, G.; Kiricsi, I. Catalytic detoxification of C_2-chlorohydrocarbons over iron-containing oxide and zeolite catalysts. *Colloids Surf. Physicochem. Eng. Asp.* 2005, 265 (1), 171-177.

81. Sanchis, R.; Dejoz, A.; Vázquez, I.; Vilarrasa-García, E.; Jiménez-Jiménez, J.; Rodríguez-Castellón, E.; López Nieto, J. M.; Solsona, B. Ferric sludge derived from the process of water purification as an efficient catalyst and/ or support for the removal of volatile organic compounds. *Chemosphere* 2019, 219, 286-295.

82. Domínguez, M. I.; Sánchez, M.; Centeno, M. A.; Montes, M.; Odriozola, J. A. 2-propanol oxidation over gold supported catalysts coated ceramic foams prepared from stainless steel waste. *J. Mol. Catal. Chem.* 2007, 277 (1), 145-154.

83. Domínguez, M. I.; Barrio, I.; Sánchez, M.; Centeno, M. Á.; Montes, M.; Odriozola, J. A. CO and VOCs oxidation over Pt/SiO_2 catalysts prepared using silicas obtained from stainless steel slags. *Catal. Today* 2008, 133-135, 467-474.

84. Kwak, I.-H.; Lee, E.-H.; Kim, J.-B.; Nam, S.-C.; Ryi, S.-K. Hydrolysis of HFC-134a using a red mud catalyst to reuse an industrial waste. *J. Ind. Eng. Chem.* 2024.

85. Zhu, F.; Chen, K.; Huang, X. Performance study of KOH-modified calcium carbide slag for carbonyl sulfur removal. *J. Chem. Eng.* 2023, 74 (06), 2668-2679.

CHAPTER 7

Combination of Artificial Intelligence and Solid Waste-Based Materials

Zijie Xiao, Bowen Yang, Yingzi Zeng, Hongtao Shi, and Xiaochi Feng

7.1 INTRODUCTION

The swift expansion of urban areas and economic prosperity has led to a considerable upsurge in global solid waste. Projections indicate that by 2025, solely urban regions will produce 2.2 billion tonnes of solid waste.[1] The Solid Waste-based Materials (SWMs) demonstrate superior applicability and environmental friendliness as the structural and functional materials.[2] Recycled structural materials are normally derived from the solid waste generated by isogenous materials after engineering and construction processes, undergoing reconfiguration to regenerate their original functions and properties.[3] Functional materials are frequently sourced from inorganic or biomass solid waste and then subjected to specialized processing, such as calcination and transition metal loading, to create materials with distinct physical and chemical properties.[4]

The recycling process has a significant impact on the structure and properties of materials. During the recycling process, raw materials are collected and reprocessed, involving multiple stages from collection and sorting to processing and manufacturing, posing a direct influence on the structure and properties of materials. Primarily, the collection and sorting stages in the recycling process can lead to heterogeneity in the original materials.[5] Recycled materials often come from various sources and may contain different types of impurities or additives, not only making the structure of recycled materials more complex and introducing variability in material properties, but also providing an external source for the synthesis of composite materials.[6] Subsequently, the processing steps, such as crushing, melting, and reshaping, in the recycling process play a crucial role in influencing the material structure.[7] These processes can affect the crystal structure, size, and boundary distribution, thereby impacting important properties such as mechanical performance, electrical conductivity, and durability. Among them, melting and reshaping may result in crystal rearrangement, influencing the strength and plasticity of the SWMs.[8, 9] Moreover, the manufacturing technology and conditions in the recycling process also influence material properties. For instance, processing parameters such as temperature, pressure, and time can affect the crystallinity, molecular arrangement, and defect formation in recycled materials, subsequently influencing their mechanical properties and chemical stability.[10-12] Generally, the impact of the recycling process on the structure and properties of materials is multifaceted. From the heterogeneity introduced during collection and sorting to changes in crystalline structure during processing and manufacturing, along with the influence of manufacturing process conditions, these factors collectively determine the final performance of recycled materials.[13] Therefore, in-depth research and optimization are necessary for different recycling processes to achieve precise control and improvement of material structure and properties.

The construction of SWMs is modelled after the conventional process of material preparation using ordinary raw materials, where the paradigm shift involves substituting solid waste as the preparative material, thereby not only achieving resource conservation and mitigation of environmental pollution, but also improving material properties through doping or by way of assistance. Therefore, parameter variations in the recycling process significantly impact the structure and properties of materials. In our previous work, the introduction of N and S heteroatoms for the preparation of Co-C composite material results in a simultaneous transformation of the Co and the structures of carbon matrix.[14, 15] Under the circumstance of catalysing the degradation of pharmaceutical pollutants with PeroxyMonoSulfate (PMS), the heteroatom-doped material alters the catalytic activation mechanism by modifying the reactive species. As a result, the modified material exhibits significantly enhanced efficacy, and the use of different reactive species leads to the conversion of degradation byproducts, thereby influencing the environmental impact of the treatment outcomes. Due to the presence of hetero-elements not included in the raw materials, the introduction of solid waste recycling into the material synthesis process can be considered as introducing doping with elements, when it is feasible that the modification to the material leads to differences in its physical and chemical properties. Metal oxides showed an influencing trend to element doping, the extent of S substitution for O induces the structure of Co_3O_4 into Co_9S_8 as the material, altering the catalytic mechanisms and causing augmentation of the efficiency in organic pollutant degradation.[16] The structure of a material influences its functionality, and changes in structure will manifest as differences in functionality through reaction mechanisms, and the relationship could be traceable and predictable.

The rapid advancements in uncovering novel materials are credited to the refinement of dependable quantum-mechanical techniques for predicting crystal structures. The characteristics of a material are notably influenced by its structure, underscoring the critical role of structure prediction in the realm of computational materials discovery.[17] While prediction was once viewed as a daunting challenge, the progression of state-of-the-art computational tools has facilitated the anticipation of structures for numerous innovative and progressively intricate materials.[18] The core challenge in prediction is to propose corresponding laws to establish a projection between the measurable and easily obtainable system attributes and their properties. Quantum chemistry provides us with such a decision rule, linking the wave functions to the properties through the Schrödinger equation, and it is proposed to replace the rather tedious rule based on the Schrödinger equation with a regression model based on similarity.[19] Machine Learning (ML) demonstrates strong applicability in tasks related to high-dimensional data such as classification and regression. Focused on extracting knowledge and insights from extensive databases, ML learns from past computations to generate reliable and reproducible decisions and outcomes. As a result, it has played a crucial role in various fields, notably in areas like speech recognition, image recognition, bioinformatics, information security, and Natural Language Processing (NLP).[20-23] ML methods can propose a general formal system, enabling the rapid discovery of relationships between structure and properties, thereby efficiently and accurately predicting a diverse set of unrelated properties of material systems.[24] Quantitative Structure-Activity Relationship (QSAR) is an approach used to guide the exploration of relationships between structure-structure, structure-function, or function-function of the materials. Besides, different materials constitute distinct catalytic systems, and contrasting the reaction mechanisms under different catalytic systems can also involve QSAR to describe how the structures of materials influence change in catalytic mechanisms and performance.[25] The Density Functional Theory (DFT) method excellently captures the intrinsic properties of materials, with the assistance of regression methods such as ML, it enables accurate predictions of properties and limitations in the design of materials for lithium-ion batteries, hydrogen production and storage, superconductors, photovoltaic materials, and thermoelectric materials, providing valuable insights for material design and invention.[26] Solid waste serves as impurities to introduce variable parameters in the synthesis of the SWMs, which allows for the description of the impact of solid waste on material synthesis through statistical relationships or logical connections. Consequently, the types and quantities of solid waste are proposed to have an influence on the optimal processes for material synthesis.

In this review, we conducted a comprehensive discussion on the synthesis of SWMs used in structural and functional materials, intervention methods for solid waste raw materials, preparation availability, and application scenarios, and explored the process of solid waste recycling and regeneration, discussing its availability, economic viability, and current challenges in preparing SWMs. We provided a review of the physical and chemical properties, as well as functionalities, of various types of SWMs, and specifically conducted an analysis and review based on catalytic properties, particularly focusing on physical properties or adsorption as the front-end reaction properties of catalytic actions. We analysed the regulation of catalytic performance based on catalytic mechanisms, thereby discussing material modification strategies for SWMs used in catalysis, to comprehensively consider the application of models at three levels, namely decision-tree, composite neural network models, and graph neural networks, in predicting the structure and function relationships of SWMs. Finally, we presented a forward-thinking perspective on the integration of SWMs and AI-technology, spanning from solid waste recycling to the regeneration process of SWMs, and advancing to the role of AI-technology in guiding material modification and preparation. In this review, we aim to utilise AI-technology as an auxiliary role in various aspects of the regeneration process, understanding the challenges and breakthroughs in the preparation, efficacy, and application of SWMs, to propose considerations and suggestions for the future improvement of AI technology in all aspects of SWMs

7.2 STRUCTURING OF SOLID WASTE-BASED MATERIAL

In the process of solid waste regeneration, waste resources are utilised in the production of materials with distinct structural and functional properties. Structural performance primarily concerns the macroscopic mechanical characteristics and deformability of materials, whereas functional performance focuses on specific attributes such as photoelectronic properties, magnetism, and catalytic effects. The manufacturing process of materials significantly influences their structure and properties, leading to markedly different outcomes based on varying methodologies. Consequently, critical manufacturing processes decisively impact both the structure and properties of materials. Furthermore, based on the origins from solid waste, the collection and pre-processing of waste raw materials directly impacts the properties of the materials along with the normal manufacturing process.

7.2.1 Recycled Structural Materials

Recycled structural materials are produced using recycled or SWMs as the origin, often resulting in the integrated or bulk materials with considerable mechanical strength (Figure 7.1). Given the need to diminish the demand for natural resources and significantly mitigate environmental impacts, the recycled structural materials exhibit great potential, contingent upon cost feasibility. The manufacturing process involves the collection, treatment, and reprocessing of solid waste to render it suitable for the fabrication of SWMs. Commercial structural materials usually demand high structural integrity, and as a consequence, the incorporation of recycled or waste materials in their entirety into finished structural materials proves challenging. Hence, solid waste is frequently subjected to pre-processing and subsequently mixing with other raw materials during manufacturing to enhance the structural properties of the resulting SWMs.

Cement can be considered to be in one of the fundamental categories within structural materials, where its strength, cost, and feasibility of production are among the primary concerns in the manufacturing process of cementitious SWMs. The ash generated from the incineration of various

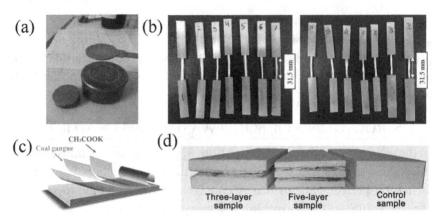

FIGURE 7.1 (a) Concrete sample with the addition of Sugarcane Bagasse Ash (SBA),[28] (b) Fibre roving specimens for tensile testing from the dissolution experiment (left) and from a roll of virgin Johns Manville StarRov 086 (right),[39] (c) Coal gangue sample construction, and (d) samples with different content of plastic waste and EPS.

agricultural waste materials can serve as an admixture for cementitious mortar. The SiO_2 content in the combustion ash of most agricultural waste is comparable to that of slag-based binders. However, when compared to slag-based binders, incorporating agricultural waste ash into fly ash-based binders can result in improved acid resistance and significantly reduce the drying shrinkage of slag-based binders.[27] Additionally, X-Ray Fluorescence (XRF) results indicate that the incorporation of ash from bio-waste significantly increases the presence of active inorganic compounds, thereby enhancing the strength and toughness of cement.[28] The use of bio-waste ash as a partial substitute for cement is considered a potentially more environmentally friendly approach. At certain substitution levels, the mechanical properties of cement exhibit minimal to no compromise and may even experience enhancement. The utilisation of sludge ash presents a commendable case of hazardous waste reclamation. However, the partial replacement of silicate cement with Sewage Sludge Ash (SSA) leads to a similar decrease in the functional availability and cohesion of mortar.[29] Particularly at substitution rates exceeding 30%, there is an extended solidification time, consequently diminishing their processability.

Due to its homogeneous amorphous state, glass is easily recycled through melting, making the recycling of waste glass a well-established process. Glass fibres can serve as micro-fillers in aluminium-based ultra-lightweight concrete, with advancements stemming from the bonding properties exhibited by urban-waste glass, therefore, cement can be replaced by some portion of recycled glass fibre.[30, 31] Compared to bio-waste ash, glass fibres can further enhance the toughness of structural materials.[32] By extracting valuable resources from landfills and transforming them into beneficial raw materials, global carbon dioxide emissions can be reduced. The current research on the recycling process of stainless steel and aluminium waste primarily focuses on the mechanical and electrochemical effects, including the adjustment of precipitation-free zones around crystal boundaries, casting microstructures, processing parameters, and the resulting mechanical, functional, and chemical properties. The fabrication of structural metallic materials mainly focused on the control of impurities, which emphasizing the significance of the calcination process.

The recycling of plastic waste involves reprocessing shredded plastic through melting and reshaping, commonly used for repurposing in packaging and structural engineering.[33] The recycling process of packaging plastics primarily encounters challenges in waste management, including sorting, cleaning, and energy consumption.[34] Conversely, the recycling process of structural plastics still faces various tests concerning mechanical strength.[35] In the assembly process of construction, recycled plastics can serve as alternatives to conventional aggregates, substituting as clay-fired paper bricks, offering fire-resistant and waterproof qualities to fireboards.[36-38] Using green recycled plastics as substitutes for traditional building materials helps in conserving limited natural resources, maximizing the reduction of recycled resources in minimizing the environmental impact of the concrete and plastics industries, simultaneously lowering their life-cycle costs.

7.2.2 Recycled Functional Materials

Although bio-waste is more environmentally friendly compared to non-biodegradable waste, it cannot be directly regenerated as is the case with plastics and metals. The World Bank has estimated that by 2025, municipal solid waste (only a part of bio-waste) from urban areas worldwide may reach 2.2 billion tonnes per annum, and waste generation rates might double over the next two decades in developing countries.[40, 41] Agricultural waste in particular, far exceeds the accumulation rate of plastics, metals, or other construction materials.[42] The treatment of bio-waste is generally carried out through biological methods involving microbial fermentation and degradation, as well as chemical methods involving incineration.[43] Typically, biomass waste is harmless to the environment, except for a few types of waste, such as sludge, which pose environmental hazards. However, excessive accumulation of biomass waste can occupy significant land or other resources. Therefore, converting it into solid waste materials is desirable. The synthesis of carbon-based materials mainly involves high-temperature calcination, which not only has low equipment requirements to minimize costs, but also facilitates the construction of most material functionalities. Plastics, manufactured fibres, and other polymer waste are also crucial feedstocks for synthesizing carbon-based materials, particularly in the carbonization process, where the generation of nanocarbon materials with particular structures is facilitated.[44, 45] The functional materials prepared from bio-waste as raw materials have low production costs, high availability, and have potential for industrial applications. Resource utilisation of bio-waste is beneficial for alleviating environmental burdens, expanding resource sources, and simultaneously enabling the application of the products as functional materials in a broader range of fields.

Bio-waste can serve as raw materials for crystalline carbon, and obtaining various nanocarbon materials through carbonizing and further processing, with the various material establishment showed in Figure 7.2. Zero-dimensional nanocarbon materials have extremely small dimensions, typically exhibiting a spherical or near-spherical structure. The unique electrical, optical, and magnetic properties of these materials arise from their nano-scaled characteristics, where quantum effects and surface effects often lead to properties and behaviours distinct from those of larger-scale materials.[46] Utilising bio-waste as a raw material for the synthesis of zero-dimensional carbon materials presents an opportunity to achieve waste reduction and product augmentation. Uchimiya et al.[47] investigates the widespread occurrence of pyrogenic carbon in soil and examines the interactions between engineered carbon nanoparticles (nC60-stir) and natural pyrogenic carbon. One-dimensional carbon materials are popular in current research also because of their superior mechanical and electrical properties, which facilitate the utilisation of bio-waste for the production of this recycled material. Liu et al.[48] successfully converted waste PolyPropylene (PP) through plastic pyrolysis techniques into Carbon NanoTubes (CNTs) containing Fe. They employed chemical vapour deposition to prepare carbon nanotube-coated Fe-Mo/MgO nanomaterials (CNT@Fe-Mo/

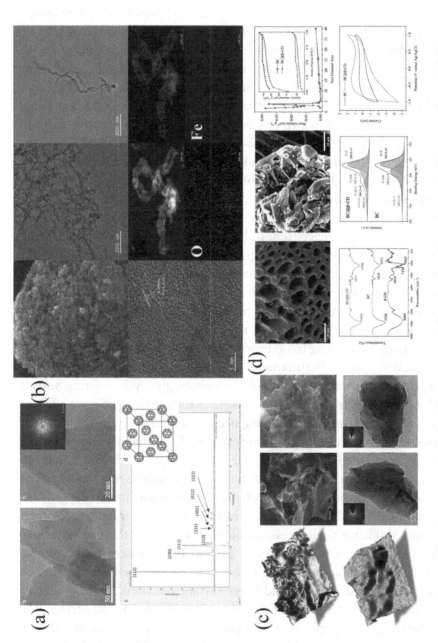

FIGURE 7.2 The characterisation of solid waste-based functional carbon materials (a) 0-dimensional,[47] (b) 1-dimensional,[48] (c) 2-dimensional,[49] and (d) 3-dimensional[52] materials.

MgO) using PP as a precursor. This material demonstrated efficient and low-energy inactivation of pathogenic bacteria without the need for additional energy or reactive oxidants. Two-dimensional carbon materials offer considerable prospects for applications through their considerable extensibility and conductivity, to mitigate the pollution associated with solid waste, enabling sustainable waste-to-resource recovery, upgrading, and remanufacturing. Bio-waste such as coconut shells, pine cones, and palm kernels, and so forth, serve as excellent sources of carbon-based functional materials, which can be utilised to serve as effective reduced Graphene Oxide (rGO), with applications ranging from fuel cells and capacitor electrodes to Volatile Organic Compound (VOC) adsorbents.[49-51]

The process of generating amorphous carbon from bio-waste is more feasible and low-cost. Among the various kinds, biochar, which stands out for its abundant raw materials, expansive surface area, well-developed pore structure, and cost-effectiveness, has shown substantial potential in the removal of water pollutants.[52] In the pathways of thermochemical conversion, both pyrolysis and gasification prove to be effective and economical techniques for generating a diverse array of products, encompassing bio-oil, combustible producer gas, stable biochar, liquid fuels, and specialized chemicals.[53] From a commercial standpoint, pyrolysis is the established technology for yielding bio-oil and biochar with high efficiency and rapid production rates.[54] Furthermore, microwave radiation technology has emerged as a promising method for biomass pyrolysis, offering several advantages over traditional pyrolysis.[55] Microwave energy proves efficient for selective feedstocks, as heating occurs internally within the material, resulting in reduced energy wastage during the overall heating processes. According to the International Biochar Initiative (IBI), the most accepted and standardized definition of biochar is, "It is a carbon-rich solid material obtained from biomass thermochemical conversion under oxygen-free conditions" (IBI 2012).[56] Parameters such as heating rate, temperature, vapour residence time, pressure, biomass type, and the like, influence the distribution of product yields. The biochar produced from biomass pyrolysis was noted for having a comparatively reduced specific surface area, diminished porosity, and fewer surface functional groups. Certain applications are constrained due to the biochar's low specific surface area ($< 200\,m^2/g$) and porosity.[57] Activation serves as a method to enhance the physical attributes and absorption capacity of biochar.[58, 59] The procedure involved in augmenting the specific surface area and pore density is referred to as activation. Broadly, there are two categories of activation techniques: physical and chemical. Amorphous carbon possesses abundant surface area, porosity, and functional groups with high affinity, enabling efficient adsorption of heavy metal ions and organic compounds on its surface, facilitating the removal of pollutants from wastewater and soil.[60] The catalytic effects of biochar on aquatic environmental contaminants are often achieved through the input of photoelectric energy or the presence of oxidants. The proportions of heterogeneous non-metal elements, and metals present in the bio-waste used as a raw material will decisively influence the reaction performance and even switch the reaction mechanism.[61, 62] Due to its abundant surface area, porosity, and functional groups with high affinity, biochar can readily form complexes with metals. By incorporating metal sites, biochar is accessible to enhance catalytic activity, significantly improving its catalytic capabilities.[63]

Bio-waste and plastic waste, due to their predominantly organic composition, can be utilised for the preparation of carbon-based functional materials. Furthermore, as shown in Figure 7.3, substantial amounts of metal, electronic products, and construction material waste, characterised by their predominantly inorganic composition rich in silicon and aluminium, may be considered for the production of non-carbon-based functional materials. Silicon or aluminium-based functional materials, owing to their thermal stability, corrosion resistance, mechanical strength, and unique catalytic and photoelectronic properties, represent a promising class of functional materials for regeneration from inorganic solid waste.[64-67]

Photovoltaic devices and semiconductor waste contains a significant amount of silicon, and can be used to prepare high-performance silicon-based anodes for lithium-ion batteries, through extraction and compounding with other metal elements. Moreover, the chemical etching techniques can create a porous structure within nanoscale silicon particles, facilitating the adhesion of metal particles and effectively suppressing the volume expansion issue during charge-discharge processes.[68] Combining a carbon-based matrix with recycled silicon from solid waste can serve as a buffer to dissipate the stress caused by the volume expansion and contraction of silicon during cycling, and also reduce the surface area exposed to the electrolyte, to maintain a favourable electron pathway.[69] During the regeneration process, introducing a porous structure to silicon or aluminium-based materials through chemical treatment is a strategic approach for constructing an excellent foundation for composite materials, and similarly

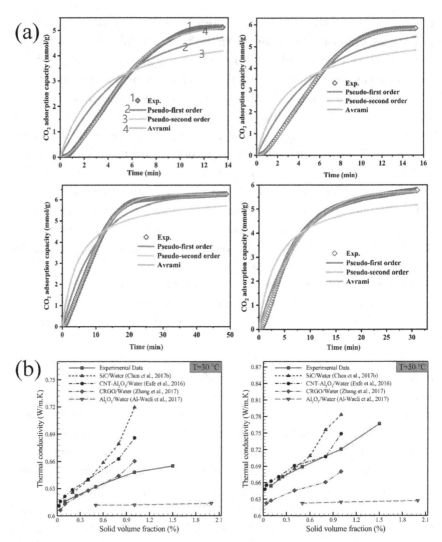

FIGURE 7.3   (a) The CO_2 adsorption of aluminium SWMs[70] and (b) thermal conversion of the silicon SWMs.[64]

promising prospects for utilising as for the adsorption process. Yan et al.[70] accomplished the synthesis of disordered hierarchical porous silica-based support materials from industrial waste blast furnace slag. The distinctive micro-meso-macro hierarchical pore characteristics of the supporting material selectively accommodate polyethyleneimine in smaller pores while retaining some larger pores. Mittal et al.[71] prepared aluminium-based SWM for pollutant adsorption. The encapsulation of smelting industrial waste Aluminium Dross (AD) into Calcium Alginate (Ca-Alg) beads were achieved, transforming it into a high-performance composite (Ca-Alg-AD), which can be employed for the adsorption or absorption, removal, and recovery of phosphates from aqueous environments.

7.3 EFFECTIVENESS OF SOLID WASTE-BASED MATERIAL

SWMs exhibit distinctive physical and chemical functionalities upon modification, which enables solid waste not only to be utilised for consumption but also to enhance the performance of materials. For structural materials, mechanical strength, durability, and safety are noteworthy concerns. Introducing solid waste as a material modification method can retard corrosion and oxidation, counteract stress concentration and temperature-humidity variations, resist aging, and withstand other environmental pressuress.[72] The bottlenecks faced by functional materials extend beyond the current performance realization and maintenance, and breaking through the performance limitations is a focal point of materials research. Introducing solid waste into material modification can enhance the optical, electrical, magnetic, and catalytic properties of materials, thereby increasing the application potential of SWMs.[73] The efficient digestion and reduction of solid waste form the foundation of the SWMs, while the enhancement of functionality constitutes a significant objective and research significance in material development.

7.3.1 Physical Properties

The significance of physical properties for structural materials far exceeds that of functional materials, as macroscopic applications rely on the mechanical performance of materials, and the consumption of structural materials is often much greater than that of functional materials (Figures 7.4a and b). Construction materials often find themselves as one of the primary destinations for solid waste recycling. However, recycled construction materials often fall short in terms of structural and mechanical strength when compared to unused materials.[74] Guo et al.[75] employed solid waste materials as binder precursors, and aluminium powder as a foaming agent to prepare foam concrete, discovering that an appropriate reaction temperature was identified as a necessary condition to accelerate the early hydration reactions, thereby forming a fresh mixture conducive to pore closure. A higher water-to-binder ratio facilitated bubble diffusion, resulting in a more uniform pore distribution. Similarly, Shi et al.[76]

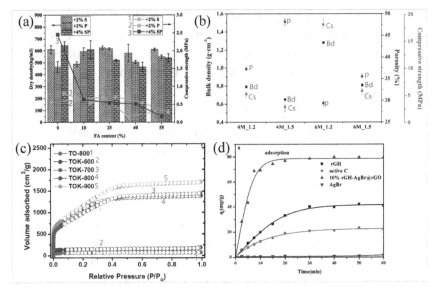

FIGURE 7.4 (a) The effect of stabilizer in the foam concrete from solid waste and the densities[75] and (b) bulk density, porosity, and compressive strength of alkali activation of ceramic-stone-porcelain and PAVAL.[81] The Nitrogen adsorption-desorption isotherms of (c) alkali-activated carbon materials for CO_2 adsorption[82] and (d) three-dimensional graphene hydrogel-AgBr@rGO for adsorption/photo-catalysis synergy of bisphenol A.[83]

employed ash and waste glass to regenerate foam glass ceramics through three stages of sintering densification, expansion, and bubble coalescence, with the resulting product deemed suitable for application in the field of thermal insulation in building structures. Plastic waste bricks with plastic content of 20% and 25% can similarly substitute a portion of the original materials while maintaining their inherent structural strength.[77] The maximum strength reached above 52 MPa, and the plastic bricks penetrate approximately 7 mm after complete solidification in 32 minutes. The water absorption rate of the waste plastic bricks is about 4%. This performance is even superior to the strength values of other masonry bricks, indicating that incorporating crushed plastic waste significantly enhances the performance of the bricks.

For the preparation of structural materials, sintering temperature is one of the most crucial process parameters, and structural materials based on solid waste sintering temperature becomes even more critical, as it plays a significant role in recrystallization and the reshaping of the material structure.[78] For building materials with partial content substitution, the admixture ratio is equally crucial. The appropriate ratio can maximize the consumption of solid waste conversion products while not compromising or even enhancing the strength and functionality of the material.[79, 80] The mechanical strength of functional materials may not be sufficient to meet specific demands in certain scenarios and requires modification for enhancement. The introduction of solid waste should lead to a positive change in the mechanical strength of the material rather than adversely affecting its utility.

7.3.2 Adsorption

Biochar is considered a significant destination for bio-waste, as this processing aims to achieve the harmless disposal and reduction of waste, with considerable commercial value of the resulting products. The most common application of biochar is as an adsorbent, determined by its high specific surface area, porous structure, and affinity for organic molecules.[59] Each year, a substantial amount of bio-waste is regenerated and processed into biochar for adsorption purposes, making the performance of biochar as a solid waste material a crucial indicator for anticipated improvement.[84]

The removal of heavy metal ions from aqueous solutions has been thoroughly investigated through various methods, including ion exchange, adsorption, (co-)precipitation, membrane separation, biological preconcentration, electrocatalytic/photocatalytic reduction-precipitation, and oxidation–reduction, among others.[85-87] In addition, the performance of biochar loaded with metals can be significantly enhanced, and the adsorbed metals can be sintered in situ without the need for additional treatment. Increasing the specific surface area and the number of pores plays a crucial role in adsorbing metals. Furthermore, a substantial portion of the effectiveness relies on chelation. Functional groups containing N and O exhibit stronger electronegativity, leading to a more pronounced chelation effect and a substantial improvement in the adsorption process for metal ions.[88]

Biochar serves as an exceptionally efficient adsorbent for organic pollutants, particularly demonstrating outstanding performance at extremely low concentrations.[89] Hydrophobic partitioning, hydrogen bonding, and π–π electron donor–acceptor interactions constitute the three mechanisms governing the adsorption of pharmaceuticals.[90] As the pyrolysis temperature increases, the surface hydrophobicity, zero-point charge value, and surface area of biochar all increase, facilitating the binding of organic molecules onto the biochar.[91] Yang et al.[92] fabricated phosphoric acid-modified biochar@TiO$_2$ (PMBC@TiO$_2$) composite materials and applied them for the removal of sulfadiazine from wastewater. The spontaneous and endothermic adsorption of sulfadiazine was primarily governed by microporous binding, hydrogen bonding, π–π interaction, and electrostatic attraction. Sulfadiazine adsorbed on the surface could be rapidly photocatalytically degraded under visible and UV light irradiation. The presence of conjugated structures and oxygen-containing groups facilitated the efficient separation of e$^-$-h$^+$ pairs and the generation of free ·O$_2^-$ and ·OH radicals, ultimately enhancing the adsorption capacity of the biochar composite material. Introducing heteroatoms (such as S, N, P, B, and the like) into biochar has a significant impact on enhancing its photocatalytic degradation efficiency. Heteroatom doping can be achieved through in-situ synthesis or post-modification strategies. Doping biochar with metals or metal oxides is an effective strategy to reduce the catalyst band gap, enhance visible light absorption, improve the generation and separation of photogenerated electron-hole pairs, and generate more free active radicals, thereby effectively enhancing the photocatalytic performance of biochar-based catalysts.[93]

As shown in Figures 7.4c and d other carbon-based materials, such as reduced Graphene Oxide (rGO), CarbonNanoTubes (CNT), and other activated carbon materials, also serve as excellent adsorbents, which can be regenerated from solid waste, contributing to waste reduction and resource recovery.[82, 83, 94] Non-carbon-based SWMs also exhibit excellent adsorption capabilities, which feedstock sources for these materials include but are not limited to plastics, rubber, and glass.[95-97] Although non-carbon-based SWMs can achieve effective adsorption, their regeneration capability and post-adsorption treatment are not as effective as carbon-based materials, especially biochar. In terms of cost control and maximizing benefits, preparing bio-waste into biochar for pollutant adsorption is the most suitable and rational choice, and the practical significance of regenerating other materials into different materials is more substantial. For most materials, adsorption is not the sole endpoint in terms of functionality. After achieving adsorption, many materials, including SWMs, often demonstrate additional functionalities to fulfil specific objectives. For instance, biochar may form in situ metal-loaded composites after re-adsorption. Functional materials frequently utilise adsorption to immobilize reactants, facilitating subsequent catalytic reactions and other processes. Enhanced adsorption capabilities play a crucial role in achieving better overall performance for materials involved in subsequent tasks.

7.3.3 Catalysis Activity

Functional materials with catalytic activity constitute a significant category within SWMs, and both in applications individually for catalysis and synergistically combined with other functionalities they performed well, with SWMs exhibiting substantial effectiveness in enhancing the conversion efficiency of solid waste at low levels of material and feedstock consumption, to reach maximum potential. Synthesis of catalysts is constrained by factors such as cost, availability, and processing compatibility of raw materials. Resourcefully utilising solid waste as a more convenient and cost-effective source for raw material preparation can contribute to the advancement and widespread adoption of synthetic processes, which enhances specific aspects of product performance while concurrently achieving reduction and control of environmental hazards.[73] This method aligns with the requirements of efficiency, high quality, and environmental sustainability.

The recovery of electrochemical equipment poses a significant challenge, and the reutilisation of discarded batteries can effectively mitigate the environmental hazards associated with their disposal. New components needed for other energy storage devices can depend on SWMs to achieve optimal performance. Lu et al.[98] utilised mango seed waste as a precursor and synthesized N and O co-doped porous carbon through high-temperature carbonization coupled with subsequent KOH activation. At a current density of $1\,A\,g^{-1}$, the resulting material exhibited an exceptionally high specific capacitance of $402\,F\,g^{-1}$ and retained 102.4% of the initial capacitance after 5000 cycles. The biochar-derived activated carbon electrode material features low cost, a facile synthesis process, and outstanding electrochemical performance, providing an economical and viable strategy for obtaining porous carbon materials for energy conversion and storage systems.

Chemical synthesis is also a promising application of SWMs, especially considering the metal sources provided by solid waste materials, which are sufficient to provide SWMs with active sites, thereby serving as catalysts for synthetic reactions. Cui et al.[99] achieved the structural modulation of a novel porous carbon-based solid acid catalyst from biomass pyrolysis byproducts, biomass tar, through sulfonation and Al-Ti metal loading. This process successfully established an inexpensive and efficient heterogeneous catalytic system for the production of 5-HydroxyMethylFurfural (HMF). The chemical synthesis of small molecules is primarily reliant on metal catalysts, which the availability can be achieved through two approaches within SWMs: the regeneration of metal-based catalysts and the loading of carbon-based catalysts. Yang et al.[100] reported findings from DFT computations suggest that the chemisorption mechanism operating on the $CuMn_2O_4$ (100) surface is accountable for the binding of reactants, potential intermediates, and resultant products onto the catalyst. The surface Cu atom, with a coordination number of 2, plays a pivotal role in the NH_3-SCR reaction, functioning as the active site for the adsorption of both NH_3 and NO. On the other hand, these metal-based active catalysts for small molecule synthesis can be easily prepared as SWMs by incorporating solid waste materials, thereby achieving the regeneration and functionalization of solid waste.[101, 102]

Biochar has the potential to serve as an electrode for electrochemically breaking down organic pollutants in water.[103] The composite of nZVI-BC demonstrated a degradation efficiency of 96.2% for NonylPhenol (NP) within 120 minutes when 0.4 g/L of nZVI-BC and 5 mM persulfate were present.[104] In this process, ZVI is synthesized by reducing ferrous iron with sodium borohydride, utilising low-cost rice husk to shape it into a carbon-based substrate under a nitrogen atmosphere. In our previous study, Co single atoms and corresponding nanoparticle catalysts were prepared on coconut shell biochar substrates.[61] We found that they activated persulfates to generate interface oxygen transfer reactions and solution-phase oxidation reactions, respectively. We successfully extended the comparative cases to heterogeneous reactions, confirming that the interface oxygen transfer reaction mediated by the catalysts exhibited higher stability and greater oxidation capacity. Additionally, the influence of the catalyst material's morphology and microscopic size on the oxidation process was elucidated. The catalytic degradation examples have been adequately described in the previous text.

The investigation of the mechanism in synthetic reactions can guide the development of catalysts, enhance suitable active metal sites and substrates, and inform the selection of appropriate preparation processes, including the choice of waste materials. Abu-Ghazala et al.[105] utilised Aluminium Industry Waste (AIW) as a precursor to prepare a novel environmentally friendly CaO/Al_2O_3 nanocatalyst for efficient biodiesel production from Waste Cooking Oil (WCO) at optimized reaction temperatures. The study explores the efficiency of the catalyst in the low-temperature conversion of Waste Cooking Oil (WCO) to biodiesel. The potent alkaline nature of Al_2O_3, enhanced by the electronic effect on the carrier Al, promotes hydrogen extraction to form methanol, and the efficient support of aluminium with specific surface morphology, porosity, and high roughness is crucial as well. Pu et al.[106] utilised natural manganese Ferrite Ore (FO) and industrial waste Lithium-silicon Powder Residue (LSP) as raw materials to design a low-cost Mn-Fe/SAPO-34 catalyst through Solid-State Ion Exchange (SSIE). The coexistence of iron and manganese in FO facilitated the dispersion of Fe, resulting in abundant Mn^{4+} and Fe^{3+}. This promoted the adsorption and activation of NH_3 beyond that of the 1.2Mn–Fe/SAPO-34 catalyst.

The concept of utilising waste to create catalysts is a primary objective for SWMs in degradation, clarifying and exploring the degradation mechanisms, is beneficial for the improvement and modification of catalysts. Zhang et al.[107] fabricated C/BiOBr composite materials through hydrothermal treatment of biochar obtained from the pyrolysis of food waste. By selecting an appropriate carbonization temperature and adjusting the carbon content in the C/BiOBr composite, a significant enhancement in photocatalytic performance can be achieved. The smooth structure and surface properties (oxygen functional groups and persistent free radicals) prepared at 400°C play a crucial role in enhancing photocatalytic activity, facilitating the primary active species of h^+ and $\cdot O_2^-$, therefore it contributes to the photodegradation process.

Advanced oxidation processes effectively remove pollutants by generating various reactive species. Among them, Fenton and Fenton-like technologies produce a wide range of active substances to achieve selective degradation of pollutants. Fenton and Fenton-like oxidation technologies rely on various radical and non-radical reactive substances, including sulfate radicals ($SO_4^-\cdot$), hydroxyl radicals ($\cdot OH$), halogen radicals ($\cdot X$), organic radicals, singlet oxygen ($^1O_2$), holes and electrons (h^+ and e^-), and high-valent metal ($M_{(n+2)^+}$) sites (Figure 7.5).[108-112] Many pollutants exhibit varying sensitivity to the degradation by different reactive species; thus, diverse oxidative degradation pathways need to be employed for the selective oxidation of different pollutants. The regulation of the mechanism of the activated persulfate system should be the goal for the controlled adjustment of the composition and structure of catalyst materials.[15] Our previous research has revealed that altering the process of regenerating

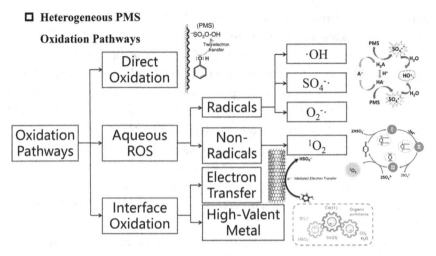

FIGURE 7.5   The oxidation pathways of heterogeneous PMS activation.[113-117]

waste biochar affects the morphology of loaded metals, thereby influencing the variation in active species generated during the catalytic production of PMS.[61] This ultimately leads to changes in pollutant degradation efficiency and the ecological risk of by-products. Investigating the reaction mechanisms applied is beneficial for guiding the preparation of catalysts and other functional materials, significantly aiding in improving the recovery efficiency of waste regeneration and the ultimate performance of the products.

7.4 RELATIONSHIP MODEL BETWEEN STRUCTURE AND EFFECT OF SOLID WASTE-BASED MATERIALS

Solid waste-based materials are considered green alternatives for several catalytic reactions, due to their intrinsic properties such as acid/alkaline resistance, biocompatibility and restorability. However, owing to their various chemical parameters and complex structural of solid waste-based materials, it is very difficult to predict their performance before actual production. Moreover, finding out the relationship between various indexes and the performance of materials and trying to predict these is very helpful to guiding the construction and production of materials. If the inherent correlation cannot be clearly identified, it will substantially obstruct the understanding of the intrinsic active centre working and make the rational design of functional solid waste-based materials much more challenging, thus limiting the further use. Thanks to the rapid development of chemical computing, the qualitative characterisation of materials is more convenient, and the emergence of various ML models makes this work possible.[118-120] As an artificial intelligence technology, ML has received wide attention as a robust and versatile technique for dealing with large amounts of data. ML has excellent potential in managing environmental remediation. This chapter mainly introduces the three basic models and their corresponding different construction ideas.

7.4.1 Decision-Tree Modelling

The Decision-Tree (DT) modelling is widely used in the prediction of material properties, and its basic flow is shown in Figure 7.6a.[121] Among them, Random Forest Regressor (RFR) and XGBoost classifier (XGB) are two widely used algorithms. The main difference between them is that RFR is an integrated learning algorithm based on DTs, while XGB is a gradient descent algorithm based on DTs.[122]

RFR is based on the DTs, while it is an ensemble learning method for classification, regression and other tasks that operates by constructing a multitude of decision trees at training time, which corrects for decision

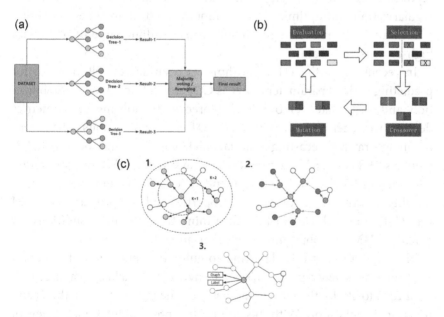

FIGURE 7.6   (a) Decision-tree modelling, (b) genetic algorithm, (c) graph neural network.

trees' tendency for overfitting to their training set. In a random forest, each decision tree operates independently and is trained by randomly selecting features and samples. Ultimately, the results of all individual trees are aggregated through voting or averaging to obtain the final prediction. Consequently, random forest exhibits strong generalization capability.

XGB, which is a scalable boosting tree and a variant of the GBDTs, also employs numerous DTs and uses the weighted quantile search to aid in parallel and distributed computing. It might take time to optimize its numerous hyperparameters, but it possesses greater capability in excavating complex patterns from general datasets. Moreover, XGB constructs decision trees sequentially where previous tree predictions are taken into account at each step to enhance overall predictive performance. During the construction of each decision tree, XGB incorporates regularization techniques and gradient descent optimization methods which contribute to its improved robustness and generalization ability.[123]

Both RFR and XGB are ensemble learning algorithms; however, RFR offers simplicity along with effective prevention of overfitting issues making it suitable for addressing small to medium-sized classification and regression problems.[124] On the other hand, XGB introduces more advanced regularization and optimization techniques resulting in superior prediction accuracy thereby being well-suited for handling large datasets with high-dimensional features.[125]

In recent years, the DT algorithm has been successfully applied for predicting various parameters such as CO_2 absorption, metal immobilization, and waste water removal.[126-128] Moreover, by utilising chemometrics derived from chemical calculations, the DT algorithm exhibits great potential in accurately forecasting material efficiency.[129,130] Palansooriya et al.[131] employs RFR approaches to predict the Heavy Metal (HM) immobilization efficiency of biochar in biochar-amended soils. This study presents new insights into the effects of biochar properties and soil properties on HM immobilization, which can help determine the optimum conditions for enhanced HM immobilization in biochar-amended soils.

The application of this algorithm to guide the generation of materials has been extensively reported in several environment fields.[132] Paula et al.[133] used RFR to clarify the correlation between the preparation methods and functional application. With this approach, a noncomputational review of scientific articles for materials with a huge parameter space such as Carbon

Functional Materials (CFM) is largely obsolete. In addition, the results provide a powerful platform for data-oriented experimental design of CFMs produced from biomass. Moreover, a systematic study using DT modelling was conducted by Wang et al.[134] to reveal the remarkable capability of XGB for accelerating bio-based catalyst development. Two nonradical-enhanced solid waste-based materials with different active sites were prepared based on the XGB results, which serves as a proof of concept for applying ML in the synthesis of tailored biochar for persulfate activation. Furthermore, through effective utilisation of previous or unsuccessful data, decision tree algorithms can continuously guide the design of functional materials.[133] To summarize, the feasibility and potential of introducing DT modelling into the rational solid waste-based material synthesis to expedite the exploration period and boost catalytic capability are worth being thoroughly investigated.

7.4.2 Genetic Algorithm Coupled with Back Propagation Neural Network Modelling

There are a lot of parameters that need to be set during the initialization of traditional neural networks. If they are not defined, they are usually assigned randomly.[135] This method will not only slow down the convergence rate of the model, but also cause the instability of the model prediction effect, and cause the model to fall into the problem of local minimum value.[136] By using a genetic algorithm to optimize the initial conditions of the network, the global search ability of the model can be increased, and the prediction accuracy of the model can be further improved.[137]

The Genetic Algorithm (GA), first proposed in the 1970s, is inspired by the evolution of organisms in nature.[138] Based on the evolutionary process in Mendelian genetics and the theory of natural selection in Darwinian biological evolution, the optimal solution of the problem is searched by simulating the natural evolutionary process.[139] The genetic algorithm is a kind of evolutionary algorithm, which uses the idea of survival of the fittest in nature for reference, and transforms the problem solving process into the process of finding the most suitable gene in biological evolution.[140] From the beginning of the generation, the fitness of genes in individuals is gradually increased through the calculation of fitness, selection, crossover and other processes, so as to obtain the optimal solution to the problems faced.[141, 142] The schematic diagram of the genetic algorithm is shown in Figure 7.6b.

The operation of the genetic algorithm starts from encoding the chromosomes of each individual in a population, and calculates the fitness of each individual according to the structure of the chromosome and the information contained in it, so as to allocate different reproduction opportunities for each individual.[143] The higher the fitness of an individual presents, the closer to the optimal solution we need, so it can obtain greater reproduction opportunities.[144] Individuals with lower fitness will have less reproductive opportunities and will be selected out. The new generation of individuals will be formed through the cross-over of the previous generation of individuals in order to achieve the purpose of obtaining the local optimal solution in the population.[145] At the same time, in the process of crossover, the data contained in different chromosomes will also vary to a certain extent, so that the population can complete the global search in a large data space to get close to the global optimal solution we want to obtain.[146] Later, with the increase of the number of iterations, due to the increasing fitness of each generation of individuals, the continuously evolving population is also closer to the optimization goal, and finally the optimal solution can be obtained to meet the needs of the problem.[147]

With this model, traditional neural networks can be optimized while facilitating global network searches. Additionally, coupled chemical calculations can effectively predict material degradation efficiency. Xiao et al.[141] established QASR model according to GA-BPNN, which not only increases our understanding of contaminant degradation in PMS treatment systems but also highlights a novel QSAR model to predict the degradation performance in complicated heterogeneous AOPs. By utilising GA-BPNN, the iteration rate and accuracy of a degradation rate prediction model can be effectively increased, and guide the design and production of solid waste-based materials. Sigmund et al present a widely applicable GA-BPNN approach that excellently predicted the conventionally used Freundlich isotherm fitting parameters $logK_F$ and n. With this approach, they provided a method to select the appropriate sorbent for a given contaminant based on the ability to predict sorption.

7.4.3 Graph Neural Networks Modelling

Graph Neural Network (GNN) is a kind of framework that has used deep learning to learn graph structure data directly in recent years. Its excellent performance has attracted great attention and in-depth exploration by scholars. By formulating certain strategies in the nodes and edges of the graph, GNN converts the graph structure data into a standardized and standard representation and feeds it into a variety of neural networks for training, achieving excellent results in tasks such as node classification, edge information propagation, and graph clustering. With this model, we can input chemical structure information directly into the model in the form of a molecular graph. The network automatically optimizing feature extraction schemes to establish associations between molecules and solid waste-based materials, and achieving excellent predictive performance without requiring feature engineering steps.[148] Its basic structure is shown in Figure 7.6c. The capability to automatically optimize the molecular representations allows GNNs to simplify Quantitative Structure–Activity Relationship (QSAR) modelling and achieve high performances in general tasks. Therefore, GNNs may revolutionize the descriptor-based QSAR paradigm and can potentially facilitate the constructing models for predicting the effort.

Graph Attention Networks (GAT) introduce an attention mechanism in GNN that allows each node to assign different weights based on the characteristics of its neighbours.[149] By assigning different weights, each node can be weighted according to the importance of its neighbouring nodes when updating features, so as to better capture the complex relationship between nodes.[150] Compared with traditional GNN, GAT has the following advantages:

1. Suitable for inductive tasks: Traditional graph convolutional networks are good at handling transudative tasks.[151, 152] That is, the same graph data is used for training and testing, but inductive tasks cannot be completed. GAT, on the other hand, requires only first-order neighbour node information, so it can process a wider range of graph data and implement inductive tasks.[153]

2. Reasonable learning of nodes: Traditional graph convolutional networks use the same weight in convolution operations for different neighbours of the same node, while GAT allows learning different weights for different neighbour nodes through the attention mechanism.[154, 155] This allows GAT to capture important relationships between nodes more precisely, improving the performance of the model.[156]

With the addition of attention mechanisms, GAT is able to make more flexible use of graph data for more efficient, flexible, and accurate modelling.[157,158] This model is suitable for a wider range of graph structures and tasks, and has better representation ability, so it has achieved remarkable results in the field of molecular graph recognition.[159] Xiong et al.[148] labelled aromatic atoms accurately according to attention weight parameters and accordingly verified the reliability of the developed models. They demonstrate that a new GNN architecture achieves state-of-the-art predictive performances on a variety of data sets and that what it learns is interpretable. The feature visualization for GNN suggests that it automatically learns nonlocal intramolecular interactions from specified tasks, which can help us gain chemical insights directly from data beyond human perception. Wang et al.[160] developed a characterisation method with a predefined molecular fingerprint-based GAT. This applicability domains characterisation method has been shown to be superior to some conventional applicability domains characterisation methods for QSAR models constructed on large data sets using various machine learning. Wang et al.[161] developed the GAT model together with the applicability domain characterisation, which may serve as efficient tools for screening chemicals. And the modelling methodology can be applied to other physicochemical, environmental, behavioural, and toxicological parameters of chemicals that are necessary for their risk assessment and management. The effective utilisation of GAT in molecular image recognition will facilitate the establishment of a correlation between degradation efficiency and molecular structure. Swift and efficient classification unveils the specific types of molecules that different materials are suited for, thereby guiding material design or screening for diverse pollutants.

7.5 PREDICTIONS FOR THE FUTURE OF SOLID WASTE-BASED MATERIALS

The preparation of SWMs involves multiple domains, including waste collection and pre-processing, recycling processes, and applications. Environmental management is poised to enter a new phase with advancements in technology and computational power, with the reduction of environmental harm and the concept of waste-to-waste to be integral throughout the entire production and waste recovery processes of SWMs. The recycling of solid waste constrains the sourcing of raw materials for SWMs, and the judicious implementation of recycling and sorting facilities, societal policies, heightened public awareness, and advancements in relevant technologies render it not only environmentally beneficial but also economically profitable. The preparation process of SWMs is heavily reliant on the progress in materials science and a crucial shift in mindset, further conservation of resources and enhancement of efficiency can be achieved by applying materials to suitable scenarios. AI-technologies stand out as one of the most esteemed and transformative technologies for the future of SWM preparation. AI-technologies not only guide the preparation of materials in SWMs but also lend a helping hand in their application scenarios, thereby enhancing the spatial utilisation of materials.

7.5.1 Source and Collection

The preparation of SWMs requires an increase in the yield and proportion of waste raw materials that can be used for reproduction. Therefore, the classification and collection of solid waste play a crucial role in the entire recycling industry and process. Managing waste poses a significant challenge for city authorities in developing countries, primarily attributed to the escalating waste generation, the strain on municipal budgets resulting from the high costs associated with waste management, and a lack of comprehensive understanding regarding various factors influencing different stages of waste management and the necessary linkages to ensure the effective functioning of the entire handling system.[162]

AI technology has significantly economized human resources and redundant costs in both the process and management aspects, making advancement in propelling the recycling of solid waste and its subsequent utilisation in reproduction towards greater feasibility and broader application. Waste management processes involve intricate operations and non-linear parameters due to the interconnection of multiple processes and the significant variability of demographic and socio-economic factors that impact the overall systems. Additionally, achieving satisfactory performance in solid waste management systems without compromising other health and environmental factors poses a considerable challenge.[163] The emerging field of AI is considered highly suitable for application in solid waste management. AI technology focuses on designing computer systems and programs capable of emulating human characteristics such as problem-solving, learning, perception, understanding, reasoning, and awareness of surroundings. AI models, including ANN, expert systems, GA, and Fuzzy Logic (FL), possess the ability to address ambiguous problems, configure intricate mappings, and make predictions.[164] A comprehensive improvement approach can have multiple and simultaneous effects on waste recycling. Optimization of the routes for solid waste transport vehicles, the structural modification of vehicles, and advancements in loading methods collectively contribute to the more efficient recycling of waste, which leads to an improvement in the collection of secondary raw materials through a circular economy approach.[165] Typically, the cutting-edge solutions are confined to statistical methods that lack predictive capabilities and offer a restricted perspective of real-world scenarios. As a result, their efficacy in addressing everyday business challenges such as overflowing containers and subpar service quality is relatively low. Deriving appropriate formulations from historical data sources of waste collection can improve the entire collection process, and simultaneously utilising GPS information from tracking devices and data collected by ultrasonic sensors can enhance the waste collection services provided by businesses.[166]

The utilisation of intelligent approaches can improve the efficiency of waste recycling. The circular economy concept, which involves recycling and subsequent reuse of material flows, seeks to redesign artificial systems to harmonize economic and environmental well-being, with the goal of increasing recycling rates.[167] Chemical recycling converts plastic packaging waste into chemical products, reducing the necessity for their initial production from fossil feedstocks.[168] Because chemical recycling yields products that are chemically identical to the ones being replaced, it is expected to decrease the demand for the Earth's finite fossil resources and alleviate greenhouse gas emissions.[169] Unlike past mechanical recycling of plastic packaging, which led to noticeable performance losses, chemical recycling avoids these issues, enabling products to circumvent incineration or eventual disposal in landfills after shorter usage cycles.[170] Apart from economic interventions, policies are pivotal in efficiently encouraging waste recycling, exerting significant influence on waste generation at the residential level, particularly in the context of food waste. The formulation of beneficial policies is driven by factors such as resource conservation, food security, and the environmental and economic implications of food waste.[171] Analysing various factors contributing to food waste offers valuable insights into the most effective policy approaches for achieving sustainable food waste management. Among several key factors, waste management, quality of life, environmental policy, natural resource utilisation, and population growth are noteworthy. The moderate role of environmental policy significantly promotes the development of industries such as renewable electricity in waste management, quality of life, and natural resource utilisation.[172]

7.5.2 Potential Reformation on Structure and Function

The production process of SWMs presents significant opportunities for enhancement, and the development of these technologies and novel materials is closely intertwined. Ren et al.[173] advocate for the exploration of the correlation between the characteristics of solid waste and the properties of geopolymers. Through the regulation of the composition and reactivity of solid waste, as well as the determination of parameters like solid temperature, liquid-solid ratio, and mixing ratio, it is feasible to attain the anticipated performance of geopolymers. Utilising cost-effective silico-aluminium-enriched solid waste and alkaline activators for the production of geopolymers, intended for applications such as adsorbents, catalysts, and other high-value uses, offers notable advantages, encompassing low-temperature solidification, reduced energy consumption, resource retrieval, straightforward preparation, and scalable manufacturing. Suffo et al.[174] showed that elevating the concentration of carbocal, a byproduct of the sugar beet industry, leads to increased rigidity in the composite material, decreased fluidity, and enhanced thermo-mechanical performance, while preserving the polymer's structural integrity. Thermo-mechanically, carbocal serves as a reinforcement for the polymer rather than functioning as a loading agent. This implies the practical potential to diminish the use of traditional thermoplastic materials, offering a valuable application for abundant waste, such as in the production of packaging items. Zhang et al.[175] identified that the cost of Magnesium potassium Phosphate Cement (MPC) primarily arises from magnesium sources and phosphate retarders. Materials such as dolomite, magnesite tailings, and the like, can replace dead-burnt magnesia. Similarly, substitutes for phosphates, such as phosphogypsum, phosphorus-containing sludge, and phosphate tailings, can efficiently replace traditional high-cost phosphates. Regarding performance, solid waste with higher Si and Al content generally does not adversely affect MPC; in fact, it can enhance the mechanical properties and water resistance of MPC when added in appropriate quantities. Consequently, transforming these materials into functional carbon substances for energy storage devices, serving as promising electrode materials crucial for energy storage, represents a significant accomplishment in the scientific community.[176] To meet the requirements of future energy storage systems, supercapacitors should possess the capability to efficiently store and generate energy at exceptionally high rates, accommodating numerous charge and discharge cycles.

7.5.3 The Roles of AI-Based Technologies in Solid Waste-Based Materials

In the preceding presentation, intelligent and automated methods have been applied to waste management and collection, providing significant assistance for the regeneration and acquisition of raw materials in SWMs. The application patterns and methods of intelligence and automation in the preparation processes of SWMs closely parallel the approaches employed in the preparation of conventional materials. The guiding principles of materials science can be summarized into four paradigms: empirical trial and error, physical and chemical laws, computer simulation, and big data-driven science.[177] New methods based on big data have expanded the application of artificial intelligence in existing materials science, encompassing various aspects. However, traditional approaches to discovering novel materials, such as empirical trial and error and DFT, often entail prolonged research and development cycles with high costs and low efficiency.[178] ML presents a highly effective means to significantly reduce computational expenses and shorten development cycles, emerging as a viable alternative to DFT calculations and even repetitive experiments.[179] As a branch of artificial intelligence, ML leverages vast amounts of data to continuously optimize models under the guidance of algorithms, facilitating rational predictions. Presently, ML has found extensive application in predicting the toxicity of nanomaterials, uncovering non-toxic nanoparticles, establishing relationships between multiple structures and single properties of nanoparticles, exploring quantum-mechanical observables in molecular systems, analysing chemical reactions involving nanomaterials, and addressing kinetic systems.[180, 181] In addition to its role in function prediction, ML surpasses human judgment in terms of accuracy and convenience when applied to material analysis tasks such as detecting metal corrosion, identifying asphalt pavement cracking, and determining concrete strength.[182] ML is frequently employed in the discovery of novel materials, several classification and regression algorithms can be utilised to forecast the chemical composition of a material based on its structure.[183] ML is frequently employed either independently or in conjunction with computer simulations, such as DFT, to streamline the calculations of intricate issues within the realm of quantum chemistry. Functioning as a data-driven approach, ML has the capability to circumvent the need to solve complex equations, like the Schrödinger equation, and can directly determine properties associated with the energy, geometry, and curvature of potential energy surfaces of molecules.[180] In future research, the expeditious development of comprehensive material databases is essential for the effective utilisation of ML in materials science. The quantity and quality of data play a direct role in influencing the accuracy of ML. By employing text mining techniques, valuable data dispersed throughout scientific literature and experimental records can significantly bolster existing databases and streamline the establishment of dedicated repositories for material information. The progress in AI-technology and the growing computational capacity have led to significant advancements in utilising machine learning to guide the preparation of SWMs. In simulations of quantum matter and quantum technologies, a fundamental concept is the many-body wave function, an exceptionally intricate mathematical entity in physics. This has given rise to the notion of neural network quantum states, which entail representing quantum states through potent function approximators for the wave function using neural networks. The application of feed-forward neural networks for wave function representation involves optimizing the network parameters through a micro-genetic algorithm, ensuring the satisfaction of the Schrödinger equation for a one-dimensional harmonic oscillator by the neural network.[184]

ML models heavily depend on the choice and quality of reference data, a crucial component shaping their reliability and applicability. In the context of ML-FF (Machine Learning Force Fields), reference data plays a pivotal role. Generating datasets in computational chemistry and physics poses challenges, as each reference point requires computationally intensive calculations, limiting data collection. Additionally, the vast configurational space of molecules, solids, or liquids complicates the identification of representative geometries. The optimal selection of reference data may need adjustment based on the specific properties of the ML model or its intended application.[185] Due to the limited extrapolation capabilities of ML methods, their predictions are reliable only within regions where training data is available. Therefore, when generating reference data, it is crucial to adequately sample all regions of the Potential Energy Surface (PES) that may be relevant for subsequent studies. For instance, in the investigation of a reaction, the dataset should encompass not only configurations corresponding to reactant and product structures but also the region around the transition state and along the transition pathway.[186] If the reaction coordinate defining the transition process is already known, a direct approach to generating reference data would involve sampling the transition path region. However, even if an ML model accurately reproduces the entire reference dataset, issues may arise when applying the model to study the reaction. Inadequate sampling of rare transition processes can lead to uncertainties in Molecular Dynamics (MD) simulations utilising the ML-FF.[187,188] The reference data may be confined to a specific subset of molecular configurations along the transition pathway, causing the model to enter the extrapolation regime between boundary states, potentially resulting in unreasonable transition pathways compared to those generated by MD simulations. Another potential concern is that, following the transition state region, a substantial amount of potential energy is often converted into internal motions like bond vibrations. Consequently, the effective temperature, defined by kinetic energy, surpasses ambient conditions significantly. Even with the use of a thermostat, the rapid increase in thermal energy may not be immediately manageable, leading the trajectory to explore high-energy configurations not included in the reference data, requiring the model to extrapolate once again.

7.6 CONCLUSION

This review comprehensively explores the synthesis, properties, and applications of SWMs. It discusses catalytic properties, material modification strategies, and utilises AI technology for predicting the structure-function relationships of SWMs. The review aims to leverage AI as an auxiliary tool in the regeneration process, addressing challenges and proposing considerations for future advancements in SWM preparation and application.

The solid waste regeneration process involves converting waste resources into materials with distinct structural and functional properties. Structural performance focuses on macroscopic mechanical characteristics, while functional performance includes specific attributes like photoelectronic properties, magnetism, and catalytic effects. Manufacturing processes significantly impact material structure and properties, with critical manufacturing processes decisively influencing both aspects. Recycled structural materials from solid waste show potential for high mechanical strength but face challenges in complete integration due to structural integrity demands. Recycling processes for glass, metals, and plastics contribute to sustainability. Carbon-based functional materials, derived from bio-waste, plastics, and metals, exhibit promise in catalysis, adsorption, and photoelectric properties, offering low-cost and readily available solutions. Nanocarbon materials, carbon quantum dots, and various carbon structures synthesized from bio-waste find applications in bioimaging, catalysis, and pollutant removal. Aluminium/silicon-based functional materials, utilising waste with predominantly inorganic composition, show promise in applications like silicon-based anodes for lithium-ion batteries and pollutant adsorbents. Utilising waste resources for functional materials represents an innovative and sustainable approach, addressing environmental concerns and promoting resource conservation.

However, due to the various chemical parameters and complex structure of SWMs, it is challenging to predict their performance before actual production. To guide the construction and production of materials, identifying the relationships between various indicators and material performance, and attempting to predict them, are crucial. This is particularly helpful for understanding the functioning of intrinsic active centres and rational design of functional solid waste-based materials. The lack of clear identification of inherent correlations significantly hinders the understanding of intrinsic active centre functioning, making the rational design of functional solid waste-based materials more challenging and limiting further utilisation. With the rapid development of chemical computing, qualitative characterisation of materials has become more convenient. The emergence of various ML models has made this work feasible. Decision-tree modelling, the genetic algorithm coupled with backpropagation neural network modelling, and graph neural network modelling are three fundamental models in machine learning for managing environmental remediation. Decision-tree modelling is widely used for material performance prediction, with RFR and XGB being commonly employed algorithms. RFR, an integrated learning algorithm based on decision trees, and XGB, a gradient descent algorithm based on decision trees, are both ensemble learning methods. RFR is simpler and more effective for small to medium-sized classification and regression problems, while XGB performs better in handling large datasets with high-dimensional features. GA coupled with backpropagation neural network modelling optimizes traditional neural networks by enhancing global search capabilities and prediction accuracy. GNN modelling, specifically GAT, directly learns graph structure data, achieving excellent performance in tasks such as node classification, edge information propagation, and graph clustering. These three models provide powerful tools for the design and performance prediction of solid waste-based materials.

In the future, the source and collection of waste materials are crucial, and AI technology significantly economizes resources in waste management, offering efficient solutions. The application of intelligent approaches enhances recycling efficiency, contributing to a circular economy. AI-based technologies also play a vital role in the reformation of SWM structure and function, optimizing production processes and material development. The integration of AI in materials science accelerates the discovery of novel materials and aids in predicting their properties. Additionally, AI technologies contribute to the potential reformation of SWM production, utilising waste materials in diverse applications, such as structural support, fillers, catalysis, synthesis, and more. The roles of AI-based technologies extend to the entire SWM life cycle, from waste collection to the preparation of materials, showcasing their transformative impact on sustainable waste management.

REFERENCES

1. Abbà, A.; Collivignarelli, M. C.; Sorlini, S.; Bruggi, M. On the reliability of reusing bottom ash from municipal solid waste incineration as aggregate in concrete. *Compos. B Eng.* 2014, 58, 502–509.

2. Safiuddin, M.; Jumaat, M. Z.; Salam, M. A.; Islam, M. S.; Hashim, R. Utilisation of solid waste in construction materials. *Int. J. Phys. Sci.* 2010, 5 (13), 1952–1963.

3. Yin, K.; Ahamed, A.; Lisak, G. Environmental perspectives of recycling various combustion ashes in cement production – A review. *Waste Manage.* 2018, 78, 401–416.

4. Tatrari, G.; Karakoti, M.; Tewari, C.; Pandey, S.; Bohra, B. S.; Dandapat, A.; Sahoo, N. G. Solid waste-derived carbon nanomaterials for supercapacitor applications: A recent overview. *Mater. Adv.* 2021, 2 (5), 1454–1484.

5. Sarc, R.; Curtis, A.; Kandlbauer, L.; Khodier, K.; Lorber, K. E.; Pomberger, R. Digitalisation and intelligent robotics in value chain of circular economy oriented waste management – A review. *Waste Manage.* 2019, 95, 476–492.

6. Zhao, J.; Yang, K.; Wang, J.; Wei, D.; Liu, Z.; Zhang, S.; Ye, W.; Zhang, C.; Wang, Z.; Yang, X. Expired milk powder emulsion-derived carbonaceous framework/Si composite as efficient anode for lithium-ion batteries. *J. Colloid Interface Sci.* 2023, 638, 99–108.

7. Vakili, M.; Deng, S.; Cagnetta, G.; Wang, W.; Meng, P.; Liu, D.; Yu, G. Regeneration of chitosan-based adsorbents used in heavy metal adsorption: A review. *Sep. Purif. Technol.* 2019, 224, 373–387.

8. Tanigaki, N.; Fujinaga, Y.; Kajiyama, H.; Ishida, Y. Operating and environmental performances of commercial-scale waste gasification and melting technology. *Waste Manage. Res.* 2013, 31 (11), 1118–1124.

9. Avasthi, I.; Lerner, H.; Grings, J.; Gräber, C.; Schleheck, D.; Cölfen, H. Biodegradable mineral plastics. *Small Methods.* 2024, 8, 2300575.

10. Pu, M.; Zhou, X.; Liu, X.; Fang, C.; Wang, D. A facile, alternative and sustainable feedstock for transparent polyurethane elastomers from chemical recycling waste PET in high-efficient way. *Waste Manage.* 2023, 155, 137–145.

11. Battistel, A.; Palagonia, M. S.; Brogioli, D.; La Mantia, F.; Trócoli, R. Electrochemical methods for lithium recovery: A comprehensive and critical review. *Adv. Mater.* 2020, 32 (23), 1905440.

12. Kim, H.; Yang, J.; Gim, H.; Hwang, B.; Byeon, A.; Lee, K.-H.; Lee, J. W. Coupled effect of TiO_{2-x} and N defects in pyrolytic waste plastics-derived carbon on anchoring polysulfides in the electrode of Li-S batteries. *Electrochim. Acta* 2022, 408, 139924.

13. Jiang, B.; Wu, M.; Wu, S.; Zheng, A.; He, S. A review on development of industrial solid waste in tunnel grouting materials: Feasibility, performance, and prospects. *Materials* 2023, 16 (21), 6848.

14. Xiao, Z.-J.; Feng, X.-C.; Shi, H.-T.; Zhou, B.-Q.; Wang, W.-Q.; Ren, N.-Q. Why the cooperation of radical and non-radical pathways in PMS system leads to a higher efficiency than a single pathway in tetracycline degradation. *J. Hazard. Mater.* 2022, 424, 127247.

15. Feng, X.-C.; Xiao, Z.-J.; Shi, H.-T.; Zhou, B.-Q.; Wang, Y.-M.; Chi, H.-Z.; Kou, X.-H.; Ren, N.-Q. How nitrogen and sulfur doping modified material structure, transformed oxidation pathways, and improved degradation performance in peroxymonosulfate activation. *Environ. Sci. Technol.* 2022, 56, 1404.

16. Zhou, X. Q.; Luo, M. Y.; Xie, C. Y.; Wang, H. B.; Wang, J.; Chen, Z. L.; Xiao, J. W.; Chen, Z. Q. Tunable S doping from Co_3O_4 to Co_9S_8 for peroxymonosulfate activation: Distinguished radical/nonradical species and generation pathways. *Appl. Catal. B: Environ.* 2021, 282, 119605.

17. Oganov, A. *Modern Methods of Crystal Structure Prediction.* Wiley Press. 2011.

18. Oganov, A. R.; Pickard, C. J.; Zhu, Q.; Needs, R. J. Structure prediction drives materials discovery. *Nat. Rev. Mater.* 2019, 4 (5), 331–348.

19. Poggio, T.; Rifkin, R.; Mukherjee, S.; Niyogi, P. General conditions for predictivity in learning theory. *Nature* 2004, 428 (6981), 419–422.

20. Joze, H. R. V.; Drew, M. S. In *Improved machine learning for image category recognition by local color constancy*, 2010 IEEE International Conference on Image Processing, 2010, 3881–3884.

21. Larranaga, P.; Calvo, B.; Santana, R.; Bielza, C.; Galdiano, J.; Inza, I.; Lozano, J. A.; Armananzas, R.; Santafé, G.; Pérez, A.; Robles, V. Machine learning in bioinformatics. *Briefings Bioinform.* 2006, 7 (1), 86–112.

22. Eminagaoglu, M.; Eren, S. In *Implementation and comparison of machine learning classifiers for information security risk analysis of a human resources department*, 2010 IEEE International Conference on Image Processing, 2010; 187–192.

23. Olsson, F. *A literature survey of active machine learning in the context of natural language processing. Digitala Vetenskapliga Arkivet.* 2009, 59.

24. Pilania, G.; Wang, C.; Jiang, X.; Rajasekaran, S.; Ramprasad, R. Accelerating materials property predictions using machine learning. *Sci. Rep. UK* 2013, 3 (1), 2810.

25. Xiao, Z.; Yang, B.; Feng, X.; Liao, Z.; Shi, H.; Jiang, W.; Wang, C.; Ren, N. Density functional theory and machine learning-based quantitative structure–activity relationship models enabling prediction of contaminant degradation performance with heterogeneous peroxymonosulfate treatments. *Environ. Sci. Technol.* 2023, 57 (9), 3951–3961.

26. Jain, A.; Shin, Y.; Persson, K. A. Computational predictions of energy materials using density functional theory. *Nat. Rev. Mater.* 2016, 1 (1), 15004.

27. Athira, V. S.; Charitha, V.; Athira, G.; Bahurudeen, A. Agro-waste ash based alkali-activated binder: Cleaner production of zero cement concrete for construction. *J. Clean. Prod.* 2021, 286, 125429.

28. Berenguer, R.; Lima, N.; Cruz, F.; Pinto, L.; Lima, N. B. D. Thermodynamic, microstructural and chemometric analyses of the reuse of sugarcane ashes in cement manufacturing. *J. Environ. Chem. Eng.* 2021, 9 (4), 105350.

29. Monzó, J.; Payá, J.; Borrachero, M. V.; Girbés, I. Reuse of sewage sludge ashes (SSA) in cement mixtures: The effect of SSA on the workability of cement mortars. *Waste Manage.* 2003, 23 (4), 373–381.

30. Yang, Q.; Fan, Z.; Yang, X.; Hao, L.; Lu, G.; Fini, E. H.; Wang, D. Recycling waste fibre-reinforced polymer composites for low-carbon asphalt concrete: The effects of recycled glass fibres on the durability of bituminous composites. *J. Clean. Prod.* 2023, 423, 138692.

31. Serelis, E.; Vaitkevicius, V. Utilisation of glass shards from municipal solid waste in aluminium-based ultra-lightweight concrete. *Constr. Build. Mater.* 2022, 350, 128396.

32. Tushar, Q.; Salehi, S.; Santos, J.; Zhang, G.; Bhuiyan, M. A.; Arashpour, M.; Giustozzi, F. Application of recycled crushed glass in road pavements and

pipeline bedding: An integrated environmental evaluation using LCA. *Sci. Total Environ.* 2023, 881, 163488.

33. Kazemi Najafi, S. Use of recycled plastics in wood plastic composites – A review. *Waste Manage.* 2013, 33 (9), 1898–1905.

34. Schyns, Z. O. G.; Shaver, M. P. Mechanical recycling of packaging plastics: A review. *Macromol. Rapid Commun.* 2021, 42 (3), 2000415.

35. Gu, L.; Ozbakkaloglu, T. Use of recycled plastics in concrete: A critical review. *Waste Manage.* 2016, 51, 19–42.

36. Aneke, F. I.; Shabangu, C. Green-efficient masonry bricks produced from scrap plastic waste and foundry sand. *Case Stud. Constr. Mater.* 2021, 14, e00515.

37. Moghaddam Fard, P.; Alkhansari, M. G. Innovative fire and water insulation foam using recycled plastic bags and expanded polystyrene (EPS). *Constr. Build. Mater.* 2021, 305, 124785.

38. Alqahtani, F. K.; Abotaleb, I. S.; ElMenshawy, M. Life cycle cost analysis of lightweight green concrete utilising recycled plastic aggregates. *J. Build. Eng.* 2021, 40, 102670.

39. Cousins, D. S.; Suzuki, Y.; Murray, R. E.; Samaniuk, J. R.; Stebner, A. P. Recycling glass fibre thermoplastic composites from wind turbine blades. *J. Clean. Prod.* 2019, 209, 1252–1263.

40. Xu, C.; Nasrollahzadeh, M.; Selva, M.; Issaabadi, Z.; Luque, R. Waste-to-wealth: Biowaste valorization into valuable bio(nano)materials. *Chem. Soc. Rev.* 2019, 48 (18), 4791–4822.

41. Srivastava, R. K.; Shetti, N. P.; Reddy, K. R.; Nadagouda, M. N.; Badawi, M.; Bonilla-Petriciolet, A.; Aminabhavi, T. M. Valorization of biowaste for clean energy production, environmental depollution and soil fertility. *J. Environ. Manage.* 2023, 332, 117410.

42. Watson, J.; Zhang, Y.; Si, B.; Chen, W.-T.; de Souza, R. Gasification of biowaste: A critical review and outlooks. *Renew. Sustain. Energ. Rev.* 2018, 83, 1–17.

43. Wong, S.; Ngadi, N.; Inuwa, I. M.; Hassan, O. Recent advances in applications of activated carbon from biowaste for wastewater treatment: A short review. *J. Clean. Prod.* 2018, 175, 361–375.

44. Kamali, A. R.; Yang, J.; Sun, Q. Molten salt conversion of polyethylene terephthalate waste into graphene nanostructures with high surface area and ultra-high electrical conductivity. *Appl. Surf. Sci.* 2019, 476, 539–551.

45. Yu, R.; Wen, X.; Liu, J.; Wang, Y.; Chen, X.; Wenelska, K.; Mijowska, E.; Tang, T. A green and high-yield route to recycle waste masks into CNTs/Ni hybrids via catalytic carbonization and their application for superior microwave absorption. *Appl. Catal. B: Environ.* 2021, 298, 120544.

46. Pan, F.; Ni, K.; Xu, T.; Chen, H.; Wang, Y.; Gong, K.; Liu, C.; Li, X.; Lin, M.-L.; Li, S.; Wang, X.; Yan, W.; Yin, W.; Tan, P.-H.; Sun, L.; Yu, D.; Ruoff, R. S.; Zhu, Y. Long-range ordered porous carbons produced from C_{60}. *Nature* 2023, 614 (7946), 95–101.

47. Uchimiya, M.; Pignatello, J. J.; White, J. C.; Hu, S.-T.; Ferreira, P. J. Structural transformation of biochar black carbon by C_{60} superstructure: Environmental implications. *Sci. Rep.* 2017, 7 (1), 11787.

48. Liu, M.; Yan, X.; Su, L.; Dong, H.; Hu, Z.; Peng, Y.; Guan, L.; Zhang, J.; Zhou, Z.; Zhu, Y.; Zhou, N. Waste plastic thermal-transformed CNT@ Fe–Mo/MgO for free radical activation and bacteria sterilization. *J. Clean. Prod.* 2022, 373, 133794.

49. Tamilselvi, R.; Ramesh, M.; Lekshmi, G. S.; Bazaka, O.; Levchenko, I.; Bazaka, K.; Mandhakini, M. Graphene oxide-based supercapacitors from agricultural waste: A step to mass production of highly efficient electrodes for electrical transportation systems. *Renew Energ.* 2020, 151, 731–739.

50. Mishra, R.; Kumar, A.; Singh, E.; Kumari, A.; Kumar, S. Synthesis of graphene oxide from biomass waste: Characterisation and volatile organic compounds removal. *Process Saf. Environ.* 2023, 180, 800–807.

51. Omenesa Idris, M.; Asshifa Md Noh, N.; Nasir Mohamad Ibrahim, M.; Ali Yaqoob, A. Sustainable microbial fuel cell functionalized with a bio-waste: A feasible route to formaldehyde bioremediation along with bioelectricity generation. *Chem. Eng. J.* 2023, 455, 140781.

52. Shi, H.; Feng, X.; Xiao, Z.; Jiang, C.; Wang, W.; Zhang, X.; Xu, Y.; Wang, C.; Guo, W.; Ren, N. How β-cyclodextrin-functionalized biochar enhanced biodenitrification in low C/N conditions via regulating substrate metabolism and electron utilisation. *Environ. Sci. Technol.* 2023, 57 (30), 11122–11133.

53. Bridgwater, A. V. Renewable fuels and chemicals by thermal processing of biomass. *Chem. Eng. J.* 2003, 91 (2), 87–102.

54. Yu, K. L.; Lau, B. F.; Show, P. L.; Ong, H. C.; Ling, T. C.; Chen, W.-H.; Ng, E. P.; Chang, J.-S. Recent developments on algal biochar production and characterisation. *Bioresour. Technol.* 2017, 246, 2–11.

55. Huang, Y.-F.; Chiueh, P.-T.; Kuan, W.-H.; Lo, S.-L. Microwave pyrolysis of lignocellulosic biomass: Heating performance and reaction kinetics. *Energy* 2016, 100, 137–144.

56. Sakhiya, A. K.; Anand, A.; Kaushal, P. Production, activation, and applications of biochar in recent times. *Biochar* 2020, 2 (3), 253–285.

57. Liu, N.; Charrua, A. B.; Weng, C.-H.; Yuan, X.; Ding, F. Characterisation of biochars derived from agriculture waste and their adsorptive removal of atrazine from aqueous solution: A comparative study. *Bioresour. Technol.* 2015, 198, 55–62.

58. Liu, W.-J.; Jiang, H.; Yu, H.-Q. Development of biochar-based functional materials: Toward a sustainable platform carbon material. *Chem. Rev.* 2015, 115 (22), 12251–12285.

59. Nguyen, T.-B.; Nguyen, T.-K.-T.; Chen, W.-H.; Chen, C.-W.; Bui, X.-T.; Patel, A. K.; Dong, C.-D. Hydrothermal and pyrolytic conversion of sunflower seed husk into novel porous biochar for efficient adsorption of tetracycline. *Bioresour. Technol.* 2023, 373, 128711.

60. Zhou, B.-Q.; Sang, Q.-Q.; Wang, Y.-J.; Huang, H.; Wang, F.-J.; Yang, R.-C.; Zhao, Y.-T.; Xiao, Z.-J.; Zhang, C.-Y.; Li, H.-P. Comprehensive understanding of tetracycline hydrochloride adsorption mechanism onto biochar-based gel pellets based on the combination of characterisation-based and approximate site energy distribution methods. *J. Clean. Prod.* 2023, 416, 137909.

61. Xiao, Z.-J.; Zhou, B.-Q.; Feng, X.-C.; Shi, H.-T.; Zhu, Y.-N.; Wang, C.-P.; Van der Bruggen, B.; Ren, N.-Q. Anchored Co–oxo generated by cobalt single atoms outperformed aqueous species from the counterparts in peroxymonosulfate treatment. *Appl. Catal. B: Environ.* 2023, 328, 122483.

62. Ido, A. L.; de Luna, M. D. G.; Ong, D. C.; Capareda, S. C. Upgrading of *Scenedesmus obliquus* oil to high-quality liquid-phase biofuel by nickel-impregnated biochar catalyst. *J. Clean. Prod.* 2019, 209, 1052–1060.

63. Zhang, C.; Dong, Y.; Liu, W.; Yang, D.; Liu, J.; Lu, Y.; Lin, H. Enhanced adsorption of phosphate from pickling wastewater by Fe-N co-pyrolysis biochar: Performance, mechanism and reusability. *Bioresour. Technol.* 2023, 369, 128263.

64. Ranjbarzadeh, R.; Moradikazerouni, A.; Bakhtiari, R.; Asadi, A.; Afrand, M. An experimental study on stability and thermal conductivity of water/silica nanofluid: Eco-friendly production of nanoparticles. *J. Clean. Prod.* 2019, 206, 1089–1100.

65. Lv, C.; Wang, H.; Liu, Z.; Zhang, W.; Wang, C.; Tao, R.; Li, M.; Zhu, Y. A sturdy self-cleaning and anti-corrosion superhydrophobic coating assembled by amino silicon oil modifying potassium titanate whisker-silica particles. *Appl. Surf. Sci.* 2018, 435, 903–913.

66. Ammarullah, M. I.; Santoso, G.; Sugiharto, S.; Supriyono, T.; Wibowo, D. B.; Kurdi, O.; Tauviqirrahman, M.; Jamari, J. Minimizing risk of failure from ceramic-on-ceramic total hip prosthesis by selecting ceramic materials based on tresca stress. *Sustainability* 2022, 14 (20), 13413.

67. Tarish, S.; Wang, Z.; Al-Haddad, A.; Wang, C.; Ispas, A.; Romanus, H.; Schaaf, P.; Lei, Y. Synchronous formation of ZnO/ZnS core/shell nanotube arrays with removal of template for meliorating photoelectronic performance. *J. Phys. Chem. C.* 2015, 119 (3), 1575–1582.

68. Li, Y.; Chen, G.; Liu, W.; Zhang, C.; Huang, L.; Luo, X. Construction of porous Si/Ag@C anode for lithium-ion battery by recycling volatile deposition waste derived from refining silicon. *Waste Manage.* 2023, 156, 22–32.

69. Sreenarayanan, B.; Vicencio, M.; Bai, S.; Lu, B.; Mao, O.; Adireddy, S.; Bao, W.; Meng, Y. S. Recycling silicon scrap for spherical Si–C composite as high-performance lithium-ion battery anodes. *J. Power Sources* 2023, 578, 233245.

70. Yan, H.; Zhang, G.; Liu, J.; Li, G.; Zhao, Y.; Wang, Y.; Wu, C.; Wu, W. Amine-functionalized disordered hierarchical porous silica derived from blast furnace slag with high adsorption capability and cyclic stability for CO_2 adsorption. *Chem. Eng. J.* 2023, 478, 147480.

71. Mittal, Y.; Srivastava, P.; Tripathy, B. C.; Dhal, N. K.; Martinez, F.; Kumar, N.; Yadav, A. K. Aluminium dross waste utilisation for phosphate removal and recovery from aqueous environment: Operational feasibility development. *Chemosphere* 2024, 349, 140649.

72. de Andrade Salgado, F.; de Andrade Silva, F. Recycled aggregates from construction and demolition waste towards an application on structural concrete: A review. *J. Build. Eng.* 2022, 52, 104452.

73. Chen, H.; Wan, K.; Zhang, Y.; Wang, Y. Waste to wealth: Chemical recycling and chemical upcycling of waste plastics for a great future. *ChemSusChem* 2021, 14 (19), 4123–4136.

74. Alhawat, M.; Ashour, A.; Yildirim, G.; Aldemir, A.; Sahmaran, M. Properties of geopolymers sourced from construction and demolition waste: A review. *J. Build. Eng.* 2022, 50, 104104.

75. Guo, R.; Xue, C.; Guo, W.; Wang, S.; Shi, Y.; Qiu, Y.; Zhao, Q. Preparation of foam concrete from solid waste: Physical properties and foam stability. *Constr. Build. Mater.* 2023, 408, 133733.

76. Shi, X.; Liao, Q.; Liu, L.; Deng, F.; Chen, F.; Wang, F.; Zhu, H.; Zhang, L.; Liu, C. Utilising multi-solid waste to prepare and characterise foam glass ceramics. *Ceram. Int.* 2023, 49 (22, Part A), 35534–35543.

77. Wahane, A.; Dwivedi, S.; Bajaj, D. Effect in mechanical and physical properties of bricks due to addition of waste polyethylene terephthalate. *Mater. Today: Proc.* 2023, 74, 916–922.

78. Lyu, H.; Hao, L.; Zhang, S.; Poon, C. S. High-performance belite rich eco-cement synthesized from solid waste: Raw feed design, sintering temperature optimization, and property analysis. *Resour. Conserv. Recycl.* 2023, 199, 107211.

79. Rodríguez Aybar, M.; Porras-Amores, C.; Moreno Fernández, E.; Pérez Raposo, Á. Physical-mechanical properties of new recycled materials with additions of padel-tennis ball waste. *J. Clean. Prod.* 2023, 413, 137392.

80. Pavlík, Z.; Vyšvařil, M.; Pavlíková, M.; Bayer, P.; Pivák, A.; Rovnaníková, P.; Záleská, M. Lightweight pumice mortars for repair of historic buildings – Assessment of physical parameters, engineering properties and durability. *Constr. Build. Mater.* 2023, 404, 133275.

81. Maldonado-Alameda, A.; Mañosa, J.; López-Montero, T.; Catalán-Parra, R.; Chimenos, J. M. High-porosity alkali-activated binders based on glass and aluminium recycling industry waste. *Constr. Build. Mater.* 2023, 400, 132741.

82. Rehman, A.; Heo, Y.-J.; Nazir, G.; Park, S.-J. Solvent-free, one-pot synthesis of nitrogen-tailored alkali-activated microporous carbons with an efficient CO_2 adsorption. *Carbon* 2021, 172, 71–82.

83. Chen, F.; An, W.; Liu, L.; Liang, Y.; Cui, W. Highly efficient removal of bisphenol A by a three-dimensional graphene hydrogel-AgBr@rGO exhibiting adsorption/photocatalysis synergy. *Appl. Catal. B: Environ.* 2017, 217, 65–80.

84. Leng, L.; Xiong, Q.; Yang, L.; Li, H.; Zhou, Y.; Zhang, W.; Jiang, S.; Li, H.; Huang, H. An overview on engineering the surface area and porosity of biochar. *Sci. Total Environ.* 2021, 763, 144204.

85. Cai, Y.; Zhang, Y.; Lv, Z.; Zhang, S.; Gao, F.; Fang, M.; Kong, M.; Liu, P.; Tan, X.; Hu, B.; Wang, X. Highly efficient uranium extraction by a piezo catalytic reduction-oxidation process. *Appl. Catal. B: Environ.* 2022, 310, 121343.

86. Wang, X.; Chen, L.; Wang, L.; Fan, Q.; Pan, D.; Li, J.; Chi, F.; Xie, Y.; Yu, S.; Xiao, C.; Luo, F.; Wang, J.; Wang, X.; Chen, C.; Wu, W.; Shi, W.; Wang, S.; Wang, X. Synthesis of novel nanomaterials and their application in efficient removal of radionuclides. *Sci. China Chem.* 2019, 62 (8), 933–967.

87. Hao, M.; Chen, Z.; Yang, H.; Waterhouse, G. I. N.; Ma, S.; Wang, X. Pyridinium salt-based covalent organic framework with well-defined nanochannels for efficient and selective capture of aqueous $^{99}TcO_4^-$. *Sci. Bull.* 2022, 67 (9), 924–932.

88. Jeon, P.; Lee, M.-E.; Baek, K. Adsorption and photocatalytic activity of biochar with graphitic carbon nitride (g-C_3N_4). *J. Taiwan Inst. Chem. Eng.* 2017, 77, 244–249.

89. Yao, L.; Yang, H.; Chen, Z.; Qiu, M.; Hu, B.; Wang, X. Bismuth oxychloride-based materials for the removal of organic pollutants in wastewater. *Chemosphere* 2021, 273, 128576.

90. Ndoun, M. C.; Elliott, H. A.; Preisendanz, H. E.; Williams, C. F.; Knopf, A.; Watson, J. E. Adsorption of pharmaceuticals from aqueous solutions using biochar derived from cotton gin waste and guayule bagasse. *Biochar* 2021, 3 (1), 89–104.

91. Ouyang, D.; Chen, Y.; Yan, J.; Qian, L.; Han, L.; Chen, M. Activation mechanism of peroxymonosulfate by biochar for catalytic degradation of 1,4-dioxane: Important role of biochar defect structures. *Chem. Eng. J.* 2019, 370, 614–624.

92. Yang, C. X.; Zhu, Q.; Dong, W. P.; Fan, Y. Q.; Wang, W. L. Preparation and characterisation of phosphoric acid-modified biochar nanomaterials with highly efficient adsorption and photodegradation ability. *Langmuir* 2021, 37 (30), 9253–9263.

93. Lu, Y.; Cai, Y.; Zhang, S.; Zhuang, L.; Hu, B.; Wang, S.; Chen, J.; Wang, X. Application of biochar-based photocatalysts for adsorption-(photo) degradation/reduction of environmental contaminants: Mechanism, challenges and perspective. *Biochar* 2022, 4 (1), 45.

94. Hayati, B.; Maleki, A.; Najafi, F.; Gharibi, F.; McKay, G.; Gupta, V. K.; Harikaranahalli Puttaiah, S.; Marzban, N. Heavy metal adsorption using PAMAM/CNT nanocomposite from aqueous solution in batch and continuous fixed bed systems. *Chem. Eng. J.* 2018, 346, 258–270.

95. Holmes, L. A.; Turner, A.; Thompson, R. C. Interactions between trace metals and plastic production pellets under estuarine conditions. *Mar. Chem.* 2014, 167, 25–32.

96. Gupta, V. K.; Nayak, A.; Agarwal, S.; Tyagi, I. Potential of activated carbon from waste rubber tire for the adsorption of phenolics: Effect of pretreatment conditions. *J. Colloid Interface Sci.* 2014, 417, 420–430.

97. Liu, S.; Chen, X.; Ai, W.; Wei, C. A new method to prepare mesoporous silica from coal gasification fine slag and its application in methylene blue adsorption. *J. Clean. Prod.* 2019, 212, 1062–1071.

98. Lu, S.; Yang, W.; Zhou, M.; Qiu, L.; Tao, B.; Zhao, Q.; Wang, X.; Zhang, L.; Xie, Q.; Ruan, Y. Nitrogen- and oxygen-doped carbon with abundant micropores derived from biomass waste for all-solid-state flexible supercapacitors. *J. Colloid Interface Sci.* 2022, 610, 1088–1099.

99. Cui, X.; Zheng, L.; Li, Q.; Guo, Y. A remarkable bifunctional carbon-based solid acid catalyst derived from waste bio-tar for efficient synthesis of 5-hydroxymethylfurfural from glucose. *Chem. Eng. J.* 2023, 474, 146006.

100. Yang, Y.; Liu, J.; Liu, F.; Wang, Z.; Ding, J.; Huang, H. Reaction mechanism for NH_3-SCR of NO_x over $CuMn_2O_4$ catalyst. *Chem. Eng. J.* 2019, 361, 578–587.

101. Mohamed, H. H.; Alsanea, A. A. TiO_2/carbon dots decorated reduced graphene oxide composites from waste car bumper and TiO_2 nanoparticles for photocatalytic applications. *Arabian J. Chem.* 2020, 13 (1), 3082–3091.

102. Sobhani, A. Hydrothermal synthesis of $CuMn_2O_4$/CuO nanocomposite without capping agent and study its photocatalytic activity for elimination of dye pollution. *Int. J. Hydrogen Energ.* 2022, 47 (46), 20138–20152.

103. Liu, Q.; Bai, X.; Su, X.; Huang, B.; Wang, B.; Zhang, X.; Ruan, X.; Cao, W.; Xu, Y.; Qian, G. The promotion effect of biochar on electrochemical degradation of nitrobenzene. *J. Clean. Prod.* 2020, 244, 118890.

104. Hussain, I.; Li, M.; Zhang, Y.; Li, Y.; Huang, S.; Du, X.; Liu, G.; Hayat, W.; Anwar, N. Insights into the mechanism of persulfate activation with nZVI/BC nanocomposite for the degradation of nonylphenol. *Chem. Eng. J.* 2017, 311, 163–172.

105. Abu-Ghazala, A. H.; Abdelhady, H. H.; Mazhar, A. A.; El-Deab, M. S. Enhanced low-temperature production of biodiesel from waste cooking oil: Aluminum industrial waste as a precursor of efficient CaO/Al$_2$O$_3$ nano-catalyst. *Fuel* 2023, 351, 128897.

106. Pu, Y.; Yang, L.; Yao, C.; Jiang, W.; Yao, L. Low-cost Mn–Fe/SAPO-34 catalyst from natural ferromanganese ore and lithium-silicon-powder waste for efficient low-temperature NH$_3$-SCR removal of NO$_x$. *Chemosphere* 2022, 293, 133465.

107. Zhang, X.; Wu, Y.; Giwa, A. S.; Xiong, J.; Huang, S.; Niu, L. Improving photocatalytic activity under visible light over a novel food waste biochar-based BiOBr nanocomposite. *Chemosphere* 2022, 297, 134152.

108. Liu, Y.; Zhao, Y.; Wang, J. L. Fenton/Fenton-like processes with in-situ production of hydrogen peroxide/hydroxyl radical for degradation of emerging contaminants: Advances and prospects. *J. Hazard. Mater.* 2021, 404.

109. Oh, W.-D.; Dong, Z.; Lim, T.-T. Generation of sulfate radical through heterogeneous catalysis for organic contaminants removal: Current development, challenges and prospects. *Appl. Catal. B: Environ.* 2016, 194, 169–201.

110. Hou, J. F.; He, X. D.; Zhang, S. Q.; Yu, J. L.; Feng, M. B.; Li, X. D. Recent advances in cobalt-activated sulfate radical-based advanced oxidation processes for water remediation: A review. *Sci. Total Environ.* 2021, 770, 145311.

111. Ding, D.; Yang, S.; Qian, X.; Chen, L.; Cai, T. Nitrogen-doping positively whilst sulfur-doping negatively affect the catalytic activity of biochar for the degradation of organic contaminant. *Appl. Catal. B: Environ.* 2020, 263, 118348.

112. Liu, B. H.; Guo, W. Q.; Wang, H. Z.; Zheng, S. S.; Si, Q. S.; Zhao, Q.; Luo, H. C.; Ren, N. Q. Peroxymonosulfate activation by cobalt(II) for degradation of organic contaminants via high-valent cobalt-oxo and radical species. *J. Hazard. Mater.* 2021, 416, 125679.

113. Zhang, Y.-J.; Huang, G.-X.; Winter, L. R.; Chen, J.-J.; Tian, L.; Mei, S.-C.; Zhang, Z.; Chen, F.; Guo, Z.-Y.; Ji, R.; You, Y.-Z.; Li, W.-W.; Liu, X.-W.; Yu, H.-Q.; Elimelech, M. Simultaneous nanocatalytic surface activation of

pollutants and oxidants for highly efficient water decontamination. *Nat. Commun.* 2022, 13 (1), 3005.

114. Zhou, S.; Yu, Y.; Zhang, W.; Meng, X.; Luo, J.; Deng, L.; Shi, Z.; Crittenden, J. Oxidation of microcystin-LR via activation of peroxymonosulfate using ascorbic acid: Kinetic modelling and toxicity assessment. *Environ. Sci. Technol.* 2018, 52 (7), 4305–4312.

115. Zhou, Y.; Jiang, J.; Gao, Y.; Ma, J.; Pang, S. Y.; Li, J.; Lu, X. T.; Yuan, L. P. Activation of peroxymonosulfate by benzoquinone: A novel nonradical oxidation process. *Environ. Sci. Technol.* 2015, 49 (21), 12941–12950.

116. Yun, E. T.; Lee, J. H.; Kim, J.; Park, H. D.; Lee, J. Identifying the nonradical mechanism in the peroxymonosulfate activation process: Singlet oxygenation versus mediated electron transfer. *Environ. Sci. Technol.* 2018, 52 (12), 7032–7042.

117. Yi, Q.; Li, X.; Li, Y.; Dai, R.; Wang, Z. Unraveling the Co(IV)-mediated oxidation mechanism in a Co_3O_4/PMS-based hierarchical reactor: Toward efficient catalytic degradation of aromatic pollutants. *ACS EST Eng.* 2022, 2 (10), 1836–1846.

118. Zhao, Y.; Wang, L.; Luo, J.; Huang, T.; Tao, S.; Liu, J.; Yu, Y.; Huang, Y.; Liu, X.; Ma, J. Deep Learning prediction of polycyclic aromatic hydrocarbons in the high arctic. *Environ. Sci. Technol.* 2019, 53 (22), 13238–13245.

119. Zhang, K.; Zhong, S.; Zhang, H. Predicting aqueous adsorption of organic compounds onto biochars, carbon nanotubes, granular activated carbons, and resins with machine learning. *Environ. Sci. Technol.* 2020, 54 (11), 7008–7018.

120. Yuan, X.; Suvarna, M.; Low, S.; Dissanayake, P. D.; Lee, K. B.; Li, J.; Wang, X.; Ok, Y. S. Applied machine learning for prediction of CO_2 adsorption on biomass waste-derived porous carbons. *Environ. Sci. Technol.* 2021, 55 (17), 11925–11936.

121. Kotsiantis, S. B. Decision trees: A recent overview. *Artif. Intell. Rev.* 2013, 39 (4), 261–283.

122. Rokach, L. Decision forest: Twenty years of research. *Inform. Fusion* 2016, 27, 111–125.

123. Sheridan, R. P.; Wang, W. M.; Liaw, A.; Ma, J.; Gifford, E. M. Extreme gradient boosting as a method for quantitative structure–activity relationships. *J. Chem. Inform. Model.* 2016, 56 (12), 2353–2360.

124. Li, Z. Extracting spatial effects from machine learning model using local interpretation method: An example of SHAP and XGBoost. *Comput. Environ. Urban Syst.* 2022, 96, 101845.

125. Ma, M.; Zhao, G.; He, B.; Li, Q.; Dong, H.; Wang, S.; Wang, Z. XGBoost-based method for flash flood risk assessment. *J. Hydrol.* 2021, 598, 126382.

126. Zhu, J.-J.; Kang, L.; Anderson, P. R. Predicting influent biochemical oxygen demand: Balancing energy demand and risk management. *Water Res.* 2018, 128, 304–313.

127. Zhu, X.; Wang, X.; Ok, Y. S. The application of machine learning methods for prediction of metal sorption onto biochars. *J. Hazard. Mater.* 2019, 378, 120727.

128. Zhu, J.-J.; Anderson, P. R. Performance evaluation of the ISMLR package for predicting the next day's influent wastewater flowrate at Kirie WRP. *Water Sci. Technol.* 2019, 80 (4), 695–706.

129. Sigmund, G.; Gharasoo, M.; Hüffer, T.; Hofmann, T. Deep learning neural network approach for predicting the sorption of ionizable and polar organic pollutants to a wide range of carbonaceous materials. *Environ. Sci. Technol.* 2020, 54 (7), 4583–4591.

130. Zhong, S.; Zhang, K.; Bagheri, M.; Burken, J. G.; Gu, A.; Li, B.; Ma, X.; Marrone, B. L.; Ren, Z. J.; Schrier, J.; Shi, W.; Tan, H.; Wang, T.; Wang, X.; Wong, B. M.; Xiao, X.; Yu, X.; Zhu, J.-J.; Zhang, H. Machine learning: New ideas and tools in environmental science and engineering. *Environ. Sci. Technol.* 2021, 55 (19), 12741–12754.

131. Palansooriya, K. N.; Li, J.; Dissanayake, P. D.; Suvarna, M.; Li, L.; Yuan, X.; Sarkar, B.; Tsang, D. C. W.; Rinklebe, J.; Wang, X.; Ok, Y. S. Prediction of soil heavy metal immobilization by biochar using machine learning. *Environ. Sci. Technol.* 2022, 56 (7), 4187–4198.

132. Liu, X.; Gharasoo, M.; Shi, Y.; Sigmund, G.; Hüffer, T.; Duan, L.; Wang, Y.; Ji, R.; Hofmann, T.; Chen, W. Key physicochemical properties dictating gastrointestinal bioaccessibility of microplastics-associated organic xenobiotics: Insights from a deep learning approach. *Environ. Sci. Technol.* 2020, 54 (19), 12051–12062.

133. Paula, A. J.; Ferreira, O. P.; Souza Filho, A. G.; Filho, F. N.; Andrade, C. E.; Faria, A. F. Machine learning and natural language processing enable a data-oriented experimental design approach for producing biochar and hydrochar from biomass. *Chem. Mater.* 2022, 34 (3), 979–990.

134. Wang, R.; Zhang, S.; Chen, H.; He, Z.; Cao, G.; Wang, K.; Li, F.; Ren, N.; Xing, D.; Ho, S.-H. Enhancing biochar-based nonradical persulfate activation using data-driven techniques. *Environ. Sci. Technol.* 2023, 57 (9), 4050–4059.

135. Jin, G.; Feng, W.; Meng, Q. Prediction of waterway cargo transportation volume to support maritime transportation systems based on GA-BP neural network optimization. *Sustainability* 2022, 14 (21), 13872.

136. Liang, Y.; Jiang, K.; Gao, S.; Yin, Y. Prediction of tunnelling parameters for underwater shield tunnels, based on the GA-BPNN method. *Sustainability* 2022, 14 (20), 13420.

137. Boutaba, R.; Salahuddin, M. A.; Limam, N.; Ayoubi, S.; Shahriar, N.; Estrada-Solano, F.; Caicedo, O. M. A comprehensive survey on machine learning for networking: Evolution, applications and research opportunities. *J. Internet Serv. Appl.* 2018, 9 (1), 16.

138. Zhu, Y.; Xu, J.; Zhang, S. Application of optimized GA-BPNN algorithm in English teaching quality evaluation system. *Comput. Intell. Neurosci.* 2021, 2021, 4123254.

139. Katoch, S.; Chauhan, S. S.; Kumar, V. A review on genetic algorithm: Past, present, and future. *Multimed. Tools Appl.* 2021, 80 (5), 8091–8126.

140. Yan, H.; Zhang, J.; Zhou, N.; Li, M. Application of hybrid artificial intelligence model to predict coal strength alteration during CO_2 geological sequestration in coal seams. *Sci. Total Environ.* 2020, 711, 135029.

141. Xiao, Y.; Cao, Y.; Zhong, K.-Q.; Yin, L.; Deng, J. Optimized neural network to predict the experimental minimum period of coal spontaneous combustion. *Environ. Sci. Pollut. Res.* 2022, 29 (19), 28070–28082.

142. Aziz, R. M.; Mahto, R.; Goel, K.; Das, A.; Kumar, P.; Saxena, A. Modified genetic algorithm with deep learning for fraud transactions of Ethereum smart contract. *Appl. Sci.* 2023, 13 (2), 697.

143. Wang, W.; Li, G.; Ye, J.; Li, G.; Tang, Y.; Fang, S. Optimization of preparation of cerium-loaded intercalated bentonite by response surface method and genetic algorithm-back propagation neural network and its application in simultaneous removal of ammonia nitrogen and phosphorus. *Chem. Lett.* 2022, 51 (8), 886–890 .

144. Asadi, E.; Silva, M. G. d.; Antunes, C. H.; Dias, L.; Glicksman, L. Multi-objective optimization for building retrofit: A model using genetic algorithm and artificial neural network and an application. *Energ. Build.* 2014, 81, 444–456.

145. Kou, G.; Xiao, H.; Cao, M.; Lee, L. H. Optimal computing budget allocation for the vector evaluated genetic algorithm in multi-objective simulation optimization. *Automatica* 2021, 129, 109599.

146. Slowik, A.; Kwasnicka, H. Evolutionary algorithms and their applications to engineering problems. *Neural. Comput. Appl.* 2020, 32 (16), 12363–12379.

147. Vidal, T.; Crainic, T. G.; Gendreau, M.; Prins, C. A hybrid genetic algorithm with adaptive diversity management for a large class of vehicle routing problems with time-windows. *Comput. Oper. Res.* 2013, 40 (1), 475–489.

148. Xiong, Z.; Wang, D.; Liu, X.; Zhong, F.; Wan, X.; Li, X.; Li, Z.; Luo, X.; Chen, K.; Jiang, H.; Zheng, M. Pushing the boundaries of molecular

representation for drug discovery with the graph attention mechanism. *J. Med. Chem.* 2020, 63 (16), 8749–8760.

149. Wang, Y.; Wang, J.; Cao, Z.; Barati Farimani, A. Molecular contrastive learning of representations via graph neural networks. *Nat. Mach. Intell.* 2022, 4 (3), 279–287.

150. Wu, Z.; Ramsundar, B.; Feinberg, E. N.; Gomes, J.; Geniesse, C.; Pappu, A. S.; Leswing, K.; Pande, V. MoleculeNet: A benchmark for molecular machine learning. *Chem. Sci.* 2018, 9 (2), 513–530.

151. Mansouri, K.; Cariello, N. F.; Korotcov, A.; Tkachenko, V.; Grulke, C. M.; Sprankle, C. S.; Allen, D.; Casey, W. M.; Kleinstreuer, N. C.; Williams, A. J. Open-source QSAR models for pKa prediction using multiple machine learning approaches. *J. Cheminform.* 2019, 11 (1), 60.

152. Tang, W.; Chen, J.; Hong, H. Development of classification models for predicting inhibition of mitochondrial fusion and fission using machine learning methods. *Chemosphere* 2021, 273, 128567.

153. Zang, Q.; Mansouri, K.; Williams, A. J.; Judson, R. S.; Allen, D. G.; Casey, W. M.; Kleinstreuer, N. C. In silico prediction of physicochemical properties of environmental chemicals using molecular fingerprints and machine learning. *J. Chem. Inform. Model.* 2017, 57 (1), 36–49.

154. Huang, K.; Zhang, H. Classification and regression machine learning models for predicting aerobic ready and inherent biodegradation of organic chemicals in water. *Environ. Sci. Technol.* 2022, 56 (17), 12755–12764.

155. Ciallella, H. L.; Zhu, H. Advancing computational toxicology in the big data era by artificial intelligence: data-driven and mechanism-driven modelling for chemical toxicity. *Chem. Res. Toxicol.* 2019, 32 (4), 536–547.

156. Chuang, K. V.; Gunsalus, L. M.; Keiser, M. J. Learning molecular representations for medicinal chemistry. *J. Med. Chem.* 2020, 63 (16), 8705–8722.

157. Jiang, D.; Wu, Z.; Chang-Yu, H.; Chen, G.; Liao, B.; Wang, Z.; Shen, C.; Cao, D.; Wu, J.; Hou, T. Could graph neural networks learn better molecular representation for drug discovery? A comparison study of descriptor-based and graph-based models. *J. Cheminform.* 2021, 13 (1).

158. Zhang, K.; Zhang, H. C. Predicting solute descriptors for organic chemicals by a deep neural network (DNN) using basic chemical structures and a surrogate metric. *Environ. Sci. Technol.* 2022, 56 (3), 2054–2064.

159. Stokes, J. M.; Yang, K.; Swanson, K.; Jin, W. G.; Cubillos-Ruiz, A.; Donghia, N. M.; MacNair, C. R.; French, S.; Carfrae, L. A.; Bloom-Ackerman, Z.; Tran, V. M.; Chiappino-Pepe, A.; Badran, A. H.; Andrews, I. W.; Chory, E. J.;

Church, G. M.; Brown, E. D.; Jaakkola, T. S.; Barzilay, R.; Collins, J. J. A deep learning approach to antibiotic discovery. *Cell* 2020, 180 (4), 688.

160. Wang, Z.; Chen, J.; Hong, H. Developing QSAR models with defined applicability domains on PPARγ binding affinity using large data sets and machine learning algorithms. *Environ. Sci. Technol.* 2021, 55 (10), 6857–6866.

161. Wang, H.; Wang, Z.; Chen, J.; Liu, W. Graph attention network model with defined applicability domains for screening PBT chemicals. *Environ. Sci. Technol.* 2022, 56 (10), 6774–6785.

162. Guerrero, L. A.; Maas, G.; Hogland, W. Solid waste management challenges for cities in developing countries. *Waste Manage.* 2013, 33 (1), 220–232.

163. Vitorino de Souza Melaré, A.; Montenegro González, S.; Faceli, K.; Casadei, V. Technologies and decision support systems to aid solid-waste management: A systematic review. *Waste Manage.* 2017, 59, 567–584.

164. Abdallah, M.; Abu Talib, M.; Feroz, S.; Nasir, Q.; Abdalla, H.; Mahfood, B. Artificial intelligence applications in solid waste management: A systematic research review. *Waste Manage.* 2020, 109, 231–246.

165. Nowakowski, P.; Szwarc, K.; Boryczka, U. Combining an artificial intelligence algorithm and a novel vehicle for sustainable e-waste collection. *Sci. Total Environ.* 2020, 730, 138726.

166. Melakessou, F.; Kugener, P.; Alnaffakh, N.; Faye, S.; Khadraoui, D. Heterogeneous sensing data analysis for commercial waste collection. *Sensors* 2020, 20 (4), 978.

167. European Parliament the Council of the European UnionDirective (EU) 2018/852 of the European Parliament and of the Council of 30 May 2018 amending Directive 94/62/EC on packaging and packaging waste. *European Parliament.* 2018, 150, 141–154.

168. Closing the plastics loop. *Nat. Sustain.* 2018, 1 (5), 205–205.

169. Geyer, R.; Kuczenski, B.; Zink, T.; Henderson, A. Common misconceptions about recycling. *J. Ind. Ecol.* 2016, 20 (5), 1010–1017.

170. Shen, L.; Worrell, E.; Patel, M. K. Open-loop recycling: A LCA case study of PET bottle-to-fibre recycling. *Resour. Conserv. Recy.* 2010, 55 (1), 34–52.

171. Thyberg, K. L.; Tonjes, D. J. Drivers of food waste and their implications for sustainable policy development. *Resour. Conserv. Recy.* 2016, 106, 110–123.

172. Shah, S. A. R.; Zhang, Q.; Abbas, J.; Tang, H.; Al-Sulaiti, K. I. Waste management, quality of life and natural resources utilisation matter for renewable electricity generation: The main and moderate role of environmental policy. *Util. Policy* 2023, 82, 101584.

173. Ren, B.; Zhao, Y.; Bai, H.; Kang, S.; Zhang, T.; Song, S. Eco-friendly geopolymer prepared from solid waste: A critical review. *Chemosphere* 2021, 267, 128900.
174. Suffo, M.; Mata, M. d. l.; Molina, S. I. A sugar-beet waste based thermoplastic agro-composite as substitute for raw materials. *J. Clean. Prod.* 2020, 257, 120382.
175. Zhang, Q.; Cao, X.; Ma, R.; Sun, S.; Fang, L.; Lin, J.; Luo, J. Solid waste-based magnesium phosphate cements: Preparation, performance and solidification/stabilization mechanism. *Constr. Build. Mater.* 2021, 297, 123761.
176. Utetiwabo, W.; Yang, L.; Tufail, M. K.; Zhou, L.; Chen, R.; Lian, Y.; Yang, W. Electrode materials derived from plastic waste and other industrial waste for supercapacitors. *Chin. Chem. Lett.* 2020, 31 (6), 1474–1489.
177. Wei, J.; Chu, X.; Sun, X.-Y.; Xu, K.; Deng, H.-X.; Chen, J.; Wei, Z.; Lei, M. Machine learning in materials science. *npj Comput. Mater.* 2019, 1 (3), 338–358.
178. Ruoff, R. S.; Tse, D. S.; Malhotra, R.; Lorents, D. C. Solubility of C-60 in a variety of solvents. *J. Phys. Chem.* 1993, 97 (13), 3379–3383.
179. Wu, W.; Sun, Q. Applying machine learning to accelerate new materials development. *Mater. Today Commun.* 2018, 48 (10), 107001.
180. Sanchez-Lengeling, B.; Aspuru-Guzik, A. Inverse molecular design using machine learning: Generative models for matter engineering. *Science* 2018, 361 (6400), 360–365.
181. Wang, M.; Wang, T.; Cai, P.; Chen, X. Nanomaterials discovery and design through machine learning. *Small Methods.* 2019, 3 (5), 1900025.
182. Butcher, J. B.; Day, C. R.; Austin, J. C.; Haycock, P. W.; Verstraeten, D.; Schrauwen, B. Defect detection in reinforced concrete using random neural architectures. *Comput.-Aided Civ. Infrastruct. Eng.* 2014, 29 (3), 191–207.
183. Oliynyk, A. O.; Antono, E.; Sparks, T. D.; Ghadbeigi, L.; Gaultois, M. W.; Meredig, B.; Mar, A. High-throughput machine-learning-driven synthesis of full-Heusler compounds. *Chem. Mater.* 2016, 28 (20), 7324–7331.
184. Sugawara, M. Numerical solution of the Schrödinger equation by neural network and genetic algorithm. *Comput. Phys. Commun.* 2001, 140 (3), 366–380.
185. Unke, O. T.; Chmiela, S.; Sauceda, H. E.; Gastegger, M.; Poltavsky, I.; Schütt, K. T.; Tkatchenko, A.; Müller, K.-R. Machine learning force fields. *Chem. Rev.* 2021, 121 (16), 10142–10186.
186. Sugiyama, M.; Krauledat, M.; Müller, K.-R. Covariate shift adaptation by importance weighted cross validation. *J. Mach. Learn. Res.* 2007, 8 (5).

187. Smith, J. S.; Isayev, O.; Roitberg, A. E. ANI-1: An extensible neural network potential with DFT accuracy at force field computational cost. *Chem. Sci.* 2017, 8 (4), 3192–3203.
188. Herr, J. E.; Yao, K.; McIntyre, R.; Toth, D. W.; Parkhill, J. Metadynamics for training neural network model chemistries: A competitive assessment. *J. Chem. Phys.* 2018, 148 (24).

Printed in the United States
by Baker & Taylor Publisher Services